W0049773

Remediation of Soil and Groundwater

NATO ASI Series

Advanced Science Institutes Series

A Series presenting the results of activities sponsored by the NATO Science Committee, which aims at the dissemination of advanced scientific and technological knowledge, with a view to strengthening links between scientific communities.

The Series is published by an international board of publishers in conjunction with the NATO Scientific Affairs Division

A Life Sciences **B Physics**	Plenum Publishing Corporation London and New York
C Mathematical and Physical Sciences **D Behavioural and Social Sciences** **E Applied Sciences**	Kluwer Academic Publishers Dordrecht, Boston and London
F Computer and Systems Sciences **G Ecological Sciences** **H Cell Biology** **I Global Environmental Change**	Springer-Verlag Berlin, Heidelberg, New York, London, Paris and Tokyo

PARTNERSHIP SUB-SERIES

1. Disarmament Technologies	Kluwer Academic Publishers
2. Environment	Springer-Verlag / Kluwer Academic Publishers
3. High Technology	Kluwer Academic Publishers
4. Science and Technology Policy	Kluwer Academic Publishers
5. Computer Networking	Kluwer Academic Publishers

The Partnership Sub-Series incorporates activities undertaken in collaboration with NATO's Cooperation Partners, the countries of the CIS and Central and Eastern Europe, in Priority Areas of concern to those countries.

NATO-PCO-DATA BASE

The electronic index to the NATO ASI Series provides full bibliographical references (with keywords and/or abstracts) to more than 50000 contributions from international scientists published in all sections of the NATO ASI Series.
Access to the NATO-PCO-DATA BASE is possible in two ways:

- via online FILE 128 (NATO-PCO-DATA BASE) hosted by ESRIN, Via Galileo Galilei, I-00044 Frascati, Italy.

- via CD-ROM "NATO-PCO-DATA BASE" with user-friendly retrieval software in English, French and German (© WTV GmbH and DATAWARE Technologies Inc. 1989).

The CD-ROM can be ordered through any member of the Board of Publishers or through NATO-PCO, Overijse, Belgium.

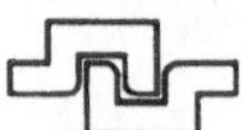

Series 2: Environment – Vol. 17

Remediation of Soil and Groundwater

Opportunities in Eastern Europe

edited by

E.A. McBean
Conestoga-Rovers & Associates Limited,
Waterloo, Ontario, Canada

J. Balek
ENEX
Tabor, Czech Republic

and

B. Clegg
Conestoga-Rovers & Associates Inc.,
Chicago, Illinois, U.S.A.

Kluwer Academic Publishers

Dordrecht / Boston / London

Published in cooperation with NATO Scientific Affairs Division

Proceedings of the NATO Advanced Research Workshop on
Remediation of Soil and Groundwater as a Technical, Institutional and
Socio-Economic Problem: Opportunities in Eastern Europe
Prague, Czech Republic
November 6–10, 1995

A C.I.P. Catalogue record for this book is available from the Library of Congress.

ISBN-13: 978-94-010-6629-7 e-ISBN-13: 978-94-009-0319-7
DOI: 10.1007/978-94-009-0319-7

Published by Kluwer Academic Publishers,
P.O. Box 17, 3300 AA Dordrecht, The Netherlands.

Kluwer Academic Publishers incorporates the publishing programmes of
D. Reidel, Martinus Nijhoff, Dr W. Junk and MTP Press.

Sold and distributed in the U.S.A. and Canada
by Kluwer Academic Publishers,
101 Philip Drive, Norwell, MA 02061, U.S.A.

In all other countries, sold and distributed
by Kluwer Academic Publishers Group,
P.O. Box 322, 3300 AH Dordrecht, The Netherlands.

Softcover reprint of the hardcover 1st edition 1996

Table of Contents

Part V: Summary Aspects

PREFACE

A NATO Advanced Research Workshop (ARW) was conducted from November 6-10, 1995 in Karlovy Vary, Czech Republic. This ARW was focused on the technical, institutional, and socio-economic implications of soil and groundwater remediation in central and eastern Europe. The five-day duration of the meeting provided an excellent forum for the forty-one delegates to discuss, on both formal and informal bases, the existing situations in central and eastern Europe with regard to a wide range of issues. As well, the meeting period included technical site visits to Chodos and Vresova, in the Czech Republic. The discussed issues included (i) development of an understanding of the extent of existing environmental hazards, (ii) the remediation methodologies currently being employed, (iii) the existing exposure risks to humans and the environment, and (iv) the alternative procedures for dealing with the distribution of costs for any necessary remediation, while creating incentives to reduce future pollution generation. With the complications and difficulties of dealing with these issues, there was never a shortage of points for discussion.

We set out the objectives to examine the extent of soil and groundwater remediation as the situation currently exists in the various parts of central and eastern Europe. Some of the attention was focused on the institutional and socio-economic issues - how are the problems being handled currently, how might they be structured, and how will the remediation costs be paid?

The workshop began by reviewing the existing conditions in an array of countries through the delivery of country reports. The foci of discussion were toward the contaminants that are the primary issue, the remediation technologies currently being utilized, the necessary financing being obtained and the potentials being considered. As the results of the discussions amply demonstrated, the extent of contamination as a result of military activities is substantial (particularly from the handling, storage and transport of fuel and lubricants). As well, the lack of adequate (if any) maintenance of pipelines, underground storage tanks, and airfields, has contributed to an enormous spatial distribution of petroleum hydrocarbon-related contamination. The losses of jet fuel acceptable to the military were 0.1% but estimates of the losses amount to magnitudes of several percent. For example, a 5 m thick petrol layer was identified in Estonia, spread over an area of 6 km^2. Monitoring activities to identify the extent of problems are severely limited by the lack of available financing.

The availability of financing for remediation is strongly related to the degree of privatization ongoing within any one country. However, even when financing is available, a complicating issue occurs when the ownership of the land is still in dispute; little expenditure is ongoing when there is uncertainty as to ownership. Interest by individual countries in remediating soil and groundwater problems is strongly linked to the desire for involvement in the European Union. In the absence of acceptable environmental quality standards being in place and enforced, membership in the European Union won't be accepted.

Overall, a major recommendation from the ARW is for the prioritization of site remediation to be established on the basis of human health problems, since the number of sites needing remediation far exceed the financial opportunities. As well, any financial support should have a follow-up component to ensure that the allocated budget has been efficiently utilized for the planned purpose. Support must be provided for the examination of the processes involved in assessing the merits of soil and groundwater remediation, such as mathematical modeling, risk analysis and software development, to improve the understanding of the effectiveness of specific types of remediation and guide future remediation activities.

The resulting papers contained in this volume represent the synthesis of the presentations and the discussions that took place during the meeting.

Part I contains introductory comments that summarize the overall problem of soil and groundwater pollution, and provides discussions of the intricacies of creating institutions and a socio-economic infrastructure that can deal with historical problems and avoid the creation of future problems.

Part II contains the country reports detailing the existing situation in the various countries.

Part III presents a series of papers indicating the characteristics and advantages/disadvantages of various remediation technologies appropriate for soil and groundwater pollution.

Part IV presents a series of case studies detailing experiences with utilization of the remediation technologies and some summary comments on the possible courses of action to address future challenges in Central and Eastern Europe.

In all, forty-one invited experts, representing a number of different disciplines as well as both NATO and Cooperating Partner countries from the region, participated in the ARW. This book provides some interesting perspectives on the extent of the problems. It is apparent that openness in discussions existed, as professionals talked freely about all aspects of the problems. As will be apparent in the following pages, indications of understanding the nature of the problems was developed and the challenge now is to figure out appropriate measures of addressing these problems. The necessity exists to solve the risk associated with environmental quality both as it currently exists and during any remediation. Also the money has to be found, a nontrivial task. In some countries there is almost no experience with remediation and the rate of privatization (which will raise some of the needed revenue) is very slow.

The problems are definitely substantial, but we are the cleaners and thus the lowest caste. We are invited to clean and restore the areas once polluted by industry, agriculture and the military. After making the area acceptable again for living, we have to leave for another filthy place. This does not mean that our work is less important than that of economists, politicians or army officers. Our attempts to keep this world clean and habitable are important elements in contributing toward global prosperity.

The Workshop organizers express their sincere appreciation to Z. Horicka and J. Krecek for their efforts during the progress of the meeting

The editors wish to acknowledge the guidance provided by Dr. L. Veiga da Cunha concerning the organization, format and focus of the workshop. We also greatfully acknowledge the financial assistance obtained from the NATO Advanced Research Workshop funding. We would like to extent our appreciation to the workshop participants whose expertise, dedication and hard work made the workshop a success.

The editors would also like to recognize the efforts of the following individuals for their diligence and care in the preparation of this document: Martin Draeger and Karen Showalter.

The designations employed and the presentation of materials throughout the publication do not imply the expression of any opinion whatsoever on the part of NATO and/or the Editors concerning the legal status of any country, territory, city, or of its authorities, or concerning the delimitation of any frontiers, boundaries and pollution in respective areas.

E. A. McBean

J. Balek

B. Clegg

Part I:

Overview Characterization

I.1

REMEDIATION OF SOIL AND GROUNDWATER IN CENTRAL AND EASTERN EUROPE: AN OVERVIEW ASSESSMENT

EDWARD A. MCBEAN

Conestoga-Rovers & Associates Limited

651 Colby Drive

Waterloo, Ontario, Canada N2V 1C2

Introduction

The development of remediation plans for contaminated soil and groundwater systems is receiving greatly-increased attention throughout the world. The diminishing availability of land and the public's concern for, and nonacceptance of, contaminants in soil and groundwater systems are all contributing toward the increasing pressures for the remediation of environmental problems. However, remediation efforts require sizable expenditures, and have varying degrees of effectiveness in different environments and for different types of environmental contaminants. The result is that while environmental problems are substantial, the funds to implement the cleanup are limited, making for slow progress. Thus, the recently-formed governments in central and eastern Europe face huge obstacles; with little available funding, they must create institutions and assess the socio-economic implications of dealing with historical problems, and simultaneously avoid the creation of future problems.

E.A. McBean et al. (eds.) Remediation of Soil and Groundwater, 1–8.

Clearly, any attempt to describe problems with such enormous dimensions in this single paper means that only a few features will be considered. Given this apology from the outset, some of the background to the problems of soil and groundwater as they exist in central and eastern European countries are described and guidance is provided to some of the subsequent papers in this book.

The Origin of Some of the Environmental Problems

Ecologically speaking, the harmful aftermath of the Cold War includes the former military sites due to aspects such as fuel leakages (e.g. airplane fuel spills, military firing grounds, heating oils and pipeline leaks). For example, acceptable levels of losses of jet fuel as far as the military was concerned were 0.1%. However, the actual fuel losses have been estimated at the several percent level. The result is that when a military airfield used more than 10,000 tonnes of jet fuel annually, this percentage represents extensive jet fuel leakage to the environment. When compounded with the knowledge of the number of military installations in the various countries (e.g. military installations occupied 100,000 ha or approximately 1.5 percent of Latvia), the resulting extent of environmental contamination of soil and groundwater is substantial.

The knowledge-base does exist to look for contamination at military sites. The military certainly contributed to the environmental quality equation but the problems are attributable to a much broader venue in relation to oil products, toxic metals, pesticides and fertilizers, and so on. For example, black oil was typically used for heating purposes for the military troops. Due to leakage of the heating oil, the area surrounding former military boilers are frequently contaminated. For example, at Paldiski, Estonia, the central boiler operating on black oil (furnace oil) used 12,000 tonnes of black oil per year in 1994. The planned consumption for 1995 is 7000 tonnes, much less than what was historically used (Tammamäe, 1995). Oil pollution in the soil around the central boiler of Paldiski covers an area of approximately 6 ha. Currently, following significant rainstorms, black oil pools into storm drainage ditches such that 400 kg per day of black oil flows into Paldiski Bay. In other situations, when the tanks were full and additional tanker cars had to be returned to the fuel storage depots, oil was discharged into the soil to avoid fines for returning unused oil.

Deterioration of the soil and groundwater environment has arisen from a wide range of activities, including the handling, storage and transport of fuel and lubricants, underground storage tanks, airfields and maintenance and repair workshops. For example, cases have been found where jet fuel at military airfields was purposely leaked into the soil. The intent of this activity was to make the number of flight hours of the pilots appear larger because jet fuel "consumption" was used by the military commanders to check on flight hours by their pilots. By illegally dumping the jet fuel, the pilots could avoid having to fly. Current residents are now finding residues from these acts showing up in their water supplies (Tammemäe, 1995).

Problems may also have been identified for other reasons. In the summer of 1993 at the Keila-Joa missile base in Estonia, the military personnel spilled tens of tonnes of hazardous rocket fuel in order to sell the containers made from alloyed steel and aluminum cisterns as scrap metal (Tammamäe, 1995). Vapours from the resulting dumped fuels have been detected over an area of approximately 32 hectares. By the end of 1995, the area of contamination had reached the borders of the local gardening cooperatives which are located next to the missile base.

Environmental problems have also arisen for other reasons. In Russia, in the fifties, as Zektser (1995) points out, design engineers planned for burial of pipelines five metres underground. The soils surrounding the pipes were intended to help control leakage of oil and oil products from the pipelines. The result was that large quantities of oil products penetrated through the soil into the underlying groundwater aquifers. In addition to the oil products causing contamination of drinking water supplies, the oil products in the soil and groundwater give off fumes that are migrating through soil strata into people's basements. As well, when surface water reservoirs were later constructed in these areas, the groundwater levels increased and the oil products (since they are lighter than water) float on the elevated groundwater levels. The result has been oil emergence into the cellars of apartment houses through cracks and floor drains.

Since the decomposition of oil products is extremely slow in deep groundwater layers without oxygen being present, water in such an aquifer will stay virtually unusable for centuries. As apparent from this example of an initial action (assuming the soil would prevent loss of oil to the environment), and the ramifications of subsequent actions (the construction of reservoirs), the multi-dimensionality of migration pathways of environmental contaminants is evident. The longevity of environmental residuals, indicates the complexity of environmental problems.

Alternatively, extensive applications of fertilizers have resulted in widespread nitrate pollution of groundwater. In Hungary, more than 300 villages are supplied with bottled water because the nitrate concentration of the drinking water supplies exceeds the permissible limit. In Moldova, much of the groundwater is impacted by nitrate pollution (Gavrilitsa, 1995). In Lithuania, pesticides were sent to regions irrespective of the crops or natural conditions; now the disposal of the unwanted pesticides represents an environmental problem (Sivickis, 1995).

An Indication of the Remediation Options Being Utilized

The spatial extent of soil and groundwater remediation and the complexity of needed actions for remediation are enormous. The result is, as will be apparent when examining the individual country reports in Part II of this book, that many countries simply cannot afford to clean up all of their environmental problems.

In addressing possible cleanup options, the first need is to understand the extent of the problem. Historical research can be extremely informative, involving interviewing people having knowledge about the site, including former owners and employees and government officials. This historical research can yield a wealth of information about activities that occurred at the site, their duration, the nature of the wastes likely on the site, and the physical and chemical properties of hazardous materials that may be encountered in any remediation. It is often difficult to obtain accurate information about military sites but still much can be learned from completing an audit.

The widespread extent of pollution from oil products suggests an initial option for cleanup may be the removal of floating product. In Piestany, Slovakia, in trying to clean up the groundwater in the vicinity of an airfield, the Slovaks pumped 3000 litres of jet fuel out of the ground in twenty-four hours. This by no means is a sufficient cleanup but does suggest a commonly-utilized approach as a first step of remediation. Subsequent technologies in use are those which involve large labor components and are low in cost. Widely utilized remediation technologies used as followup include bioremediation, pump and treat schemes, and excavation/landfarming. For example, at Sliac in Slovakia, landfarming is ongoing as a means of volatilizing the contaminants from the soil to the atmosphere. Other more sophisticated remediation technologies involving higher costs will be seen in the country reports in Part II infrequently.

Nevertheless, even for the widely used technologies many of these standard remediation technologies are now showing performance problems. For example, Doty and Travis (1991) indicated that for twenty-five percent of the sites using pump and treat systems for remedial operations, the remediation has been required for at least twice as long as predicted.

From a theoretical perspective, some of the standard technologies are appealing due to their low cost and expected performance. However, once implemented at a site the complex geological nature of the subsurface system coupled with the array of sources and associated contaminants that hinder the performance of the selected methodology became apparent. The complex nature of subsurface systems and the array of contaminants contribute to the difficulty of the remediation of groundwater systems. Our inability to characterize the heterogeneous nature of subsurface systems is a large part of the problem. For example, the existence of a low permeability region with the resulting appearance of successful site remediation may be followed by slow diffusion back to the mobile regions from the low permeability zone, and continuing unacceptable levels of contamination in the mobile region.

It is clear that the physical characteristics of the porous media (e.g. permeability, porosity, capillary pressure-saturation relationships, relative permeability relationships and sorption capacity) or fractured porous media systems (e.g. fracture frequency, orientation, and aperture; matrix porosity and sorption capacity) and the type of contaminant play a significant role in the remediation selected. Most of these properties are difficult to determine at actual field sites which means that results derived from highly-controlled laboratory research studies often encounter many unexpected problems when applied to real-world situations.

Problems beyond those relating to the technical aspects may also arise. In Slovakia, they know they have a major problem with oil contamination around a railyard. However, whenever they try to put any equipment in the field to clean up the soil and groundwater, the nearby residents steal the equipment.

Some private companies are trying to use military oil terminals without renovation activities. These local companies do not understand the accompanying risks related to the extremely high cost of cleanup activities or of construction of new groundwater intakes and water supply networks.

Difficulties with Paying for Remediation

The objective in market economies which exist in North America and western Europe is to make the polluter pay. This principle cannot be utilized to describe the actions of these historical polluters in central and eastern Europe and the weakness of the economies of many of these countries makes it a challenge to identify how such activities can be controlled in the future. There is enormous difficulty in charging the remediation costs to the thinly-capitalized firms that are currently operating. Complicating factors in identifying remediation alternatives include the lack of lending and insurance institutions. An immediate and practical consequence of liability concerns is the assignment of responsibility for existing pollution. One possibility is to provide relief from retroactive liability through the provision of government funds earmarked for cleanup. This might involve the utilization of pooled funds. Pooled funds are contrary to the notion that the polluter should pay cleanup costs because they widely distribute the costs of environmental remediation. These pooled funds may represent the least economically disruptive mechanism for dealing with large retroactive liabilities which are potentially otherwise, unable to be remediated.

An approach which has been utilized widely in the US and Canada is to impose a levy on problematic industries. This, however, has led to a large proportion of the financial resources earmarked for remediation being diverted to the costs of establishing responsibility by litigation.

Some countries are in a better position to clean up their environmental problems as a result of privatization. Privatization may provide a source of revenue when new owners of land are assessed extra charges to assist in remediating some of the environmental problems. As well, other countries are experimenting with the establishment of some type of eco-fund in which fines are levied for current environmental deterioration, with the resulting money subsequently allocated to clean up environmental problems. These types of revenue sources provide the opportunity for flexibility which otherwise might not be available.

The central and eastern European countries need enforceable regulations. Some of the responsible officers with decision-making authority are exposed to, and responsible for, all types of political powerplays and corruption. For example, when the president of an industrial enterprise was visited by a member of the enforcement sector of the Ministry of the Environment, a government official who has the authority to say 'yay' or 'nay' regarding the ability of the plant to continue to function, the president was told "you need a wastewater treatment plant and my brother-in-law will build one for you".

So, where do we go from here? As the papers in this book indicate, there is a good awareness on the part of environmental professionals who understand the technologies of remediation. However, governments and industry don't have the financial resources. Thus, legal and regulatory policies must be designed to target public revenues toward the environmental hazards that pose the greatest threat. One mechanism to assist in establishing priorities to determine which sites should be remediated and to what extent that remediation is necessary, involves use of risk assessment, a topic of discussion in Part III.

The incentive and hope for improvement in the extent of environmental remediation is a result of the desire on the part of many countries to join the European Union. As long as the requirements for environmental protection are firmly in place to restrict the joining of the European Union until the pollution standards are met, important progress in meeting acceptable environmental remediation will be attained. However, the soil and groundwater contamination problems will only be resolved at great economic and political cost.

References

1. Doty, C.B., and Travis, C.C., 1991, "The Effectiveness of Groundwater Pumping as a Restoration Technology, ORNL/TM-11866, Oak Ridge National Laboratory, Oak Ridge, Tennessee.
2. Gavrilitsa, A.O., 1995, "Water Resource Problems of Moldova," Scientific Research Insititue of Water Problems and Melioration, Republic of Moldova.
3. Sivickis, J., 1995 "Country Report for Lithuania on Soil Pollution and Prospects for its Purification," Ministry of Environmental Protection of the Republic of Lithuania, Juozapovicians str., 9, 2600 Vilnius, Lithuania
4. Tammamäe, A., Metsur, M., and Kaard, A., "Individual Country Report - Estonia, Environmental Damage Caused by the Former Soviet Union Army", unpublished report, 1995.
5. Zekster, I.S., 1995 "Some Aspects of Groundwater and Soil Remediation in Russia," Institute of Water Problems, Russian Academy of Sciences, Moscow.

Part II:

Country Reports

II.1

COUNTRY REPORT: FEDERAL REPUBLIC OF GERMANY

DETLEF GRIMSKI & SILVIA REPPE
Federal Environmental Agency
Bismarckplatz 7, Berlin
Germany 74793

Introduction

The Federal Republic of Germany is located in Central Europe with a northern borderline to the North Sea, Denmark and the Baltic Sea, an eastern borderline to Poland, a southern borderline to the Czech Republic, Austria and Switzerland and a western border with to France, Luxembourg, Belgium and the Netherlands. Germany's population has increased slightly over the past few years to 81 million. At the same time, the population density is now 228 inhabitants per km^2 (in the old federal states 264 inh./km^2 and in the new federal states 144 inh./km^2). This makes Germany one of the more densely populated countries of the world.

Germany is an industrialized country. As in many industrialized countries, the pollution of soil and groundwater is a serious problem due to the inadequate handling of hazardous substances. Highly publicized incidents, such as the Georgswerder landfill in Hamburg (where dioxin-containing oils were released, or that of the former sludge landfill in Bielefeld-Brake, which has been used for residential purposes), heighened the public and governments awareness of the problem.

E.A. McBean et al. (eds.) Remediation of Soil and Groundwater, 11–43.

Present Situation of Contaminated Land

Based on the definitions given in Germany (see below), the actual number of suspected contaminated Sites which are identifiedg and registered as of August 1994 is 143,252. These sites are now in an ongoing procedure of investigation and assessment by the responsible authorities in the federal states (see below).

The identification and registration of suspected contaminated sites in Germany is not yet finished. Estimates of the presently achieved degree of registration indicate a percentage of approximately. 70-80% for the new federal states and approx. 50-60% for the old federal states. Based on several projections, a total of 250,000 suspected sites are expected, .

The total number if suspected sites would also include more than 4,000 former armament production sites and approx. 10,000 military sites. However, the above noted estimate of suspected sites does not reflect the amount of contaminated land requiring remedial action. Based on the results of final risk assessments, it is estimated that only about 10-20% of the registered suspected sites will require remedial action.

The development of registered suspected sites in Germany over the last decade is given in Table 1. Table 2 shows the latest figures for the old federal states and the new federal states.

Table 1. Development of registered suspected contaminated sites in Germany

1985	35,000	identified or estimated suspected contaminated sites as a result of a first West German national survey
1987	42,000	suspected contaminated sites identified and registered
1989	48,377	suspected contaminated sites identified and registered
1990	28,877	suspected contaminated sites identified as a result of a first survey in the new federal states (East Germany)
1994	143,252	suspected contaminated sites identified and registered as a result of a survey by August 1994

Table 2. Registered suspected contaminated sites in the old/new federal states

	Registered Suspected Contaminated Sites	
	abandoned waste disposal sites	abandoned industrial sites
Old Federal States	55,931	17,628
New Federal States	30,008	39,685
Germany	85,939	57,313

Definitions, responsibilities and main procedure

In Germany there is a differentiation between sites which are suspected to be contaminated by their previous use and those at which evidence at the site exists.

Contaminated sites (in German: "Altlasten") which means "old burdens", i.e. the burden from the past due to inadequate handling of hazardous substances either by industrial activities or by the disposition of waste.

The recently released Draft of the Federal Soil Protection Act gives the following definitions on Suspected Contaminated Sites (Altlastverdachtsflachen) and Contaminated Sites (Altlasten):

Suspected Contaminated Sites (SCS) are:

- abandoned waste disposal sites (Altablagerungen) that include closed-down waste disposal facilities as well as other estates on which wastes have been treated, stored or disposed; and
- abandoned industrial sites (Altstandorte) that include estates of closed-down facilities and other estates on which environmentally hazardous substances have been handled as far as the estates were used for commercial purposes or economic enterprises

Contaminated Sites are:

- abandoned waste disposal sites (Altablagerungen); and
- abandoned industrial sites (Altstandorte) which cause harmful changes of the soil or impart other hazards for the individual or for the general public.

According to the German Constitution, the enforcement for identification, registration, risk assessment and remediation is with the federal states. As long as federal regulations/legislation do not exist, the states have the right to proclaim their own legislation, as they have done in Bavaria (1991), Baden-Wurttemberg (1991), Hesse

(1991), Lower Saxony (1990), Mecklenburg-Western Pommerania (1992), Northrhine-Westphalia (1992), Saarland (1994), Saxony-Anhalt (1991), Saxony (1991), and Thuringia (1991). Although this legislation varies in detail depending on the administrative structures and responsibilities that exist in each state the general procedure can be described by the following steps (see Fig. 1):

- identification and registration of suspected sites
- investigation and assessment
- remediation and/or monitoring.

According to the recommendations of a report released by a state working group on waste, the following general phases are conducted.

I Identification and first evaluation phase
II Orientation phase
III Detail phase of site assessment
IV Remedial investigation
V Remediation
VI Follow-up.

During each of these phases, decisions must be made concerning priorities and necessary immediate measures. During Phase I, all known and newly-suspected sites are identified and located. Phase I generally does not involve any on-site investigation but ends with an initial assessment of possible risks based on available information.

In phase II, on-site investigations are carried out. During this phase it is generally determined whether the suspected site represents a risk and should thus be considered as an abandoned hazardous site. For the evaluation of the suspected sites, various mathematical models have been developed by the federal states.

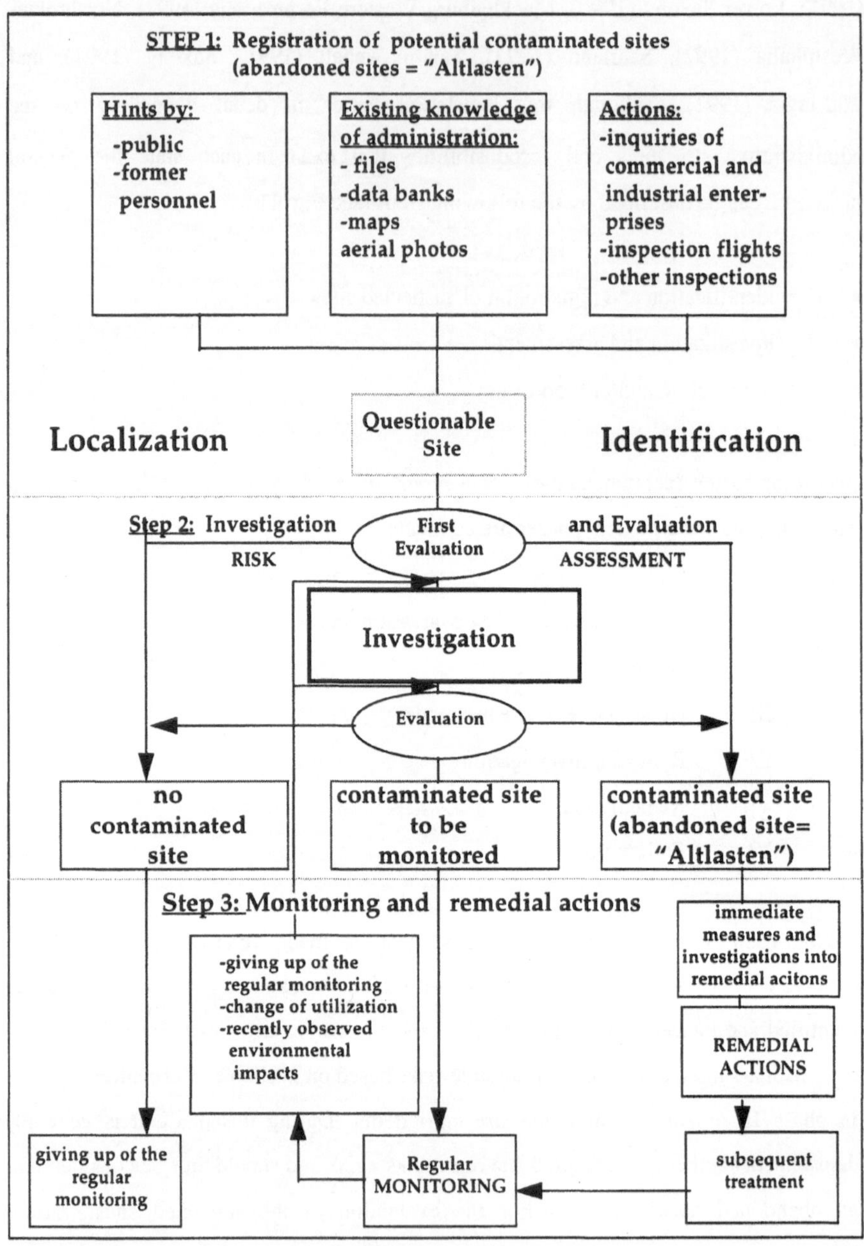

Figure 1. Scheme representing the general procedure of the German approach to the abandoned sites problem

Main Sources of Contamination

There are differences between the federal states depending on the individual industrial structure. So far, each federal state has published an individual catalogue of industrial types of use that are likely to have caused contamination in the past. Since a significant industry in one federal state may be of minor significance in another federal state. The most detailed and comprehensive data on contamination sources in Germany have been compiled from a project about sites which had been used for military purposes.

The total area that was used for military purposes is around 1 million hectares or about 2.8% of the area of the Federal Republic of Germany. These approx. 1 million hectares (ha) belonged or still belong to the following forces:

approx. 253,000 ha	German Bundeswehr
approx. 240,000 ha	former National People's Army (NVA)
approx. 200,000 ha	Allied Forces
approx. 250,000 ha	Western Group of Forces of the former Soviet Union (WGT)

The expected environmental problems of the sites formerly used by the WGT in East Germany have been the main focus of concern. Thus, in agreement with the Federal Minister of Finance (BMF) and the Federal Minister for the Environment, Nature Conservation and Nuclear Safety (BMU) ordered an initial investigation plus 20 risk assessments of the 1,026 WGT bases in early 1991.

The objectives of this project were:

- to identify, analyze and document the suspected contaminated sites on the WGT properties;

- to identify and initiate immediate measures in cooperation with the responsible authorities to ward off acute hazards;

- to initiate preliminary risk assessments for suspected sites in order to establish priorities for subsequent actions;

- to provide the basis for risk assessment for selected sites in order to obtain a representative basis for economic appraisal and effective remediation.

Similar to the situation at industrial waste sites, the suspected contamination at military bases is linked to the specific former use of each respective location. The following kinds of utilization could be distinguished: garrisons, military exercise grounds (training areas, firing ranges), airfields, telecommunication facilities, maintenance and repair workshops, stores, bunkers and hospitals.

Each of these sites have a specific potential contamination. The reasons for the contamination of soil and groundwater due to military use could be to:

- careless handling, storage and transport of fuel and lubricants
- lack of maintenance of pipelines, underground storage tanks and air fields
- a careless attitude at maintenance and repair workshops (no underground sealing)
- desolate condition of wastewater systems
- unorganized waste disposal areas, especially: household waste, hazardous waste, scrap yards, hospital waste, ammunition
- self-support of the military forces by keeping pets ie. leaching of liquid manure and fecal wastes.

In relative terms, the following main contaminents have been identified:

- mineral oil products 57 %
- waste and hazardous waste 38 %
- explosives and ammunition 5 %.

At present (as of October 1995) about 33,750 suspected contaminated sites have been recorded on 1052 real estates properties. Moreover, 9,862 immediate measures have been initiated on 524 sites. Furthermore 2,944 burials of waste and earth movements have been identified.

The preliminary results of a data assessment are provided by Figures 2-4.

Figure 2 shows the results of an initial assessment for a number of contaminated sites allocated to the main protective goods. Altogether, about 1100 properties have been assessed totally or in part.

About 24,000 suspected contaminated areas of these properties of varying size and intensity have been recorded. Figure 3 shows that mineral wastes such as building rubble, ash or contaminated excavation material, represent almost one-half of the recorded amount of water type.

Similarly, larger quantities of metal scrap and domestic waste and mineral oil products were recorded.

Figure 4 gives an overview on the amount of contaminants relative to the land use and type of waste. Troop training areas, barracks and airfields are particularly heavily polluted.

INITIAL ASSESSMENT ON MEMURA

BASIS: 28,422 SRPA

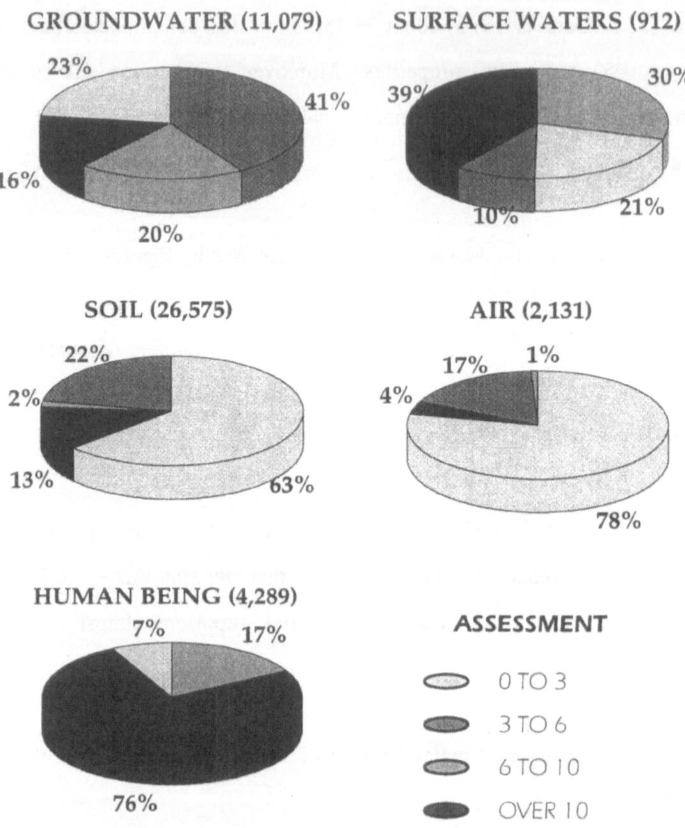

CIRCULAR AREA IS PROPORTIONAL TO THE NUMBER OF SRPA

Figure 2. Initial Assessment

CONTAMINATION PROFILES

According to quantity and number of cases

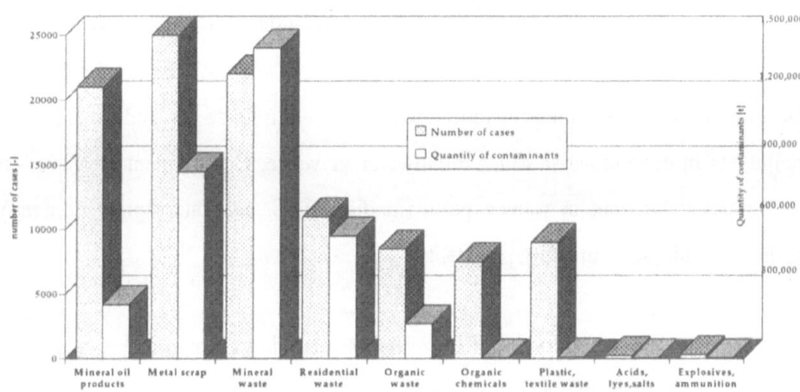

Number of properties : 1,052,

Number of SRPA : 33,750

Figure 3. Contamination Profile

According to quantity of contaminants, related to area of property

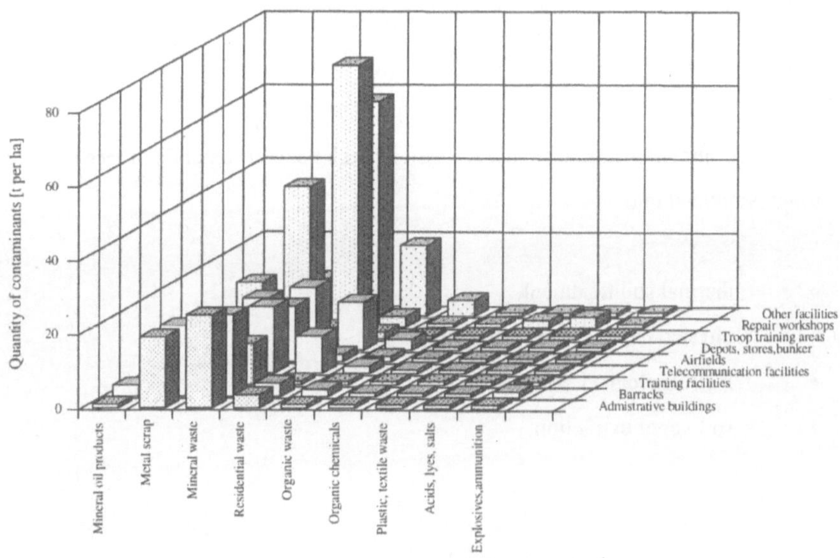

Figure 4. Contamination Profile

Remediation Technologies

In general, there is a distinction between decontamination and containment techniques. Which of these measures is more suitable depends on the individual situation.

Decontamination measures include measures to eliminate or reduce the mass of pollutants in contaminated soil, groundwater or waste. Containment measures include techniques to prevent or reduce pollutant discharges, e.g. interrupting contamination pathways without eliminating pollutants.

Figure 5. provides an overview of remediation procedures which makes distinctions on the basis of the place of application, on-site, off-site or in situ.

On-site application refers to treatment in the general areat of the contamination following excavation. Off-site application is used when the contaminated material is treated elsewhere after excavation and transport. Treating the contaminated material without excavation, that is to say under more or less natural storage conditions is referred to as in-situ application.

Decontamination techniques

The soil decontamination techniques applied in the Federal Republic of Germany can be roughly divided into:

- thermal soil treatment,
- physico chemical soil treatment,
- microbiological soil treatment,
- soil vapor extraction.

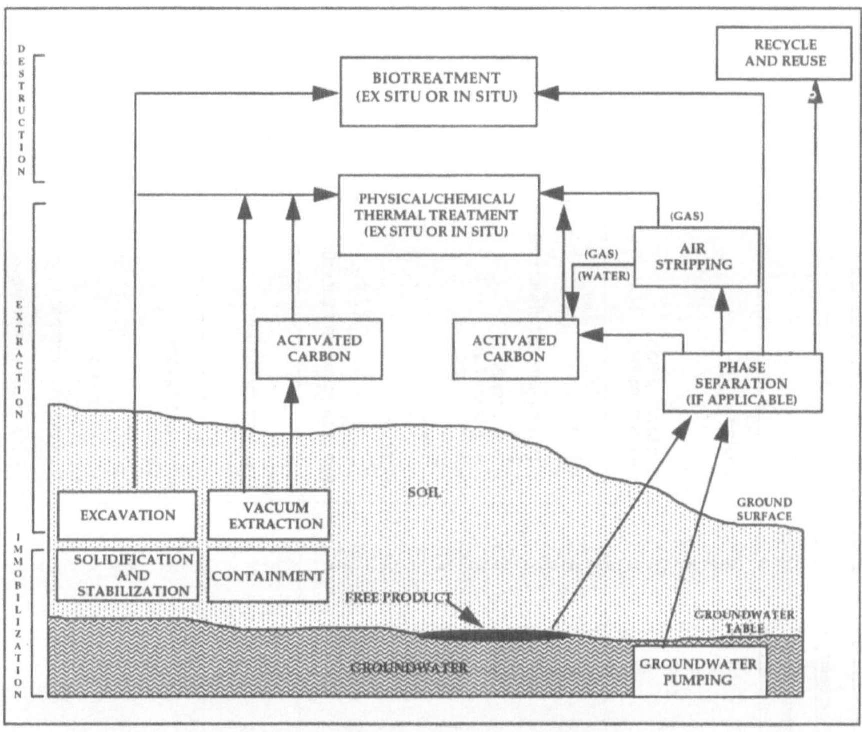

Figure 5. Overview of remediation measures

These techniques are applied either on-site or off-site in mobile or stationary treatment facilities or, in the case of microbiological treatment and soil vapor extraction, in situ. Table 3 provides an overview on the soil treatment facilities available in the Federal Republic of Germany and known at the Federal Environmental agency.

The following section briefly outlines some principles of the various soil remedy techniques and fields of application.

Table 3. Operating stationary soil treatment plants in the Federal Republic of Germany

Federal State	Site	Treatment technology			Operator
		thermal	chem./ phys.	biological	
Lower Saxony	Ahnsen			X	biodetox
	Baltje-Hörne			X	GRT
	Bardowick			X	GRT
	Barenburg		X		GAA
	Ganderkesee			X	Umweltschutz Nord
	Northeim			X	Umweltschutz Mitte
North Rhine-Westphalia	Deponie Haus Forst			X	Trienokens
	Dortmund		X		Züblin
	Emmerich		X		Entsorgung, Umweltschutz und Recycling
	Essen-Vogelheim			X	Hochtief Umwelt
	Gladbeck / Brauck			X	Umweltschutz Ruhr
	Herne	X			Hochtief Umwelt
	Münster			X	BSM
	Münster		X		Fa. Greitens
	Rheine		X		RGR
	Siegen (Zentraldeponie Fludersbach)			X	Fa. Kölsch
	Werl			X	Kreis Soest
	Wesel			X	Terra Entsorgung und Recycling
Rhineland-Palatinate	Germersheim			X	IMA
	Marbach			X	Umweltschutz Südwest
	Mutterstadt			X	Zeller
Saxony	Altenbernsdorf			X	Dierichs & Hagedorn
	Borna	X			Broerius
	Böhlen			X	ContraCon
	Espenhain			X	ESBO
	Freiberg			X	PD Umweltschutz
	Freiberg			X	Bauer + Mourik
	Grumbach			X	Umweltschutz Grumbach
	Leipzig			X	gefus
	Niederau			X	Gröbener Deponie Betriebsges.

Table 3. Operating stationary soil treatment plants in the Federal Republic of Germany

Federal State	Site	Treatment technology			Operator
		thermal	chem./ phys.	biological	
Baden-Württemberg	Schlat (Kreis Göppingen)			X	Leonhard Weiss
	Tettnang			X	Bodenreinigung Oberschwaben
Bavaria	Marktoberdorf			X	BORAG
	München		X		Gebr. Huber
	München-Freimann		X		AB-Umwelttechnik
Berlin	Berlin		X		afu
	Berlin		X		Hafemeister
	Berlin		X		Harbauer
	Berlin-Köpenick			X	Umweltschutz Ost
	Berlin-Tiergarten	X			BORAN
Brandenburg	Groß Kreutz			X	BRZ Großkreuz
Bremen	Bremen			X	Umweltschutz Nord
Hamburg	Hamburg-Peute		X		TerraCon
	Hamburg-Wilhelmsberg		X	X	Hansatec
	Hamburg-Veddel		X		NORDAC
	Hamburg-Waltershof			X	Umweltschutz Nord
Hesse	Neu-Isenburg			X	hutec
Mecklenburg-Western Pomerania	Carpin			X	Lobbe
	Langhagen			X	Umweltschutz Nord
	Poppendorf / Rostock			X	M. B. R.

Table 3. Operating stationary soil treatment plants in the Federal Republic of Germany

Federal State	Site	Treatment technology			Operator
		thermal	chem./ phys.	biological	
Saxony	Oelzschau			X	PD Umweltschutz
	Pohritzsch			X	S.D.R.
	Pohritzsch			X	S.D.R.
	Rodewisch			X	Umweltschutz Grumbach
	Schildau			X	Dierichs & Hagedorn
	Seifersbach			X	Dierichs & Hagedorn
	Zschopau			X	Bodenbehandlungszent. Zschopau
	Zschortau			X	Anlagenbau Umweltprojekt
	Zwickau			X	Umweltschutz Zwickau
Saxony-Anhalt	Bad Lauchstadt			X	Umweltschutz Mitte
	Coswig		X		AB-Umwelttechnik
	Magdeburg			X	GRT
	Halle			X	MUEG
Schleswig-Holstein	Flensburg			X	GSU
	Kiel		X		Preussag
	Lägerdorf		X		AB-Umwelttechnik
Thuringia	Großbreitenbach			X	SGDA
	Merkers			X	SGDA
	Wormstedt			X	Dr. Schilling

Thermal Soil Treatment

Thermal soil treatment processes are based on the principle of releasing pollutants from a soil in the gaseous phase by appling heat. The elimination of pollutants is via either a pyrolytic process or incineration

In general, two types of combustion processes can be distinguished:

- thermal treatment by direct combustion of the material (incineration),
- thermal treatment by indirect combustion of the material (pyrolysis).

Thermal treatment of contaminated soil has the highest clean up performance and the highest energy costs among all the typically considered remedial techniques. In some cases it is the only feasible solution because of the ability to cope with a wide range of soil types and contaminants. It is applicable for various soil types, including soils which contain humus, peat, loam and clay. Organic contaminants namely halogenated and nonhalogenated volatiles and semivolatiles, PCBs, pesticides, organic and inorganic cyanides can be eliminated. Except for cadmium and mercury, heavy metals cannot generally be eliminated. All other heavy metals remain in the soil or, if operated in the high temperature range, are included in glass or clay bodies.

Tables 4. and 5. show the general thermal technique applications relative to the grain size, specific soil type and pollutant.

Table 4. Suitability of thermal processes depending on the soil/grain size

Soil/Grain Size Mineral Content	*suitable*	*to some extent*	*not suitable*
gravel	X		
medium sand	X		
fine sand	X		
silt	X		
clay	X		
loam	X		
building rubble	X		
sludge		X	
fine grain residues from soil washing	X		

Table 5. Suitability of thermal procedures depending on the pollutant

Type of Pollutant	*suitable*	*to some extent*	*not suitable*
mineral oil hydrocarbons	X		
polychlorinated aromatic hydrocarbons	X		
BTEX	X		
phenol compounds	X		
cyanides	X		
chlorinated compounds[1]	X		
organic compounds	X		
N, S, O, P[2]		X	
Hg, Cd, As, So, Zn		X	
any other heavy metals[3]			X

1 These compounds include all organic chlorinated compounds such as PCB's, PCP's, CF, chlorinated pesticides and PCDD/PCDF. Other halogenated compounds can, in general, be treated but at present they are not significant in practical terms.

2 The NSOP compounds include all organic compounds that are not halogenated; however, they contain at least one of the four elements N, S, O, P. These include for example pesticides and TNT.

3 Non-volatile heavy metals.

Chemical Physical Soil Treatment

One of the most important physico chemical processes to treat contaminated soils in Germany has been the extraction of contaminants by soil washing. This technique which has been used on an industrial scale in several facilities. In the soil by means of a liquid medium like water, organic solvents, water-surfactants, acid or bases as the washing fluid. According to the type of energy that is used for separation of the contaminants from the surface of the individual grains there are different processes like high pressure washing, counter current extraction wet washing drums and use of hydro-cyclones.

The application of soil washing processes mainly depends on the content of extremely fine grains in the soil. Fine-grained soil types (e.g. loam, clay, loess) can only be effectively washed with these processes to a limited extent since soil particles with a small diameter contain the largest pollutant load relative to their mass and thus, the energy spent for washing increases inversely with grain diameter, i.e. depending on the process, the energy necessary for the separation of pollutants may not be provided economically. In the Federal Republic of Germany, plant operators refer to a ceiling of 25 to 30 percent in weight for components smaller than 0.02 mm.

Tables 6 and 7 show the possible applications depending on the size of the grain, the type of soil and the specific pollutant:

Table 6. Suitability of soil washing processes depending on the type of soil

Soil/Grain Size Mineral Content	*not suitable*	*to limited extent*	*suitable*
gravel			X
medium-grained sand			X
fine grained sand			X
silty sand			X
clay		X	
loam		X	
rubble		X	
sludge	X		
ash	X		

Table 7 Suitability of soil washing processes depending on the pollutant

Type of Pollutant	*not suitable*	*to some extent*	*suitable*
mineral oil hydrocarbons			X
polychlorinated aromatic hydrocarbons		X	
highly volatile chlorinated hydrocarbons			X
aromatic hydrocarbons			X
polychlorinated biphenyls			X
chlorinated compounds			X
dioxins and furans			
cyanides		X	
heavy metals		X	

Microbiological Soil Treatment

Biological processes employed to clean up contaminated soils are aimed at breaking down organic compounds, particularly by activating and optimizing bacteria that are already present, but also by adding specially-developed bacteria strains and by optimizing the conditions for their survival. Microorganisms have a high detoxification potential for the degradation of organic contaminants like aliphatic and aromatic hydrocarbons, with some limitation for polycyclic aromatic hydrocarbons, while many heavy chlorinated hydrocarbons are hardly biodegradable. There are certain practical limitations to the use of biological processes, and the use of this in situ process is still relatively limited when compared with the on site process.

The range of application of microbiological procedures is limited to soils with organic contaminants. In individual cases, suitability depends on the bioavailability of pollutants in the soil, their microbial degradability and certain soil qualities (homogeneity, content of fine grains).

Table 8 shows the microbial degradability provided that these substances are bioavailable.

Table 8. Microbial degradability of substances relevant for decontamination provided that they are bioavailable

Class of Substances	*in principal, easily degradable*	*in principal difficult to degrade*
aliphatic hydrocarbons, mineral oil, hydrocarbons and their derivates	+	
monocyclic aromatic (e.g. BTX aromatics) and heterocyclic (e.g. pyridin, chinolin) hydrocarbons	+	
polycyclic aromatic hydrocarbons	+[1]	
highly volatile halogenated - in particular chlorinated - hydrocarbons	+	+[2]
alicyclic chlorinated hydrocarbons and their derivates	+	
polychlorinated dibenzodioxins and dibenzofurans (PCDD and PCDF)		+
pesticides and their derivates		+
heavy metals		not degradable

Some low-chlorinated congeners are in principle degradable/halogenable. For highly chlorinated congeners degradation cannot be confirmed at present.

[1] up to 4 ring PAHC

[2] ring 5 and 6 PAHC

Soil Vapor Extraction Procedures

Soil vapor extraction procedures are utilized for the remediation of soils that are contaminated with highly volatile compounds. They are primarily applied in the unsaturated zone of the soil in in situ processes. The contaminated soil vapor is extracted by a low pressure generated by a vacuum well. At the same time, evaporation is promoted by increasing the flow of air along the pollutant molecules.

Soil vapor extraction can be combined with air sparging of the saturated zone. The pressurized air introduced into the groundwater causes a stripping process and thus converts the volatile pollutants to gas. The gaseous pollutants are collected and extracted by a soil vapor extraction device and decontaminated with activated carbon.

Containment Techniques

The containment techniques applied in the Federal Republic of Germany include:

- surface sealing,
- vertical sealing (compacted wall),
- immobilization measures,
- passive hydraulic and pneumatic measures,
- additional horizontal sealing systems.

The following section provides a brief overview of the containment techniques.

Surface Sealing Systems

The Federal Republic of Germany has comprehensive experience with the application and efficiency of surface sealing systems. They have mainly been developed and optimized within the framework of research projects to remediate existing dumps and landfills and have been frequently used in practice for this purpose.

Surface sealing systems seal off contaminated soil from atmospheric influence. Thus, they prevent infiltration of rain water in the contaminated areas and leaching/migration of pollutants into groundwater.

Surface sealing systems consist of the actual sealing component and surface water and gas drainage. Moreover, a recultivation layer is required. In case of existing depositions, the sealing component is, in general, a so-called combined sealing which consists of a mineral sealing element with a salient plastic layer sealing.

In the case of existing contaminated sites, for which containment measures have been increasingly applied recently, only one of the above sealing elements has been applied, i.e. either only a plastic layer or a mineral layer. There is no best available technology in the Federal Republic of Germany defined yet for the surface sealing of existing contaminated sites.

Vertical Sealing Systems/Compacted Walls

As in the case of surface sealing systems, compacted walls have been developed and optimized mainly within the framework of research projects to remediate existing contaminated sites. However, compacted walls have also been frequently applied in practice for existing dump sites.

Compacted walls are designed to seal off the contaminated soil from the surroundings and thus prevent lateral leaching of pollutants. They are inserted up to a depth where a natural compact layer is present; in combination with the surface sealing, they encapsulate the contaminated soil.

The requirements for the technique and the mass of the compacted wall are determined for each individual case depending on the chemical and hydraulic strain. The most common techniques for compacted walls are sheet walls, thin walls and diaphragm walls.

Immobilization Measures

In the case of immobilization measures, the pollutants contained in the soil are contained in such a way that leaching into the environment is prevented or reduced. Although various immobilization processes are available (compaction, sealing, canopy) only processes for pollutant stabilization by solidification are relevant for the decontamination of existing contaminated sites and applied in practice in the Federal Republic of Germany. In the process, the contaminated material is mixed with a bonding agent and water filled into molds or installed and compacted on the spot or pressed into molds and palletized. This process is aimed at manufacturing a waterproof and solid monolith as final product of the solidification process.

The Federal Republic of Germany applies solidification processes for both the containment of existing contaminated sites (solidification of contaminated soil) and for existing depositions (solidification of waste/waste batch).

There is no best available technology for solidification procedures with qualitative requirements for the solidified material. The following criteria are used to determine the suitability of solidification procedures:

- mixed contamination of soil,
- low to medium organic contamination,
- low to high inorganic contamination,
- fine-grain material.

The suitability is tested in advance.

Passive Hydraulic and Pneumatic Measures

Hydraulic measures have been the best available technology in hydraulic engineering and foundation engineering for a long time. These measures are to prevent contact between the contaminated material and the groundwater. This is achieved by changing the hydrodynamic conditions of the groundwater, for example by lowering the groundwater level with negative wells or reversing the groundwater flow with positive wells. Hydraulic measures are used for existing deposition and hazardous waste sites.

In contrast to that, pneumatic measures are only relevant for existing depositions with a problematic gaseous phase. They imply installing a gas drainage to control the disposal of gas and thus to prevent the gas from migrating into the surrounding environment.

Additional Horizontal Sealing Systems

Additional horizontal underground bottom sealing under waste disposal sites or contaminated industrial sites has not yet been put in practice on a large technical scale in the Federal Republic of Germany. The systems available for this process have mainly been developed from mining processes, they are very complicated and cost-intensive. Their application will therefore be limited to the most problematic cases.

Remediation of Contaminated Industrial Sites Belonging to the Enterprise of the Treuhandanstalt:

A special program of the federal Government deals with contaminated industrial sites belonging to enterprises of the Treuhandanstalt (THA), which is a federal institution responsible for the transfer of former public-owned enterprises of the GDR into private property.

According to the so-called "exemption clause" of the Unification treaty, investors in the new federal states, who are acquiring installations used for commercial purposes or within the framework of economic undertakings are not liable for any damage caused prior to 1 July 1990 as a result of the operation of the installation. Exemption may be granted after the interest of the purchaser, the general public and environmental protection have been weighed. Liability due to claims based on civil law remains unaffected. The purpose of the exemption clause is to prevent the contaminated sites problem from deterring potential investors in the new federal states.

According to the "Administrative Agreement of financing the clean-up of contaminated sites" adopted by the Federal Government and the new federal states on December 22, 1992 the remediation costs are shared by the Federal Government (THA (60%) and by the new federal states (40%), on the condition that the enterprises had been exempted from the liability for environmental damages by the new federal states.

A total number of 1 billion DM per year for a period of 10 years are committed for these 60:40 projects. In addition to these 60:40 projects, the remediation costs for so-called "large projects" (projects with remediation cost over 100 million DM) will be shared by the Federal Government (THA) 75% and the new federal states 25%. Nineteen projects with an estimated cost of 6 billion DM will be financed under this 75:25 regulation. These projects comprise the highly polluted areas of the chemical industry ("chemical triangle" of Halle, Leipzig and Bitterfeld), the shipbuilding industry at the Baltic Sea and other "hot spots". Currently 19 large projects have been decided upon by the responsible working group representing the Federal Government, the THA and the new federal states (Table 9).

Another already fixed large project is the clean-up of lignite mining open casts. For this large project, initially 1.5 billion DM over a five years period are available.
For the clean-up of the Wismut sites (uranium mining), 13 billion DM are available by the Federal Government. Summing up the public funds for clean-up in East Germany, a total amount of at least 35 billion DM will be available over the next 10 years.

Table 9. Large projects according to the administrative agreement on financing of cleanup

Federal State	Name of large projects	Industry/branch
Berlin	1. Region "Industriegebiet Spree"	Gas production, chemical and textile industry
Brandenburg	2. Region "Kreis Oranienburg"	Foundry works,chemical industry
	3. Standort "Brandenburg"	Metalprocessing,chemical industry
Mecklenburg-Western Pommerania	4. Standort "Wismar"	Shipbuilding industry
	5. Standort "Rostock"	Shipbuilding industry
	6. Standort "Stralsund"	Shipbuilding industry
Saxony	7. SOW AG Boehlen	Chemical industry
	8. Saxonia AG	Iron and steel works and processing
	9. Lautawerk GmbH i.L.	Aluminum works
	10. Dresden-Coschuetz/Gittersee	Tire works, uranium ore processing (WISMUT)
	11. Bitterfeld/Wolfen	Chemical industry, Anhalt film industry
	12. Buna AG	Chemical industry
	13. LEUNA-WERKE AG	Chemical industry
	14. Hydrierwerke ZeitzGmbH	Refinery, lignite tar processing
	15. Unternehmen derehem. Mansfeld AG	Copper and silver foundry, brass processing
	16. Magdeburg-Rothensee	Industrial area
	17. Erdîl Erdgas Gommern GmbH	Oil and natural gas extraction
Thuringia	18. Kali ThÅringen	Potash mining
	19. Verwaltungs- und Verwertungsgesellschaft mbH "Rositz"	Tar processing works

Research and Development

On behalf of the Federal Ministry for the Environment, Nuclear Safety and Nature Conservation (BMU) the Federal Environmental Agency is funding R&D projects on special aspects of the contaminated sites problem. Some of the basic projects are listed as follows:

- Continuation of the gathering of basic toxicological data for approximately. 80 substances in contaminated sites and of 17 priority substances from former armament production sites;
- Derived effect levels for pollutant-specific parameters and substances based on toxicological data;
- Continuation and testing of models for the risk assessment (UMS-model for contaminated sites, MAGMA model for military and armament production sites); and
- Continuation and testing of models for setting priorities (PRISAL) and for cost estimation (KOSAL).

R&D projects concerning the remediation of contaminated sites in the Federal Republic of Germany are mainly funded by the Federal Minister for Science, Education, Research and Technology (BMBF). In general, these projects are focused on the development and implementation of innovative clean-up technologies. Several clean-up technologies have been developed in the past years, notablyEspecially thermal, chemical-physical and biological processes were investigated to be used as remediation techniques.

The BMBF alone has spent 220 million DM on research and development in the field of contaminated sites in the past decade. Furthermore, there are special funding sources provided by the federal states and other institutions like the German Research Community (Deutsche Forschungsgemeinschaft) or the Federal Foundation on the Environment (Deutsche Bundesstiftung Umwelt). In total about 200 projects with estimated costs of approx. 300 million DM have been funded in the past decade in the Federal Republic of Germany.

So far. research funding in Germany has made enormous contributions to develop remediation technologies to the degree they have now reached. For the future, the following main aspects of further research and development have been identified:

Identification and Investigation

- further research and development needs are modest
- research aspects have to be adopted to cost-effective approaches, procedures and methods for site investigation

Decontamination Technologies

- existing on-site and off-site technologies have reached a high standard
- so far, research needs are mainly identified in the field of optimization and assessment of existing technologies
- in the field of in-situ remediation, research is ongoing for biological technique (meanwhile it is discussed whether technologies using electrochemical properties for the decontamination of soil shall be considered more intensively in the future)

Safeguarding Measures

- safeguarding measures like surface liner systems and slurry walls have reached a high standard in Germany based on the experiences concerning the remediation of old waste disposal sites
- future research in this field will be concentrated on the adoption of available barrier systems to the specific conditions of old industrial sites, in particular under consideration of long-term effectiveness, monitoring measures and possible reuse options
- under consideration of the coming Soil Protection Act efforts have to be undertaken in order to identify significant parameters of safeguarding measures to assess whether they are on the same value as decontamination measures

- as immobilization/solidification measures are emerging in Germany future research on the long-term effectiveness and on adequate agents for these measures is needed

Quality Assurance/Quality Control

- in the field of quality assurance, in particular with respect to the collection, storage, transport and preparation of samples, research on analysis, documentation and control are still ongoing and must be intensified in the future

Federal Soil Protection Act

At present the Federal Government is preparing a federal Act for the Protection of Soil and the Remediation of Contaminated Sites. The responsibility for the development of the draft and its submission to the other ministries of the Government is with the Ministry for the Environment, Nature Conservation and Nuclear Safety, supported by the Federal Environmental Agency. At the moment the draft law is still in the process of final agreement between the various ministries.

As the title of the Soil Protection Act indicates, it will be divided in two parts. One part will address the prevention of soil pollution in the future, i.e. avoiding the creation of new contaminated sites by setting precaution levels for the concentration of harmful substances in the soil. These precaution levels will indicate the concentrations of substances in the soil requiring preventive measures as harmful changes in the soil cannot be excluded in the long term.

The other part of this law will address the remediation of contaminated land.

It will set the frame of reference for standards and requirements on the identification, registration, assessment and remediation of contaminated sites.

The main objective of this part of the law is "warding off of dangers" arising from contaminated sites as they are defined in the Draft (see above).

However, the Soil Protection Act will not contain detailed regulations in this area. According to the legislative structure in Germany, the law as a rule, gives the framework regulations that have to be completed in more detail by sublegal regulations such as decrees or technical instructions. So far, the obligation of the Federal Government to elaborate decrees and technical instructions are:

- the tools and standards for identification and registration;
- the sampling and analysis;
- the soil screening levels and action levels;
- the remediation measures; and
- the monitoring procedures

All of which is given by the draft law, and represents one of the main approaches to solving the contaminated land problem in Germany.

Concluding Remarks

There is no general solution to the problem of the remediation of contaminated soil. Every technique has advantages and disadvantages with respect to its applicability to the removal of a wide range of contaminants and for different soil types.

Existing technologies for the remediation of soil have reached a high standard in Germany.

Federal programs and action currently concentrate on the Legislative level with the drafting of federal statutes for former military sites and the clean-up of industrial enterprises belonging to the Treuhandanstalt. However, due to the economic situation combined with a lack of financial resources for the clean-up of contaminated land, future efforts will concentrate on the development of reliable low-cost and cost-effective remediation technologies.

References

1. Ministry for the Environment, Nature Conservation and Nuclear Safety federal Act of the Protection of Soil and the Remediation of Contaminated Sites Draft September 1995.
2. Franzius, Volker, Perspectives of remediation of contaminated land in Germany Presentation on SCI Meeting "International Perspectives on Contaminated Land", London, September 1994
3. Grimski, Detlef, Soil remediation Centers, 1994
4. Penning, Jutta, Forschungsbedarf im Bereich Altlasten Presentation on VEGAS Workshop, Stuttgart, October 1994
5. Dr. Volker Franzius and Detlef Grimski, Recent Developments in Contaminated Land Remediation in the Federal Republic of Germany: Current Programs and Future Research, Land Contamination & Reclamation, Vol. 3, Number 1, 1995

II.2

REMEDIATION OF SOIL AND GROUNDWATER IN THE CZECH REPUBLIC

JAN ŠVOMA

Aquatest Sg a.s, Geologicka 4, Praha 5, 15200

Czech Republic

Introduction

The Czech Republic (CR) like other Eastern European countries inherited a significant burden of pollution in soils, rocks and groundwater. The main ecological damages are related to nitrates, oil products, chlorinated aliphatic hydrocarbons and PCBs, in less degree by toxic metals: As, Cd, Cr, Hg, Pb and radioactive substances. Almost 40 % of drinking water from the mains in the Czech Republic does not meet Czech standards for drinking water CSN 75 7111/1991 (1). The cost of remediation is estimated up to 50 bio CZK (15).

The shortage of surface and groundwaters in the former Czechoslovakia and the number of oil spills forced the ministries of defense, transport and oil industries to start protective measures for groundwater quality protection. Large-scale remediation of groundwater contaminated with oil hydrocarbons started in 1970: airport in Prague (aviation jet fuel), oil refinery in Bratislava (crude oil), and the airfield near Pilsen (aviation jet fuel). The polluted areas reached up to ten square kilometers and the thickness of oil layer floating on groundwater table varied from one to 600 cm.

E.A. McBean et al. (eds.) Remediation of Soil and Groundwater, 45–57.

The Czechoslovak government introduced, besides funds for groundwater clean-up and enactment aiming at the mitigation of current groundwater pollution, measures for preventive water protection involving groundwater quality monitoring. Legal protection of water was anchored in the Water Act (No 138) in 1973. On the contrary, soil pollution was cleaned up with exceptions only. (e.g. in the remediation,. the cross-country pipeline spills near Havlickuv Brod and east of Prague in late 1970's.

A systematic soil remediation started in the connection with the ecological damages assessment of Soviet military bases in Czechoslovakia in 1990 (8, 13).

The second, more significant wave of soil pollution assessment is underway now in the connection with property transfers in the privatization process. However, there is great discrepancy between the quantity of soils and groundwater which should be remediated and the availability of money. The legal aspects of soil pollution were put into practice not before the nineties (1,4,13). The environmental Law (17) was approved in December 1991.

The Old Burden in the Czech Republic

There are several environmentally damaged areas in the Czech Republic: Capital of Prague, Middle Bohemian region (Melnik, Kladno), Pilsen region, Chomutov-Teplice region, Liberec region, Hradec Královć (Pardubice) region and Ostrava region

Damage has been caused by numerous constituents e.g. SO_2 and other harmful gases, fly ash, land and air traffic (1).

Soil and groundwater quality is mostly endangered in the following facilities and areas:

- former Soviet Army bases, mainly polluted with oil and chlorinated hydrocarbons, live ammunition, tear gases toxic metals, etc.
- Czech army bases - airfields and fuel depots with oil hydrocarbons and toxic metals

- oil delivery systems - depots, cross country pipelines, petrol pumps
- uranium mines and piles
- airports
- hazardous waste dumps

The extent and intensity of pollution when superposed by hydrogeological importance of local aquifers and their hydrogeological vulnerability identify the most endangered areas and sites concerning the aquifers' quality such as

- The former SA base Ralsko-Hradcany located in the Bohemian Cretaceous basin of groundwater and sandstones are intensively polluted with aviation jet fuel, aromatics, TCE and PCE. The thickness of the oil layer floating on the groundwater table reached 8 m. The amount of hydrocarbons (HC) in soil is estimated up to 10 000 t, the amount of chlorinated hydrocarbons (CHC) dissolved in water is about 1 t (7).
- The former SA base Mladá--Milovice where Soviet Headquarters were situated. It is situated in close vicinity of the waterworks supplying Prague and its surroundings with drinking water. The cretaceous fractured aquifer and Quarternary ones in sandy sediments are significantly polluted with petrol, Diesel oil, jet fuel, rocket fuel, aromatics, PAHs, PCBs, DCE, DCA, TCE, PCE, and to a minor degree by Pb, Zn, Cr and Cd. Combat gases and live ammunition in soils and waste dumps were also located. Three square km of the groundwater table was covered by an oil layer up to 3 m in thickness. The volume of soil containing 1g/kg HC or more rocketed to 300 000 m^3. Concentrations of CHC dissolved in groundwater reach units and even tens of ppm.
- Mining of radioactive raw material in Cenomanian Cretaceous aquifer at Straz pod Ralskem. Here 4 mil. of nitric and sulfuric acid were injected into sandstones in the process of uranium extraction and endangered water quality in the important overlying Turonian aquifers (6).

- Several factories in middle Moravia are situated in the vicinity of significant waterworks tapping. Plumes containing high concentrations of HC, CHC and toxic metal are approaching hygienic protection zones of the waterworks .

Review of Remedial Technologies in the Czech Republic

Soils and Rock Unsaturated Zone

Soils polluted with oil hydrocarbons are mostly treated by bioreclamation for its technical simplicity, versatility and low cost when employing ex situ methods following excavation of polluted material. Bioreclamation is performed in biobeds or composts where nutrients, oxygen and oil-consuming microorganisms are added. Non-biodegradable organics are treated thermally or by solidification and stabilization by lime or cement. Toxic metals are removed from soils by acid washing and than treated chemically or landfilled. Using ecological safe organic solvents for flushing pollutants out from the solid matrix is frequently done.

Soil venting has a dominant role among in situ methods. Unfortunately removing pollutants by vacuum extraction which is the principle of venting methods is limited to light, volatile organics such as aromatics, chlorinated solvents, petrol and aviation jet fuel. The pressure gradient is in response to the vacuum created and then the gases are sorbed in activated carbon (AC) filters or combusted. Heavier oil hydrocarbons are bioremediated in situ by injecting N,P and K solutions, oxygen and special bacteria. Removal of soluble organic and inorganic pollutants is made by water flushing.

Appropriate technologies listed in Table 1 are selected much more often according to their costs and site-specifics than according to desired level of decontamination. The estimates percentage use of different methods is shown in Tables 1 and Table 2. When new incinerators are built, the ratio between ex situ clean-up methods will change.

TABLE 1. Remediation of soil and unsaturated zone

70 % removal of soil	20 % landfilling		heterogeneous materials, degradable compounds
	50 % decontamination on site	90% biological methods	CHC, thick oil organics
		5 % solidification and stabilization	pasty organics, TM, UG
		5 % flushing and washing *	PCB, TM
	30 % decontamination off site	80 % biological methods	CHC, PAHs, PCBs
		10 % solidification and solidification	pasty organics, UG, NM
		9 % thermal methods	PCBs, UG, medicines, poisons
		1 % Na - removal of PCBs from oils	
25 % in situ decontamination	8,9 % biological methods		HC
	80 % venting		VOCs, petrol, jet fuel
	1 % bioventing		heavy and medium HC
	0,1 % thermal methods		viscous oil, medium and heavy
	10 % flushing		medium and heavy HC,
5 % encapsulation and insulation	95% horizontal and vertical barriers		waste dumps
	5 % cement and/or chemical injection		

Explanations
- * auxiliary method only
- HC hydrocarbons
- CHC chlorinated aliphatic hydrocarbons
- VOCs volatile organic compounds (e.g. petrol, BTEX, DCE, TCE, PCE)
- TM toxic metals
- UG nonbiodegradable organics

TABLE 2. Remediation of groundwater

85 % Hydraulic methods	95 % depression barriers Pump and treat method	Surface gravity separators and/or in situ skimmers	PCB, BTEX, CHC
		filters AC	PAU,PCB,BTEX,TOXICANTS
		kutex,	PAU, PCB, BTEX
		vapex or	HC
		fibroil HC	
		sprinklers	VOCs, light oil product
		bioreactors	HC, PAU, BTEX
		stripping towers with AC filters	VOCs, radon
		coke filters	phenols
		ion exchangers	TM
		chemical reactors	acids, caustics, other electrolytes
	5 % elevation barriers - pollution insulation only		
10% in situ remediation	90 % bioremediation, additional method only	downwell diffusers	
	9,5 % air sparging	bioreactors	HC, PAHs
	0,1 % radioactive radiation	blowers, AC filters, combusters	VOCs, light oil products
	0,3 % denitrification	cobalt emitter	cyanides, some organics
		mixers of ethanol or molasse	nitrates
5 % impervious barriers	50 % thick slurry walls		all pollutants
	30 % thin concrete or sheet barriers		
	20 % injected concrete or chemical curtains		

Groundwater

Hydraulic "pump and treat" methods are versatile, quickly operative and technically simple. Their advantage is the safeguarding and cleanup of pollution when an hydraulic depression between pumped wells is created. This remediation of groundwater is therefore widely used. When remediating immiscible fluids, a two pump system is employed: the lower pump creates depression and the upper pump removes pollutant and polluted water. Sometimes hydraulic skimmers are used to remove free oil phase as well.

The lack of information from behind the Iron Curtain in the past forced Czech specialists to adapt or even invent their own decontamination technologies and techniques for groundwater treatment. Some of them met European standards such as Vapex filters in the seventies, Fibroil, Kutex and Chezakarb filters in the eighties for hydrocarbons removal, decontamination well for in situ stripping (air sparging) of VOCs and cobalt emitters for cyanides and some organics destruction in situ (16, 17, 20). On the other side, effective employment of new pneumatic technologies such as soil venting, bioventing and air sparging depends on importing special devices from the West (efficient vacuum pumps and catalytic incinerators) which started in 1990.

Biological methods are used either for the enhanced HC removal from water on the ground in all remediation phases or for in situ hydrocarbon removal in the late stages of aquifer cleanup when pumping alone has low efficiency.

In the case of CHC, VOCs remediation is usually completed by stripping towers with AC filters or organic vapors condensers consisted of electrically-heated carbon fibers and nitrogen coolers.

Besides the pump-and-treat method, impervious barriers are constructed. They often isolate heavy pollution inside oil refineries, chemical factories and hazardous waste disposal sites. The disadvantages of impermeable walls are high primary costs but they have low maintenance costs. Horizontal sealing of dumps (clayey or plastic caps) is mostly used. Vertical slurry walls are limited by geological structure to sand and gravel deposits and weathered mantle of cemented or igneous rocks. Injected concrete or chemical curtains can be employed irrespective to geological conditions but are too expensive to have more widespread use.

As evident from the above, groundwater remediation methods are quite comparable with the ones used in the West. However, there are some differences concerning soil decontamination where much more bioreclamation and less thermal desorption and solvent flushing (12) are employed and no use of in situ vitrification and electrokinetic soil processing.

When the actual danger for adjacent aquifer exists so-called emergency clean up pumping was started; pollutants plume isolation was and till now is the primary aim of this activity. Postponing complex remediation of soil and groundwater in that case is caused by financial constraints. It is true that finally this emergency pumping will increase the total cost of remediation there exists no other way to protect important actual sources of drinking water.

Strategy and Instruments for Ecological Burden Removal, its Costs, Reimbursement, and Protection of the Environment in General

The principle "polluter will pay" is valid for current and proven deterioration of soil and groundwater quality. Nevertheless, state budgets and several funds have paid a lot of money for environmental protection. We can distinguish three main finance sources in our country: 1 National budget by which have been paid environmental damages caused by former Soviet army bases among other items. 2 The State Environmental Fund, created mostly by pollution levies(1) e.g. for emissions into the atmosphere or hydrosphere-see Table 3. It covers part of requests for ecological projects in different components of the environment such as water, soil, wastes and air.

Table 3. Expenditure for environmental protection (bik. CK)-according (1)

Financial source	year 1991	1992	1993
National Budget	7,52	10,80	8,16
(SA bases from NB	0,047 *)	0,226	0,195)
State Envir.Fund	1,90	1,49	3,38
Fund National Property	-	?	?

*) in 1991 is not included cost of Milovice, Hradcany and Olomouc SA bases remediation . Costs were met as in previous years by ministry of defense . Environmental damage caused by SA was estimated at 1,86 to 7,18 billion CK(8).

Table 3 indicates a shift in financing ecological projects from the National budget to the State Environmental Fund. This tendency will continue(1). There is a rule that at least 60 % of money coming from region has to be invested in it.

The third source of money for ecology is the National Property Fund of the Czech Republic. It was established in connection with the privatization process and can be used for reimbursement of costs for removal of damage to the environment caused by activities of the corporations until the date of accomplishment of the privatization project. Legal backgrounds are Law No 92/1991, No 171/1991 and two principal resolutions of the Czech Government No 455/1992 and No 123/1993(11). The NFP budget can be used for remediation of groundwater, soil and harmful waste dump sites only. The established scope of burdens in question leads to a preliminary cost estimate of 30 to 50 billion CK(15).

In any case, the level of economic damage to the environment has always exceeded several times the level of outlay to protect it. Environmental protection has been grossly underfunded. This was one of the factors responsible for the stagnation of the Czech economy. The rate of economic damage to the environment in the CR is estimated at 4 to 10 % of the GDP, while the rate of investment into its protection has, in the past years, been considerably below that level. ((1) p.217). The above mentioned discrepancy between the desired sum and the obtainable money for environmental protection and remediation is obvious. The vital task is to determine the priorities of ecological problems and deteriorated sites cleanup. A new approach to priorities estimation has been started by introducing the obligatory Environment Impact Assessment (EIA) by the Act No 244/1992 following the Environmental Act No 17/1992. EIA process concerns selected new project activities: construction activities and technologies, developing programs and products (1).

Assessing environmental liabilities of companies and the old burden generally is based on the Methodological Instruction Ministry of Administration National Property and Its Privatization and Ministry of Environment. Environmental audit must be performed (ordered) by privatized company but very often new investors do it. Site investigation should provide data of pollution extend and cleanup and its costs. It is desirable to complete an audit with risk assessment which will consider the degree of urgency of the remediation and distinguish such urgent cases where a threat of further contamination of important natural resources, or danger to human exists (3). Definition of target parameters for decontamination was based, before risk assessment approach is introduced, on the recommended limits A,B,C (4) resembling those of the Neverlands. Category A corresponds with the natural background, category B is triggering further investigation and value C is starting (finishing) cleanup generally. The limits in areas significant from the point of water economy or water protection resp. are more strict: C value should meet drinking water I hygienic protection zone Drinking Water Standards (CSN 7111/1991) and decontamination concentrations in outer protection zones and in protected natural water accumulations should correspond with B values instead of C ones. Risk assessment takes into account harmful doses for human health and for land and water biota. The evaluation of pathways to receptors, involving groundwater rate, pollutants migration ability, persistence and degradability and future land use, affect the limits for cleanup starting and finishing considerably because site-specifics are considered. The risk assessment however does not mean that assessed value will be less strict than mentioned A,B,C levels in all cases; e.g. groundwater biota may be endangered by pollutant concentrations lying below drinking water standards (19).

Conclusions

There are serious environmental problems in the Czech Republic concerning land and groundwater pollution from the past and continual problems. Remediating of old burden in the privatization process depends on money from the FNP CR when selection of cleanup priorities is based on environmental audits and risk assessment. Unfortunately, some doubts exist regarding the availability of money source in future(3). Continual pollution will be remediated and/or prevented on the principle "polluter will pay" provided the taxes and fines for pollution are high enough to make protective measures more profitable than to continue in the environment deteoriation.

International co-operation helps us to introduce new approaches of pollution evaluation and remediation. The fruitful assistance of Dutch Province of Groeningen, Canadian AGRA, and NATO's Committee on the Challenges of Modern Society is noted. In the frame of the mentioned CCMS was started Pilot Study on Environmental Aspects of Reusing Former Military Lands. In this study among other countries from the East and the West, the Czech Republic participates also (8).

From the technical point-of-view introducing some innovative technologies from the west will be necessary for speeding up remediation of the most troubled sites where the combined action of high hydrogeological importance with heavy pollution by heavy metals and organic refractants in unfavorable conditions of unsaturated and/or saturated zone. Steam stripping and electrokinetic soil processing seem to be very promising in this respect.

References

1. Absolon K. et al. (1994): Environmental Year-Book of Czech Republic (1993-4). Czech Environmental Institute Prague.
2. AGRA (1995): Environment Risk Assessment Guideline Development andTraining M.S. AGRA Earth and Environmental Ltd, Regina, Saskatchewan, Canada. (Written communication to Czech ministry of defence.)
3. Allen J. F. , Kozusník R., Connor K. (1994): Management of Environmental Liabilities in the Czech Republic. Symp. Proc.: Soil and Groundwater Pollution, Risk Assessment and Legislation, 3p. Ceský Krumlov, Czech Republic, June 8-9, 1994.
4. Anon. (1992): Methods of Assessing Environmental Liabilities of Companies for the Preparation of Privatization Projects.Methodological Instruction of the Ministry for Administration of National Property and Its Privatization of the Czech Republic and of the Ministry of Environment of the Czech Republic of May 18, 1992. (in Czech)
5. Anon (1993): State of the Environment of the Czech Republic. Planeta, No 9, p 7-12. (In Czech)
6. Anon (1994): Assessment of Rock Environment in the Czech Republic. (In Czech)
7. Cerný J., Hercik F. (1994): Results of Investigation and Remediation of Pollution at HradËany Airfield. Workshop: Ecomilitary, p 34-38, Prague, June 16, 1994. (In Czech)
8. Chvojka R., Kom·r A., Švoma J. (1994): NATO CCMS Pilot Study on Environmental Aspects of Reusing Former Military Lands. M. S. Workshop, Ober Ammergau, FRG, May 24-26, 1994.
9. EPA (1992): Innovative Treatment Technologies.Semi-Annual Report, EPA-542-R-92-011, Nr 4 Washington October, 1992.
10. Landa I., Mazac O., Redlin D. (1994): Experience with Grounwater Pollution Remediation at Former SA Bases and at Privatisied Factories. Workshop Proc.: Ecomilitary, p14-33, Prague, June 16, 1994. (In Czech)

11. Mrkos J. (1994): Process of Indemnification of Privatised Property. Symp.: Soil and Groundwater Pollution, Risk Assessment and Legislation, 2p, Ceský Krumlov, Czech Republic, June 8-9, 1994.

12. NATO CCMS (1993): Demonstration of Remedial Action Technologies for Contaminated Land and Groundwater, EPA, 600/ R-93/012 b, February, 1993.

13. Nevyjel F. (1994): State Fund of Environment of the Czech Republic-a Tool for the State Environmental Policy Establishment. Planeta, Nr 3, p 40-41. (In Czech)

14. Notenboom J. (1994): Assessment of Ecological Risk of Soil and Groundwater Pollution. Symp. Proc.: Soil and Groundwater Pollution, Risk Assessment and Legislation, 6p. Ceský Krumlov, Czech Republic, June 8-9, 1994.

15. Ruzicka J. (1994): The Introduction of Risk Analysis in Environmental Protection. Symp. Proc.: Soil and Groundwater Pollution, Risk Assessment and Legislation, 3p,Ceský Krumlov, Czech Republic, June 8-9, 1994.

16. Švoma J. (1985): Contamination and Decontamination of Aquifers-Problems which Industrial Countries are Solving Today and Developing Ones Will Solve Tomorrow. In: Hydrogeology in the Service of Man, Memoires of the 18th Congress of the Intern. Assoc. of Hydrogeologists, p 7-13, Cambridge, 1985.

17. Švoma J. (1990): In Situ Treatment of Groundwater Polluted with Chlorinated Hydrocarbons. In: F. Arendt, M. Hinsenveld and W. J. Brink (eds.), Contaminated Soil ¥90, p 1143-1144. Kluwer Academic Publishers. Printed in the Netherlands.

18. Švoma J. (1993): Investigation and Decontamination of Soil and Groundwater at Former Soviet Army Bases in Czechoslovakia. In: F. Arendt, G. J. Annokkée, R. Bosman and W. J. Brink (eds.), Contaminated Soil ¥93, 747-754. Kluwer Academic Publishers. Printed in the Netherlands

19. van den Berg R. (1992): Risk Assessment of Contaminated Soil: Proposals for the Dutch Intervention Levels for Chlorinated Aromatic Hydrocarbons. Workshop Proc.: The Health Risk Assessment and Management of Contaminated Land, p 1-19. Perth, West Australia, November 25-26, 1992.

20. Vodiczka V. (1995): Review of Groundwater and Soil Remediation Technologies. Planeta, Nr 3, p 16-19. (In Czech)

II.3

COUNTRY REPORT LATVIA

DZIDRA HADONINA

Ministry of Environmental Protection and Regional Development of the Republic of Latvia

Head of Division of State Cadastres and Natural Resources

PELDU IELA 25

RIGA, LV1494, LATVIA

Some Facts about Latvia

The Republic of Latvia is situated in north-eastern Europe, on the east coast of the Baltic Sea, with Estonia on the north, Lithuania on the south, Russian Federation on the east and Belarus on the southeast (Figure 1). Latvia has a coastline of nearly 500 kilometers and total land area of nearly 65,000 square kilometers, which is larger than such countries as Estonia, Denmark, Holland, Belgium or Switzerland.

The climate in Latvia is typical of a northerly maritime region with moderate winters and moderately warm summers. Temperatures in Riga range from 17.5 C in July to -4.3 C in January. Average annual precipitation in Riga is about 617 mm.

Latvia has a population of approximately 2.5 million. Latvian nationals constitute only 54% of the total population and about 1 million or 43% Slavonic and other nationals now live in the country. Most of the non-Latvian population lives in the largest cities. The population density is about 40 people per square kilometer which is similar to the other Baltic States, but significantly less than European countries.

E.A. McBean et al. (eds.) Remediation of Soil and Groundwater, 59–77.

Figure 1. Latvia in the Baltic Region

Riga, the capital of Latvia, is the largest city with a population of 856,000. Daugavpils in the south-east has a population of 122,000 and Liepaja in the west has a population of 105,000.

The official language is Latvian and is written in the Latin script. Similar only to Lithuanian, this non-Slavic, non-Germanic language represents the Baltic branch of the Indo-European family of languages. Today, Russian and increasingly English and German are widely spoken in the country.

On May 4, 1990, the name of the Republic of Latvia was renewed, and the Declaration on the Renewal of Independence of the Republic of Latvia was adopted. On August 21, 1991, a new Constitutional Law was adopted ending the transition to independence from the USSR, and on September 6 of the same year, the Soviet State Council officially recognized the independence of Latvia. On September 17, 1991, Latvia restored its rightful place within the international community by becoming a full member of the United Nations.

Latvia is an independent republic with parliamentary democracy. Legislative powers are vested in the Saeima (Parliament), a 100 member elected body. Head of State is the President who is elected by the Saeima. He appoints the Prime Minister and the Saeima confirms the Cabinet of Ministers holding executive powers.

LATVIA: Gross Domestic Product
(at constant 1993 prices, Ls millions)

	1991	*1992*	*1993*	*1994*
Gross Domestic Product	2645.4	1723.2	1467.0	1475.5

The Ministry of the Environmental Protection and Regional Development (MoE) is responsible for the united state policy in the fields of environmental protection, regional development, nature conservation and sustainable use of natural resources, *inter alia* building, tourism, nuclear safety, etc. The ministry is responsible also for the organization of state control and environmental impact assessment, collecting of environmental data, elaboration of legislation and development of international relation, including investment activity, in the fields of environmental protection, regional development and building. Implementation of environmental policy is delegated to subordinate institutions, and to the local and regional authorities.

Short Overview of the Environmental Situation

The majority of environmental problems in Latvia are concentrated in the so-called "hot spots" namely large industrial centers, transportation crossroads or *in territories abandoned by the Russian army.* Just some of environmental problems are manifested in the country as a whole: eutrophication and degradation of water ecosystems, excess usage of several natural resources, transboundary pollution, accumulation of household and industrial waste. Several grave local problems have been created by excessive, and in many cases chaotic, urbanization.

On the other hand, at present, state administrative structures and systems of management during the previous decades allowed preservation of natural forests, meadows and swamps, where rich animal and plant populations are to be found now. Many of these species are on the edge of extinction in the western and north-western regions of Europe. Latvia can be proud of its comparatively untouched nature, vast forests, beaches without construction and low pollution background level.

It has to be mentioned that, during the last 3-4 years the total pollution load has significantly decreased mainly due to the decline of the economy. On the other hand we are still struggling with a lot of existing environmental problems.

Water quality is traditionally mentioned as the number one problem in Latvia.

According to hydrobiological and hydrochemical data, 85% of all the surface water is either slightly polluted or polluted. Eutrophication is the biggest problem, created by biogenus substances, and it is growing rapidly. The main sources of biogens are untreated municipal waste water and leakage from agricultural lands. In several places, water pollution with dangerous substances has been identified (e.g., heavy metals, chloroorganic compounds, oil products) and their accumulation. However, it should be stressed that after 1990, both the amount of wastewaters, and leakage from agricultural lands have decreased significantly. The total amount of pollution that was discharged into watercourses, mainly due to the start of operation of Riga municipal and other wastewater treatment facilities, has significantly decreased.

The situation in the Gulf of Riga and the Baltic sea should be stressed in particular. Relative isolation and significant discharge from rivers makes the anthropogenous factor in the Gulf of particular significance. Observations of phosphorous and chlorophyll concentration over the years show that eutrophication in the Gulf of Riga continues. The present results of hydrochemical and biological observations do not indicate further development of eutrophication in the Latvian zone of central Baltic. At present, the waters of the Latvian zone in the Baltic Sea are of moderate pollution with some regions of local pollution. Zones of ecological risk are coastal regions in the vicinity of river estuaries, locations of municipal and industrial waste water discharges, locations of extraction of minerals and gravel disposal sites, as well as regions around ports.

The Most Serious Problems Relevant to Ground Water and Soil Pollution

Latvia faces environmental problems with groundwater and soil pollution only in specific spots or contaminated areas.

Basically soil and groundwater are contaminated at all sites, where dumps, fertilizers storage, oil depots and the largest enterprises are located. Simultaneously all these contaminated sites should be characterized as local point - sources and, as opposed to some regions in Western Europe, regional development of groundwater contamination has not been observed.

Groundwater is weakly or not protected from surface pollution in Latvia. The most seriously polluted areas in Latvia are: storage of acid tar at Incukalns, storage of pharmaceutical waste at Olaine, Riga landfill at Getlini, landfill at Daugavpils, storage of mineral fertilizers in Iecava, Vilani, Jelgava, landfill at Jelgava, Jurmala landfill at Kudra, oil storage at Jaunmilgravis and Tukums, former Soviet army base at Spilve in Riga. The polluted areas do not exceed few hundred meters and covers quartarian layers, with infrequent pollution distributed in deeper layers.

Pollution of ground water has been identified also in all largest cities, but there it is less serious.

Due to earlier studies, the most contaminated sites (except the former Soviet army bases, where investigations have been carried out only at a few sites) have been ascertained (see Table 1).

- Incukalns, Riga District - the maximum penetration of contamination into depth;
- Olaine, Riga District - the maximum intensity of groundwater contamination;
- ''Getlini'', Riga District - the maximum extension of groundwater contamination plume.

TABLE 1. A Characterization of groundwater pollution at the most dangerous contaminated sites

Contamination Source	Parameter	Value, mg/L	Area and a depth of contamination in m
ACID TAR POOL	Dry residue	17.000	0.5 km^2
POOL in	$SO_4{}^{2-}$	8.000	>100 m
Incukalns,	$PO_4{}^{3-}$	300	
Riga District	SSAS	75	
(Southern Pool)	COD	1.200	
	Fe	1.800	
	Al	2.000	
	V	6	
	pH	1.5	
LIQUID	Dry residue	28.000	0.25 km^2
CHEMICAL	Cl-	15.000	15 m
WASTE	NH_4+ NH_3+	4.300	
POOL	Pyridine	4.500	
Olaine	Buthanol	3.200	
	COD	21.000	
SOLID	Dry residue	10.000	1.5 km^2
WASTE	Cl-	3.500	>30 m
DUMP	COD	1.400	
"GETLINI"	N_{tot}	490	
in RIGA	Cr	1.0	

There are also a lot of other highly contaminated sites:

- Jurmala City dump ''Kudra'', where well field ''Kauguri'' (15.000 m^3/day) is damaged;
- Daugavpils City dump ''Krizi'', where the proposed well field ''Ziemeli'' (39.600 m^3 /day) could be damaged; and
- area with wide soil and groundwater contamination in Jaunmilgravis, Riga, where the oil depot, lubricating factory and other enterprises are located.

The special problem for Latvia is environmental damages and problems caused by former Soviet army.

Environmental Damages and Problems Caused by Former Soviet Army in Latvia

According to our data, the former Soviet Army military units and bases of different scale and purpose were located in Latvia occupying about 100,000 ha or approximately 1.5% of Latvia's territory.

The first studies on environment pollution caused by Russian Army were started in 1992 and two military territories in Riga - Suzi and Spilve fuel depot were investigated in detail. Investigations were continued in 1993 and finished in 1994. At present the primary observations (without samplings) and initial assessment of former Russian army territories have been completed. Three hundred military sites occupying about 96,000 ha were observed. For 53 military sites out of these occupying about 57,500 ha, the realized investigations are not sufficiently detailed and have to be continued.

The above-mentioned investigations were carried out by the Ministry of Environmental Protection and Regional Development of the Republic of Latvia.

The most severe damage from the former Soviet Army to environment and Latvian economy were caused by military firing grounds, airfields, rocket bases, filling stations, fuel depots and naval ports. On Latvian territory the Russian army had firing grounds for all kinds of weapons.

Larger amounts of information about the environmental problems emanating from defense-related installations and activities of the former Soviet Army,and then the Russian Army, in Latvia has been and will be presented at various NATO/CCMS Pilot Study meetings.

In general, environmental problems caused by the former Soviet Army in Latvia are similar to those in post-socialism East-European countries.

Monitoring of the Pollution

Monitoring of Groundwater

Current monitoring systems are based on the State Budget, which allows control of only a part of soil and groundwater pollution.

Yearly monitoring includes 96 posts and water sample locations. Selected parameters in heavily polluted areas storage of acid tar at Incukalns, storage of pharmaceutical waste at Olaine, Riga landfill at Getlini, storage of mineral fertilizers in Jelgava, landfill at Jelgava, Jurmala landfill at Kudra, oil storage at Jaunmilgravis and others). The number of controlled boring wells, samples and parameters have decreased during the last years due to insufficient financing.

The guaranteed groundwater monitoring system includes 36 stations with 435 wells and covers all aquifers located in the zone of active groundwater exchange. All wells located in the above-mentioned stations have been sampled at least once per year until 1993 and include an additional 200 wells which are installed at the local sources of groundwater contamination (mainly to the shallow unconfined aquifer). This monitoring usually was carried out once per 3-5 years. Particular parameters for which monitoring was carried out are characterized in Table 2. The implementing organization is the

Latvian Geological Survey in co-operation with the Latvian Hydrometeorological Agency. Monitoring objectives are the level and quality of groundwater and quality of groundwater in heavily polluted areas. The key parameters are groundwater levels, pH, conductivity, Eh, t^{o}, O_2, dry matter, Cl-, $SO_4 2-$, HCO_3-, Ca^{2+}, Mg^{2+}, Na^{+}, Fe^{2+}, Fe^{3+}, NH_4+, NO_2-, NO_3-, detergents, phenols, oil products, heavy metals (atomic adsorption spectrophotometry).

TABLE 2. Parameters controlled in frames of groundwater monitoring of Latvia

During Field Investigations	*All Samples**	*Separate Samples***
Temperature	Na^{+}	COD
Conductivity	K^{+}	BOD
pH	Ca^{2+}	PO_4^{3-}
	Mg^{2+}	SSAS
	Cl-	N_{tot}
	$SO_4{}^{2-}$	Extractable subs.
	HCO_3-	Heavy metals
	$NH_4{}^{+}$	AHC
	NO_2	TCE
	NO_3	CL -containing organic pesticides
	Dry residue	CL -containing organic pesticides
	$KMnO_4$ demand	
	Fe	

Notes:

- * parameters controlled in framework of the national monitoring;
- ** parameters controlled mainly in framework of the monitoring of local contaminated sites. Of course, this set of main parameters and in particular cases other necessary parameters have been added.

The methodology of sampling includes pumping of wells with Grundfos-MP1 until stabilization of pH and conductivity, a sampling and if necessary preservation of samples and their laboratory analysis (usual colorimetry and titrometry, atomic adsorption, neutron activation, gas - liquid chromatography).

Monitoring of Soil

Basic monitoring yearly in 189 sample plots in farms situated in all territory of Latvia and representing all types of soil, climatic conditions and current types of practice. This was done in part since 1992 and in a full scale since 1995. Analyses of heavy metals in soil are carried out every 6 years.

The implementing institutions are Land Cadastre Center of Latvia of State Land Service in co-operation with Latvian Agricultural University, State enterprise 'Razìba' Institute of Biology, etc. Sources of pollution include agricultural (mineral fertilizers, pesticides) pollution, partly industrial solid or liquid substances. Monitoring objectives are accumulation of pesticides, heavy metals in soil and harvest, pollution of groundwater with nitrates and nitrites.

Key parameters: mechanical characteristics, content of total organic substances, pH, nutrients (N, P), cations (K^+, Ca^{2+}), heavy metals (Pb, Cd, Cu, Cr, Zn, Ni, Fe), residues of pesticides in soil. Heavy metals (Pb, Cd, Zn), residues of pesticides in harvest and concentration of nutrients in groundwater.

Currently we are mostly using analytical methods which are accepted in the former USSR, but following the process of integration into the European system, laboratories are going through the process of certification which means that we are accepting European analytical methods (ISO standards) and step by step, adapting new technology to our conditions.

Generally, the following methods are used for plant nutrient and chemical elements detection in the soil:

- pH (acidity) in KC1 (potassium chloride) extract;
- Soil organic mater - according to the Tjurin method used in the former USSR;
- Soluble P and K (phosphorus and Sulfur) in KC1 (potassium chloride) extract;
- Cu (copper) in the HC1 (hydrochloric acid) solution;
- Zn (Zinc) - in an ammonia acetate solution.

After these preparations, all elements are detected photocolorometrically.

We must take into account that in the different European countries, distinctive analytical methods for chemical elements detection are utilized.

For example, detection methods of P (phosphorus) and K (potassium) are equal in the Latvia, Estonia and Germany, but in the Denmark, Sweden and Finland they are different. Now the Latvian specialists are working upon the comparison of different methods used in the European countries and in Latvia.

Correlation factors exist for comparing Latvian and Finnish analytical results obtained by different methods. In such a way we can obtain equal results using either Latvian or Finnish soil analysis methods.

Latvian specialists are working to obtain the correlation factors between the Swedish, Danish and Latvian methods. It is possible to obtain results in the space of 2 or 3 years.

General and specific radioactivity are mostly measured in Latvia by radiometers. All results can be obtained spectrometrically from the soil, water or plant material. Detection of such radioactive components as Sr 90 (Strontium), Cs 137 (Caesium), Cs 134 , Co 60 (Cobalt), K 40(Potassium), Ru 106 (Ruthenium), can be completed in 30-40 minutes for one sample.

All spectra of heavy metals are extracted form the soil samples by inorganic acid solutions and then detected by atomic absorption spectrophotometers.

Residues of different pesticides are determined by Gas-liquid chromatography, according to the methods worked out by the scientific institutes in Latvia, former Soviet Union and in Western Europe.

Oil and oil products can be detected accordingly to the temporary methods now used in the Latvia but we are looking forward to implementing methods accepted in European countries.

Remedial Technologies Applied

Remediation of Tukums Bulk Fuel Terminal

Shortly after its registration in Latvia in late 1991, Baltec Associates Inc. approached the Environmental Protection Committee of Latvia with a proposal to perform a self-financed remediation project at a site to be selected by the Committee. It was hoped that completion of such a project would, in addition to bringing about actual environmental improvement, demonstrate western technology and transfer know-how to local interested groups.

The Tukums bulk fuel terminal site, located approximately 70 kilometers west of Riga, was eventually selected. This terminal is one of the oldest in Latvia and was at one time crossed by a Soviet army jet fuel pipeline. Over the years, routine operations and numerous pipeline bursts resulted in severe ground water contamination. Some of the pipeline releases reportedly flooded a nearby stream.

Baltec performed a subsurface investigation of the site in early 1992. The investigation included excavation of exploratory pits and installation and sampling of ground water monitoring wells. Laboratory analysis of ground water samples revealed the presence of fuel hydrocarbons at concentrations exceeding drinking water standards. Jet fuel was identified in the samples and observed floating on the water table to a thickness of one-half meter in one of the on-site wells.

Based on these findings, Baltec constructed a system of three ground water recovery trenches around the periphery and one infiltration trench in the center of the site. Fuel and contaminated ground water is pumped from the trenches to an aboveground treatment system consisting of oil-water separation, aeration, and biological treatment, and treated water is subsequently reintroduced into the zone of contamination to enhance in-situ bioremediation. It is estimated that within three to five years, the floating fuel will have been removed and the concentration of hydrocarbons dissolved in ground water will be significantly reduced.

The remediation system began operation in late 1993. The system startup was observed by government officials from Latvia, foreign diplomats, and local business leaders and scientists, and widely reported by the press.

Already, more than $280 000 has been invested in the project. It is estimated that an additional $ 50 000 per year will be invested for the next five years. These funds will be applied toward the development and installation of a biological filter for processing treatment system vapor emissions and for overall system maintenance.

Incukalns Sulfuric Acid Tar Waste Disposal Site - Mitigation of Environmental Impacts

From the mid- 1950's to the early 1980's, tens of thousands of tons of sulfuric acid-tar waste were dumped into two sand pits located near the village of Incukalns in Latvia. The northern of the two pits is more proximal to the Gauja River and was closed in the mid- 1960's by filling with sand. Since that time, infiltrating precipitation has served to leach the soluble components of the waste and transport them to groundwaters, which they have now penetrated to a depth of 60 m. The resultant contaminant plume is laterally extensive (having a length of 1.25 km and a width of 0.5 km) and is spreading in the direction of the Gauja River and the Rembergi well field, a water-taking facility for the city of Riga.

The Incukalns pits are on the national priorities list of sites requiring immediate environmental action and among the most widely-known environment ''hot-spots'' in Latvia. Notwithstanding the economic crisis confronting Latvia at present, during the last three years the Ministry of the Environment and Regional Development has found it prudent to finance hydrogeological investigations to define the extent of contamination and to develop methodologies to control the spread of contamination. These works which are being undertaken by the environmental consulting firm Baltec Associates, Inc. of Latvia included installation of groundwater monitoring wells, both in the water table and the bedrock confined aquifers, aquifer testing, ground water flow modeling to evaluate possible groundwater remedial alternatives, and development of a pilot-scale bioremediation system for application in the confined aquifer (Figure 2).

Ideally, measures should be taken to isolate or remove the bulk of the waste from the environment and subsequently focus remediation efforts on the residual soil and ground water contamination. Outside financing will be required for this undertaking.

Socio-Economic Context - Health Status of the Population

The base for estimation relationship between population health and quality of underground water and soil is not developed. No data currently exist on direct connection with health of population and pollution of groundwater or soil.

Financial Resources Available

About 20,000 USD from monitoring budget of MOE for monitoring of Groundwater for Latvian Geological Survey.

About 23,000 USD from monitoring budget of MOE for monitoring of radiation in Latvia for Environmental Data Center including monitoring of soil radioactivity in selected areas.

About 12000 USD for ambient monitoring of agricultural soils are required.

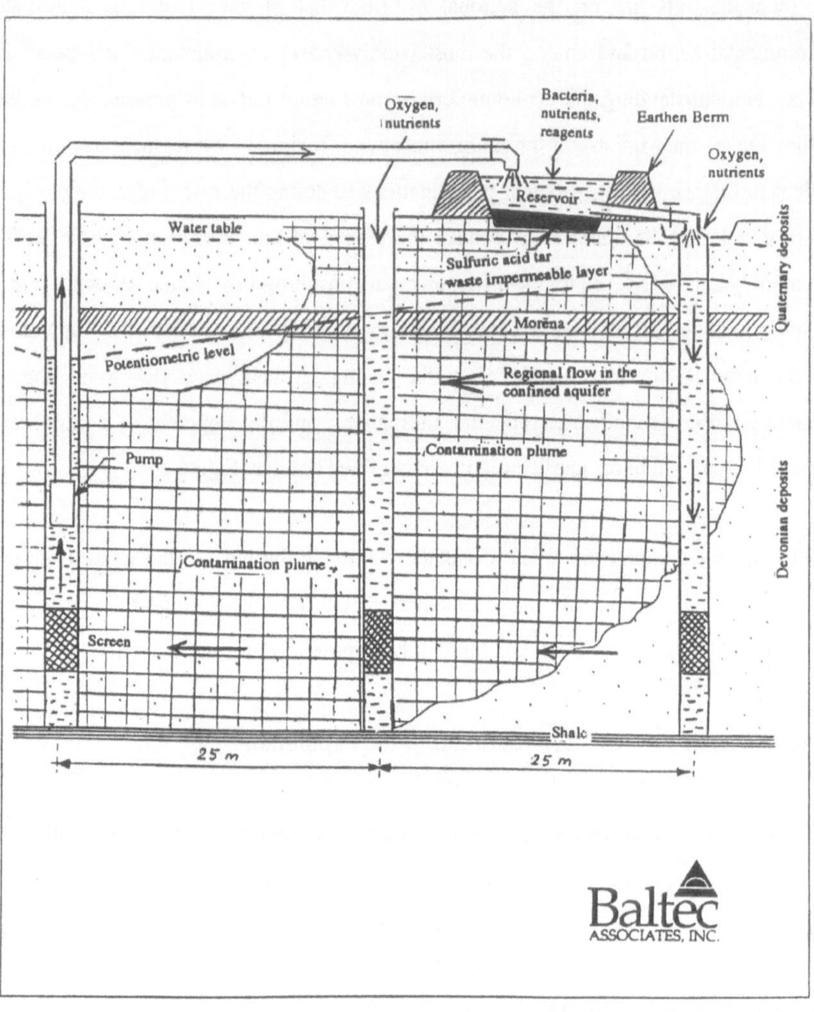

Figure 2. Proposed Bioremediation Scheme

External Co-operation Established

Project proposals: Abandoned Polluted Areas together with Germany (Bundesland Baden-Wuerttemberg, Umweltministerium).

The Danish Geological Surway is providing, assistance in analytical equipment and consulting.

Owing to the Government of Germany and the Federal Ministry for Environment, Nature Conservation and Nuclear Safety since 1992 we have good cooperation with the IABG company. Latvian specialists had the possibility to become acquainted with environmental problems in the former Soviet military territories in Germany, took part in application seminar "Data processing programs - ALADIN, MEMURA, MAGMA" and received software.

The joint Latvian - German project "*Contaminated military airfield Lielvârde Environmental Impact Assessment*" investigating the airfield Lielvârde with the German methodology is completed. The priority of this project was the identification of measures necessary to prevent the further spreading of groundwater pollution. It is the first completed joint project in the Latvian territories formerly occupied by the Soviet Army.

Norway will assist in carrying out geological examination to identify the degree of soil pollution in the territories left by the Russian army. A joint Latvian - Norwegian project "Investigation of Former Soviet Army Bases in Latvia and Identification of Environmental Damage and Problems" is proposed. An important part of the project is the transfer of appropriate philosophy, methodology and technology to Latvian colleagues. The project will be co-ordinated by the Norwegian Defense Research Establishment.

In cooperation with Canada the demonstration project of cleaning up soils polluted with rocket fuel in rocket bases Taši and Barta in Liepaja region was started in 1994 and will be carried out during the next three years. The preliminary analyses of the soil pollution have been carried out. The appropriate soil cleaning technology will be chosen from those applied in Canada, most probably the physico-chemical methods with the application of biological methods in the final phase of cleaning.

This project also includes the training of Latvian specialists in Canada. The total funding for all the project is about one million USD.

Involved in this project are:

- Environment Canada - Emergencies Engineering Division;
- Gartner - Lee International Inc.;
- Canadian-Latvian community;
- Riga Technical University.

The Latvian Ministry of Environmental Protection and Regional Development is strongly supportive of this project.

We have prepared the project proposal together with specialists from United Kingdom "Assistance to Private Farmer in Saldus Region". The principal objective of this project is to provide development, planning and management advice to private farmers who are authorized to resume farming operations on the peripherly of the former Soviet bombing range at Zvarde. The aim of the project is to have a totally positive impact on the environment, and on the social sector.

Twenty four thousand five hundred hectares of farming land and forests at Zvârde, Saldus district, was taken up by military firing range with aviation targets. Mechanical and chemical pollution, such as aviation bomb splinters, and undetonated bombs have defaced the terrain, diffuse contaminants, aircraft fuel, burning wastes, and explosives render the soil unusable. To be able to reclaim the land for cultivation, undetonated bombs have to be disposed of and soil samples analyzed, so that the necessary measures can be identified.

The most serious overall constraints are the financial resources available, the lack of information about possible future work tactics and strategies to be implemented, lack of hardware and software necessary for data processing in general and data basis creation in particular needed help in the development of monitoring systems, at different levels, in collaboration with other interested nations; needed special training courses for our specialists in European countries where laboratories are already working on the determination of soil and groundwater contaminants from an environmental standpoint.

In Latvia, technological resources are scarce. The assistance offered by other countries is of great importance for Latvia.

II.4

COUNTRY REPORT FOR LITHUANIA ON SOIL POLLUTION AND PROSPECTS FOR ITS PURIFICATION

J. ŠIVICKIS

Head of Soil Conservation Division

Ministry of Environmental Protection of the Republic of Lithuania,

Juozapaviciaus str. 9, 2600 Vilnius, Lithuania

Background

Large areas in Lithuania are impermissibly contaminated with oil products and propellants as a result of the dislocation of military bases of the former Soviet Union. There were 27 such bases in Lithuania and they occupied more than 1% of the country's territory. Cleaning them to the permitted contamination levels requires huge funds (720 m. USD) not available in Lithuania.

In 954 storage locations of the country, about 2,200 t. of pesticides that are unsuitable and prohibited from use are accumulated. These pesticides must be immediately utilized ir destroyed because cases of fire are frequent in such storage. There are large quantities of contaminated ground in the territories of these storage facilities. Investigation and cleaning of these territories also requires considerable investment.

Purification of contaminated ground to a large extent has been started in Lithuania just in 1995. There is only one purification site so far, located near the city of Klaipeda.

E.A. McBean et al. (eds.) Remediation of Soil and Groundwater, 79–85.

The biological method of purification with the help of bacteria is employed. It is considered that there must be three more such sites in Lithuania.

Legislative and normative bases for the monitoring and purification of contaminated ground is still not sufficient. Therefore the Republic of Lithuania is in need of financial, technical, and organizational assistance of economically stronger states having experience in the cleaning of soil and ground water.

The Republic of Lithuania occupies an area of 65.6×10^3 km^2, with a population of 3.7 million, or 56 people per sq. km. The largest Lithuanian cities are Vilnius (600 ths. people), Kaunas (400 ths. people), Klaipeda (200 ths. people), Siauliai (150 ths. people), and Panevezys ((130 ths. people). These cities are the largest industrial centers and, at the same time, the main polluters of soil and groundwater. Along with these industrial centers, Mazeikiai Oil Refinery, Akmene Cement Plant, Jonava and Kedainiai Fertilizer Plants, as well as road and railway transport and agricultural enterprises are among the most significant polluters.

It should be noted that the extent of pollution caused by industry and agriculture has decreased considerably since 1990 because, after the fall of the Soviet empire, unilateral economic links oriented toward the East were disrupted thus decreasing the negative effect upon the development of the Lithuanian economy. Today the Government of the Republic of Lithuania, businessmen, and industrialists are attempting to develop economic links in all directions, with Western countries in particular; however, this process is difficult and it will take much time before the Lithuanian economy has recovered.

The amount of fertilizers and pesticides used in agriculture is now one-tenth of those figures of 1988. The volume of production has decreased several times. Therefore current soil, water, and air pollution levels are considerably lower. However, road and railway transport pollutes roadside areas, railway embankments, and bus/railway station territories. The pollution of Pauoste railway station near Klaipeda, the Lithuanian port through which up to 10 million tons of oil and oil products are carried every year, is especially great. Here oil products just lay spilled on the ground. Oil is brought to Klaipeda in tank cars that frequently are not in proper order. The territory of Naftos Terminalas (Oil Terminal) company in Klaipeda, where oil is pumped from tank cars into tank vessels, is extensively polluted.

In earlier years in Lithuania, as well as in the former Soviet Union as a whole, attention previously was focused on water and air quality, leaving the problems of soil and groundwater pollution unaddressed. This determined the above-described situation in Klaipeda and other places. For instance, pollution of about 110×10^3 m^3 of soil and ground reaches 5,000-10,000 mg/kg in the territory of the Naftos Terminalas company, while in Pauoste railway station such pollution level is observed in about 22,000 m^3 of ground.

There are cases when soil and deeper ground layers (and sometimes ground water) are polluted as a result of various technological processes or emergencies, when oil products are spilled from tanks or pipelines.

No mechanism has been worked out in Lithuania so far for the requirement for plants, enterprises and firms to examine soil and ground pollution levels and to remediate polluted areas in the situations of excess concentrations. Therefore, polluted soil and ground are being isolated or cleaned only when the pollution is absolutely obvious (emergency, visually noticeable pollution of territory) or when corresponding requirements are set during the working out of construction or reconstruction designs (designs for the reconstruction of the Naftos Terminalas company and Pauoste railway station).

Until now, in emergency cases (shipwrecks, overturns of tank cars, bursting of oil product pipelines) polluted soil was manually collected and brought away to exhausted mineral resource pits for storage. For example, after the wreck of a tank vessel GLOBE ASSIMI, the mixture of sand and fuel oil collected from the beach was brought to the pit in Sventoji forest. However, this does not provide a guarantee that polluted soil so located will not do further harm to the environment (filtration of oil products to underground water; hazardous dust, vapor, etc.). Therefore it has been finally concluded that soil contaminated with oil products must be cleaned despite the high cost of this work.

The above-mentioned Naftos Terminalas Company in Klaipeda had to solve the problem of cleaning the contaminated ground; facilities for the purification of contaminated soil and water were constructed near Klaipeda by this Company. These facilities were put into operation in 1995. Biological methods are being employed for the cleaning of soil and water, using special bacteria. Contaminated soil is stored in an open concrete site, with water containing the said microorganisms poured on and regular ploughing applied. It is anticipated that the microorganisms will do their work before winter and pollutants will be eliminated to the permitted level, or 2,000 mg of oil products per kg of soil. Such conventional cleaning has been established on the condition that such soil will be used for the construction of the second line of these facilities. Monitoring will be carried out to establish whether the soil purified to the 2,000 mg/kg level will not pollute ground water and surrounding territories.

Three or four similar sites for the purification of contaminated soil to meet the needs of the country are in the planning stages in Lithuania. However, the possibilities of realization of these plans are quite limited because the economic situation of the Republic is complicated and the problem of financing such construction projects is difficult. Priority is usually given to sewage purification facilities, since the damage done to people's health and environment by contaminated water is more noticeable. Construction of the soil purification facilities of Naftos Terminalas Company cost about 3 million USD. Finding such amounts of money is a very complicated task for an economically-backward country. Soil purification is not financed on the state scale; environmental protection services just exert "pressure" on the enterprises that have polluted soil and ground water so that they bring such substances to purification sites and pay themselves for the purification. At present, the current price for the purification of 1 m^3 of contaminated soil is 66 USD.

Additional strategies for the monitoring of soil and groundwater pollution and purification work in the Republic of Lithuania are as follows:

1. Rules and standards regulating pollution levels are in preparation and soon will be completed.

2. Plants, enterprises, and firms - potential polluters of soil and groundwater - are forced to carry out investigations of pollution parameters (composition, concentration, area and depth of occurrence of pollution), i.e. engage in the monitoring of pollution.

3. Such polluters are forced to purify the soil and groundwater of excess pollution.

In the Republic of Lithuania, 277 Soviet military bases had occupied more than 1% of the country's territory, i.e. 67,000 ha. Soil and ground water in a large part of this area, particularly in the former oil product and propellant storage places, aerodromes, and tankodromes are highly contaminated with oil products and propellants. The inventory of such territories has been made. It has been established that the monetary value of the overall ecological damage done to Lithuania by the military activities of the USSR-the Russian Federation in its territory amounts to 1,728,000,000 USD, including 720,000,000 USD for the cleaning work. Compensation for this damage is the subject of interstate negotiations. Specific cleaning works are planned to be started for the former Zokniai military aerodrome where the level of ground and soil water contamination is especially high. The level of pollution may also be illustrated by the following half-humorous fact: just after the withdrawal of the Soviet army from this aerodrome, local people used to dig pits in this territory and scoop up to 1 ton of pure kerosene in one day. In these territories formerly occupied by the Soviet Army, the young Lithuanian state is to carry out large-volume and expensive cleaning work. The probability that the negotiations with Russia will result in the covering of cleaning costs is very small, and Lithuania is not capable of raising such funds due to its present economic situation.

Lithuania faces also another problem, namely, utilization of pesticides that are unsuitable and prohibited from use and purification of ground in the territories of their storage. In 954 storages of former collective farms, about 2,200 tons of such pesticides have been accumulated. Twenty cases of spontaneous fires have been recorded in such storage facilities. The ground in the territories of such storage facilities is especially contaminated with pesticides. However, it has been established during preliminary investigations that the ground in the territories of other storage facilities and near them is also extensively polluted too. A question may naturally arise why such large quantities of pesticides are accumulated in the country? This fact may be briefly explained as follows. As we know, Soviet Union was a planned economy state. Each republic was allotted, in a compulsory order, considerable quantities of pesticides irrespective of the structure of crops, natural conditions and other factors. The chemical industry had to fulfill its plans. The necessary amount of better pesticides were utilized while the rest was accumulated in storage. After the fall of the collective-farm system, most of the stores became unnecessary because farmers buy pesticides in small quantities for immediate use.

A decision was adopted by the Government to burn pesticides that are unsuitable and prohibited from use in the Akmene Cement Plant (powdered pesticides) and in Kedainiai Fertilizer Plant (liquid pesticides), although the public of Akmene and Kedainiai districts has voiced strong protest. Lack of funds prevents us from acquiring mobile pesticide burning facilities.

Lithuania lacks experience in the field of purification of ground in the territories of pesticide storage. It is believed that biological cleaning by composting should be the cheapest and most efficient method. However, we are not sure this is the situation.

To end this report several conclusions are appropriate:

1. The work of cleaning contaminated soil and ground water has been just started in Lithuania, therefore we lack experience in technological matters.

2. Due to the difficult economic situation of the country, ground purification and remediation cannot be carried out at a fast pace.

3. Financial, technical and other assistance in these matters from economically stronger and more experienced countries would be very much appreciated.

II.5

SOIL POLLUTION MANAGEMENT IN HUNGARY

G. VÁRALLYAY

Director

Research Institute for Soil Science and Agricultural Chemistry

of the Hungarian Academy of Sciences,

H-1022 Budapest, Herman O. 15. Hungary

Introduction

The accumulation of pollutants in soils, sediments and ground water due to natural factors, or as a consequence of human activities, represent serious environmental problems in some parts of Hungary. The sudden, sometimes surprising, mobilization of the (temporarily) immobile pollutants as a consequence of changes in soil properties is sometimes even more harmful (Bulla [3], Hinricksen & Enyedi [9], Kádár [11], The National Atlas of Hungary [14], National Review [15], Várallyay [28,29], Chemical Pollution [4]). It is a typical case of the Chemical Time Bomb (CTB) problem, which refers to a "chain of events, resulting in the delayed and sudden occurrence of harmful effects due to the mobilization of chemicals stored in soils and sediments in response to slow alterations of the environment" (Stigliani et al. [19]).

E.A. McBean et al. (eds.) Remediation of Soil and Groundwater, 87–112.

Contaminants

Most of the elements occurring on Earth can be found in the soil. Their quantity, quality, solubility, mobility and availability for microorganisms, plants, animals and human-beings show an extremely wide spectra. Most of these elements are essential for the living organisms, but over a certain "threshold concentration" - a great part of the same elements can be harmful, or even "toxic" for the same organisms (Stigliani et al. [20], Adriano [1], Salomons [17], Chemical Pollution [4]).

For the evaluation of the status and regime of these elements in soil the following steps are necessary (Figure 1):

(a) "Total content". The permissible total quantity of various (potentially toxic) elements in soils are summarized in Table 1, according to the standards of various countries. The data show large variability in the "total content", which depends on the determination procedure. In Table 1 data are also presented on the potentially toxic element content of various fertilizers and amendments used in agriculture. On this basis the yearly "critical loads" can be calculated, which are also included in Table 1.

(b) "Soluble content". This quantity is highly solute-specific. The "soluble content" depends on the characteristics of the given compound (e.g. solubility, electro-negativity, polarizability, rate of oxidation, ability of complex formation) and on the soil properties (e.g. soils reaction and carbonate status, texture, clay content, clay mineral association, organic matter content and quality, absorption capacity, base saturation, exchangeable cation composition, moisture content, redox potential, microbial activity, etc.). The "in situ" soluble content depends to a great extent on plant-root relationships.

Non-soluble compounds are:

- immobile: cannot be transported by the liquid flow with the result that no (or limited) possibilities of contaminating surface and subsurface water resources occur;
- not available for plants; consequently,
- not toxic for plants, animals and human-beings.

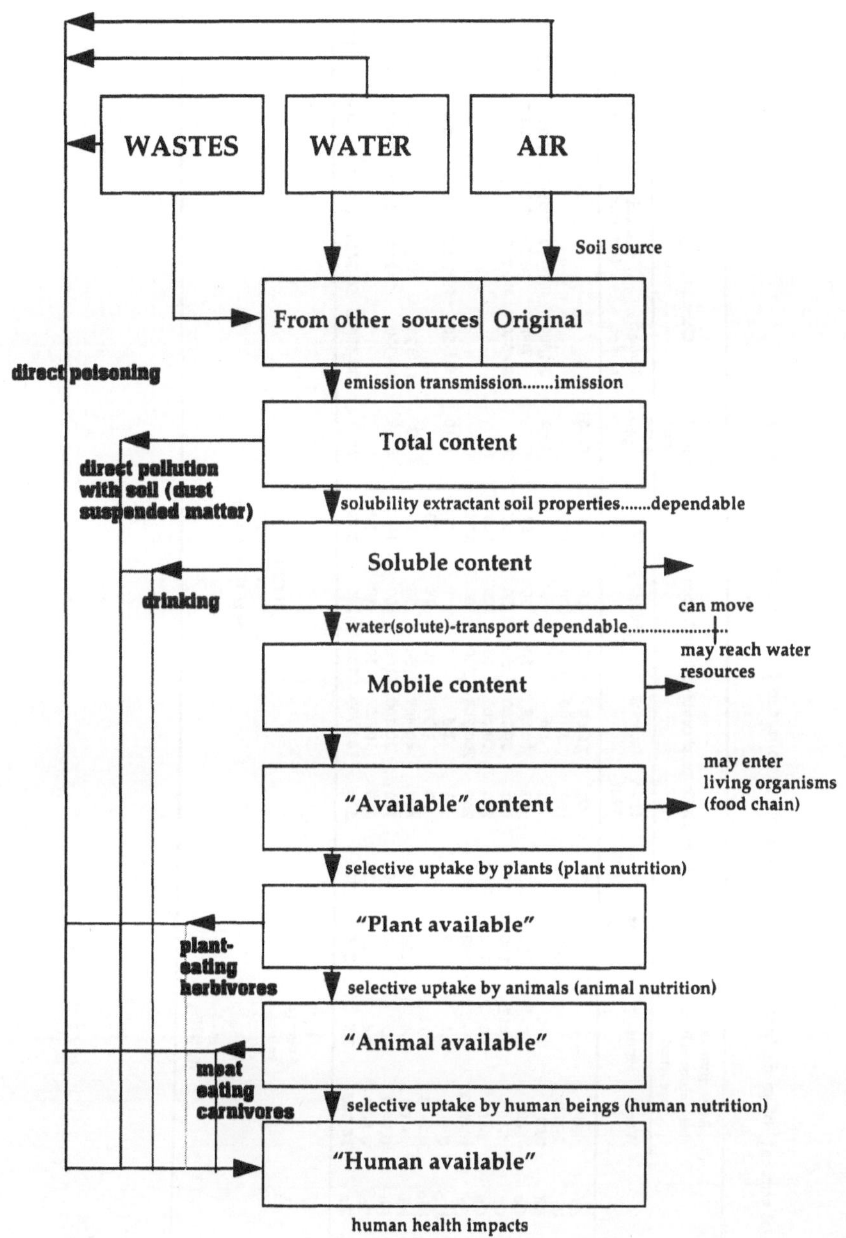

Figure 1. Sources and pathways of soil pollution

Table 1. "Total content" of some "potentially toxic" elements in soils, fertilizers and other amendements

Element		Minimum and maximum values occurred		Permissible 'total content' in soil, ppm							Critical load per year kg/ha	Content in ppm/dry matter content				
		World	Hungary	Tietjan 1983	German 1977	USA 1981	Interval	Hungary 1988				Sewage sludge	P fertilizer	N fertilizer	Lime-stone	Organic manure
Arsenic	As	1-50		15	20	60	15-60	7	10	15	0.30					
Boron	B	2-100		100	25	100	25-100	100	100	100						
Cadmium	Cd	0.01-1	0.10	5	5	6	1-6	1	2	2	0.02	2-1500	0.1-140	0.05-8.5	0.04-0.1	0.3-0.8
Cobalt	Co	1-50	2.8	50	50	80	10-80	50	50	50		2-260	1-12	5.4-12	0.4-3	0.3-24
Cromium	Cr	1-100	0.012	100	100	100	50-200	50	50	50	10	20-40000	65-245	3.2-20	10-15	6-55
Copper	Cu	2-100	7.0	100	100	200	20-200	75	100	100	10					
Iron	Fe	10-500		500	200	200	200-500	500	500	500						
Mercury	Hg	0.01-1	0.1	2	5	3	2-5	1	1	1	0.10	0.1-50	0.01-1.2	0.3-3	0.05	0.1-0.2
Molybdenum	Mo	0.2-10	0.11	10	5	20	1-30	10	10	10						
Nickel	Ni	1-100	4.4	100	50	100	40-400	50	50	50	2.0	16-5300	7-38	7-34	10-20	7.7-30
Led	Pb	0.1-10	6.5	100	100	100	20-100	100	100	100	20	50-3000	7-220	2-2.7	20-1200	6.6-15
Selenium	Se	0.1-10	0.4	10	10	20	10-20	10	10	10						
Zinc	Zn	10-300	7.5	300	300	300	100-700	200	250	300	20	700-49000	50-1450	1.0-42	10-450	15-250
			↓ Average values, deter-mined in n HNO_3					If CEC < 15	15-25	> 25 meg/100 g soil						

The solubility of various compounds shows great variability, depending on the above-listed influencing factors. This is the reason of the "Chemical Time Bomb" effect. This means that the concentration of various elements may exceed the permissible (tolerable) concentration in soil and may become "toxic" without any additional load of this element to soil, as a consequence of changing soil properties. For example, increasing acidity helps the mobilization of numerous "potentially toxic" elements (most of the heavy metals, etc.), and consequently the previously non-soluble, immobile, non-toxic element can become soluble, mobile and toxic due to acidification (Stigliani et al. [19], Chemical Pollution [4], Szabolcs [22]). With the knowledge of the existing pH-solubility relationships of the various compounds the "toxic element mobilization hazard" of soil acidification can be predicted and prevented in due time by using proper neutralization measures (e.g. liming).

(c) "Mobile content". Only the mobile fraction of the various elements (compounds) are reactive and can be transported and translocated to other places (leaching, accumulation, migration), and may reach water resources and plant root surfaces.

(d) "Plant available content". In addition to the total, soluble and mobile content of a certain element in the soil, its availability for a certain plant depends on its selective ion-uptake, which is a plant (species, sort, variety, type) dependent characteristic. Only the available fraction of a given element can be phytotoxic for plants, resulting harmful (or even lethal) changes in their metabolisms.

Non-plant available elements cannot enter the "food chain", and, consequently, cannot be toxic either for animals nor for humans. An exception is when the potentially toxic elements are not taken up by plants, but deposited and accumulated on the surface of plants (e.g. Pb on leaves, Cr and Hg on roots) and may enter the plant-eating organisms by such a mechanism.

Some plants take up, translocate and in their organs accumulate some elements which are not phytotoxic (e.g. Cd and Al accumulation in the leaves of spinach and lettuce), but can be toxic for plant-eating animals or humans.

(e) "Animal available content". The potentially toxic elements taken up by plants, or deposited to plant surfaces may enter the plant-eating (herbivore, vegetarian) organisms and may result in serious health problems or even death. In some cases the element is non-toxic (or at least not lethal) for plant-eaters, but after a quick or slow translocation it accumulates in their particular organs (e.g. in fish liver) and may cause health problems when carnivores eat that organ.

(f) "Human available content". Human-toxic elements may enter the man's organism either directly (by air, water or dust), or from plant surfaces, or through the food-chain (food prepared from plants or animals) and may result in various health problems, in some cases serious deteriorations or even death.

Consequently, the term "toxic" is not precise and not sufficiently specific. A given element over its critical concentration can be toxic for certain soils (or more exactly for their biota), for certain plants, for certain animals, and for certain people (Chemical Pollution [4], Adriano [1], Polunin & Burnett [16], Stigliani et al. [20]). The identification of these relationships is a challenging task for future multidisciplinary research. Their final result can be an imaginative "super-matrix" showing the potential "toxicity-pathways" under the given circumstances (from the total content, through solubility and mobility, up to the availability for various plants, animals and human-beings), and indicating the potential bio-deterioration, their symptoms and further consequences.

According to these criteria special attention has to be paid to the following chemicals: heavy metals; some other elements, like Al, As and F; water soluble salts; halogenated persistent organic compounds, such as pesticides, PCB's; organo-metallic compounds; oil products; nitrate, phosphate, and to a lesser extent, potassium; sulfur, NO_x and NH_3; and radionuclides.

Sources of Contaminants

The occurrence and accumulation of these elements can be due to natural sources, as:

- air (NO_x, SO_x, etc.) via wet and dry atmospheric deposition;
- water (B, Na, N, etc.) via irrigation water or groundwater;
- soil and geological deposits (P, K, Ca, Mg, Na, Fe, Al, Mn, As, Co, Cu, Ni, Se, Zn, etc.) via local weathering and soil formation processes;

or it can be the consequence of various human activities which can be defined in five major areas:

Activity	*Examples of derived pollutants*
Industrial Activities	- Zn, Cd, Pb, Hg, Fe, Mn, Cr, Ni, Cu from metallurgy
	- Pb, Al, Ga in mining waste
	- Pb, Fe, Zn, Ni, PCB's, oil, phenols and related substances on chemical waste dumps
	- Pb, Cd, oil, phenols from fuel combustion
	- Ga, Se from micro-electronics
Energy use	- Dioxins, halogenated organic compounds and related substances, NH_3, sulfur, NO_x, dust, some heavy metals from power plants
	- Radionuclides from spills by nuclear power plants
	- Radioactive fly-ash (containing Rd) after burning of coal, which is used in bricks (construction industry)
Agricultural Activities	- Cd as an impurity in phosphate fertilizer
	- P_2O_5, NO_x, K_2O in fertilizers
	- Hg in pesticides (seed production)
	- manure
	- Halogenated organic compounds, PAC's, Cu in pesticides

	- Salinizing substances (mineralization) from secondary salinization (irrigation)
Urban, domestic and commercial activities	- Dioxins, PAC's, halogenated organic compounds and some heavy metals from waste incinerators
	- NH_3, SO_2, As, dust from local heating
	- Cd, Pb, oil, PAC's from traffic
	- Heavy metals, PAC's, PCB's, halogenated organic compounds from land-fill activities
Military activities	- Cr, Ga, Ti, Se, radionuclides from the weapons industry

In addition to the increasing quantity (accumulation) of these elements the sudden (sometimes surprising) mobilization of the (temporarily) immobile pollutants as a consequence of changes in soil properties (e.g. soil acidification, salinization/alkalization, destruction of soil structure and clay minerals, decrease of organic matter content and buffer capacity, limitations in the "filter function" of the soil, etc.) is even more harmful. This time-delayed effect of potentially harmful chemical compounds is the typical "chemical time bomb" (CTB) problem (Stigliani et al. [19], Chemical Time Bombs [5, 6]).

Both accumulation and mobilization of these elements and compounds represent a serious environmental problem in numerous problem areas ("hot spots") in Hungary (Chemical Pollution [4], Molnár et al. [13], Welte & Szabolcs [33], Hinrichsen & Enyedy [9], Várallyay et al.[32]).

Pathways

The major pathways of input of chemicals in soils are as follows:

- Atmospheric deposition (wet & dry) (Downing et al. [7]);
- Vertical transport through soils to deeper layers;
- Inputs from lateral surface runoff and subsurface flow (seepage in the unsaturated zone) within soils (Várallyay, [23]);
- Inputs from groundwater (waterlogging, rising of water table), horizontal flow of groundwater) (Hinrichsen & Enyedy [9]);
- Abiotic and biotic transformation within the soil (Polunin & Burnett [16]);
- Direct inputs (from human activities, like agriculture, energy use, industrial, urban, domestic, commercial and military activities) (Welte & Szabolcs, [33]).

For atmospheric transport it is definitely true that point sources (emissions of NO_X, SO_X and other acidifying compounds from coal-, lignite- and oil-based power stations, local heating systems, etc.) produce acidification and heavy metal mobilization and that there is a problem of transboundary air pollution in Europe (Downing et al., [7]). There is also the question of local fall out from point sources (but in higher concentrations), which is in effect in the local regions (Várallyay et al. [32]).

Water transport is another problem, occurring within the various catchment areas. For example, in the Danube there apparently is much less downstream sedimentation (in the Delta Area) as might be expected, because of the construction of dams both within the tributaries and along the main stretch of the river (Stigliani et al. [20], Salomons [17], Várallyay et al. [32]). Runoff and erosion are a very important source for chemical pollution. This trend is still increasing, because of land use changes. Such a trend can affect the flow of pollutants from land to water, which problem can be solved with the application of the CTB concept.

Soil transport is also an important source of chemical contamination and potential time bombs. The vertical transport down into groundwater through percolation and the vertical transport from the groundwater to the overlying horizons by capillary action are important factors of pollutant migration as well.

Ecological/Economic Consequences

The most significant consequences of soil pollution are summarized below:

Ecological consequences	*Economic consequences*
Impact on soil (processes, properties, fertility and other soil functions, such as storage, buffer and filter function);	Higher costs of air, water and soil pollution control;
Impact on microorganisms (species spectra, diversity, activity);	Necessity of expensive complex technologies for the improvement of contaminated soil and water resources;
Impact on plants (natural vegetation and crops);	Lower biomass production;
Impact on animals and human beings (direct influences and indirect consequences through the food-chain)	Lower efficiency of applied fertilizers and pesticides.
Unfavorable influences on the "quality of life" (eg. air, drinking water supply, food quality).	

Influence of Land Use and Soil Management on Environmental Pollution

Before World War II the plant nutrient status of soils in Hungary was rather poor, due to the negative nutrient balance: more nutrients were taken up by the cultivated crops and were taken away from a given territory as yield (or biomass) than was being put back in the form of organic and green manures or fertilizers (Várallyay [28, 29], Várallyay et al. [31]).

From 1955 there was a rapid increase in fertilizer consumption. This tendency was one of the reasons of the substantial yield increase during the same period. Another consequence was that - due to the positive nutrient balance - the nutrient status of soils was significantly improved (Várallyay et al. [31], Várallyay [26]).

Despite these developments there were serious problems and inadequacies in the fertilizer application technology (improper N-P-K ratio; lack of Ca, Mg and micro-nutrient supply; limited variety of fertilizers; problems with their storage, time of application, way of distribution; etc.). The main problem, however, was an unfavorable "polarization" tendency in fertilizer application:

(a) better soil, rich farm, higher rate of fertilizer application (in spite of the lower requirements, better nutrient status of soils) , overdosage;

(b) poor soils, poor farms, lower rate of fertilizer application (in spite of the higher requirements, lower nutrient supply of soils), underdosage.

These inadequacies resulted in *environmental side-effects*, like:

- soil acidification (due to non-adequate type of fertilizer, lack of simultaneous lime application) and its consequences: mobilization of toxic elements ("chemical time bomb effect"), fixation of some of the nutritive elements, etc.;

- load of surface waters by P compounds (mainly due to surface runoff, lateral erosion and sediment transport);
- contamination of subsurface drinking water resources by nitrates (leaching);
- accumulation of harmful toxic elements in the various stages of the "food chain": in soils, plants, animals and human organs, according to their solubility, mobility and availability (Hinrichsen & Enyedy [9], Várallyay [26]).

Most of these side-effects, however, are not inevitable and uncontrollable consequences of fertilizer application, they can be prevented, or at least reduced, efficiently by precision nutrient management, based on the nutrient requirements and nutrient uptake dynamism of cultivated crops (the specific requirements of species, variety or even genotype); the nutrient status and other properties of soils; the characteristics of agroclimate and hydrology conditions of the given landsite.

All of these factors were taken into consideration in the development of the new Hungarian plant nutrition advisory system which is efficiently used on several hundreds of hectares year by year (Várallyay et al. [31]). The impacts of crop production on surface and subsurface water resources can be summarized, as follows:

(a) ***Soil erosion*** by water results in considerable soil losses in the undulating hilly regions, which means considerable losses in organic matter and plant nutrients as well. The other unfavorable consequence of soil erosion is sedimentation; silting up of waterways, canals and reservoirs (limitations in their functions; necessity of their more frequent cleaning increasing costs); and increasing hazards of waterlogging and floods in the lower parts of the watershed (Várallyay, [23]).

(b) ***P fertilization***. Because most of the P compounds have low water solubility their liquid transport and leaching is negligible (is limited to some centimeter distances). But adsorbed (fixed) P, insoluble P compounds and sometimes P fertilizer particles can be transported by surface runoff directly to surface waters. Their high P concentration may result in increasing eutrophication and its undesirable consequences: rapid silting up of canals and reservoirs and unfavorable changes in the aquatic ecosystems of shallow lakes (e.g. Lake Balaton in Hungary: recreation problems, fish disease).

(c) ***K fertilization***. Most of the potassium fertilizers are highly soluble and can be leached from the profile of light-textured soils. In heavy soils, the greater part of soil-K is fixed not only on the clay surfaces but within the lattice structure of the swelling clay minerals. Potassium has a negligible importance from the viewpoint of environmental pollution.

(d) ***N fertilization.*** The nitrate pollution of subsurface waters is one of the most important environmental problems in many countries. In Hungary more than 300 villages are supplied with bottled water because the nitrate concentration of the drinking water supplies exceeds the permissible limit. The potential sources of these high N concentrations can be the following factors:

i) Liquid manure from large, concentrated livestock farms, resulting sometimes in considerable point source N pollution of subsurface waters.

ii) Sewage waters, sewage sludges and solid wastes as a result of industrial, urban and rural development. In many settlements drinking water supply was introduced without the simultaneous establishment of canalization. This resulted in rising water table ("groundwater hills" below these villages) and an increasing hazard of N contamination of the groundwaters.

iii) Recreation and tourism, without appropriate waste water management

iv) Illegal local sources (e.g. use of "old" wells for waste disposal, etc.).

v) Irrational N fertilizer application.

Rational N fertilization cannot cause a significant N pollution, because, if we use the necessary amount of N - according to the crop's requirement - then N losses (sources of N pollution) can be efficiently reduced to a minimum level. What are the main possibilities of the N pollution of groundwaters due to N fertilization?:

- leaching of N through preferential pathways, such as cracks and biological channels (roots, earthworm channels);
- uncontrolled N application in "hobby gardens";
- improper fertilizer application, non-adequately selected for the crop requirement (nutrient uptake), soil properties and weather conditions; problems in uniform distribution, or differential distribution according to the N status of the soil; time of application; etc.

Any improvement in the technology of N fertilizer application will result in the reduction of losses (higher efficiency) and environmental hazards.

(e) ***Water soluble salts.*** Leaching of Na salts from the soil profile is favorable for the given soil (decreasing salinity), but increases the salt concentration in the drainage water. Consequently, this water cannot be used for irrigation again, and can be drained to international waterways only up to a certain quality limit prescribed by international agreements. In addition to other facts (high clay and swelling clay content, Na_2CO_3-$NaHCO_3$-type salinity high alkalinity high ESP very low permeability of the soil; lack of frost-free period after the growing season; lack of good quality water) this is the main reason why we cannot use the traditional leaching-drainage concept for salinity-alkalinity control and the only way for that is a well-functioning prediction and prevention system (Szabolcs [21]).

(f) ***Pesticides and other organic chemicals.*** Many countries from the region take part in the various international programs on these subjects ("Mapping of critical loads" (Downing et al.[7]), "Chemical Time Bomb" (Stigliani et al. [19]); "Vulnerability of soils to organic pollutants") and these research is the focus of our future scientific activities.

Susceptibility of Soils to Environmental Pollution

The susceptibility (vulnerability) of soils to environmental pollution is determined primarily by the following soil functions (Várallyay [27]):

- Soil is reactor, transformator and integrator of the combined influences of other natural resources (solar radiation, atmosphere, surface and subsurface waters, biological resources).
- Soil is an efficient "natural filter" and may prevent the deeper horizons and the subsurface waters from various pollutants, deposited on the soil surface or put into the soil.
- Soils represent a high capacity buffer media of the biosphere, which (to a certain limit) may buffer the various stresses caused by environmental factors and/or human activities representing a sharply increasing ecological threat to the biosphere.

The ability of soil to fulfill these functions ("soil value") is determined by the integrated impacts of various soil properties, which are the results of soil processes: mass and energy regimes, abiotic and biotic transport and transformation and their interactions. As any soil-related human activity influences the soil through these processes, consequently their control is the main task of soil science and soil management (Greenland & Szabolcs [8], Várallyay [24, 25], Várallyay et al. [32]).

The conceptual "flow-chart" of an efficient strategy for the control of soil processes (indicating the logical sequence of the necessary consecutive steps) is presented in Figure 2..

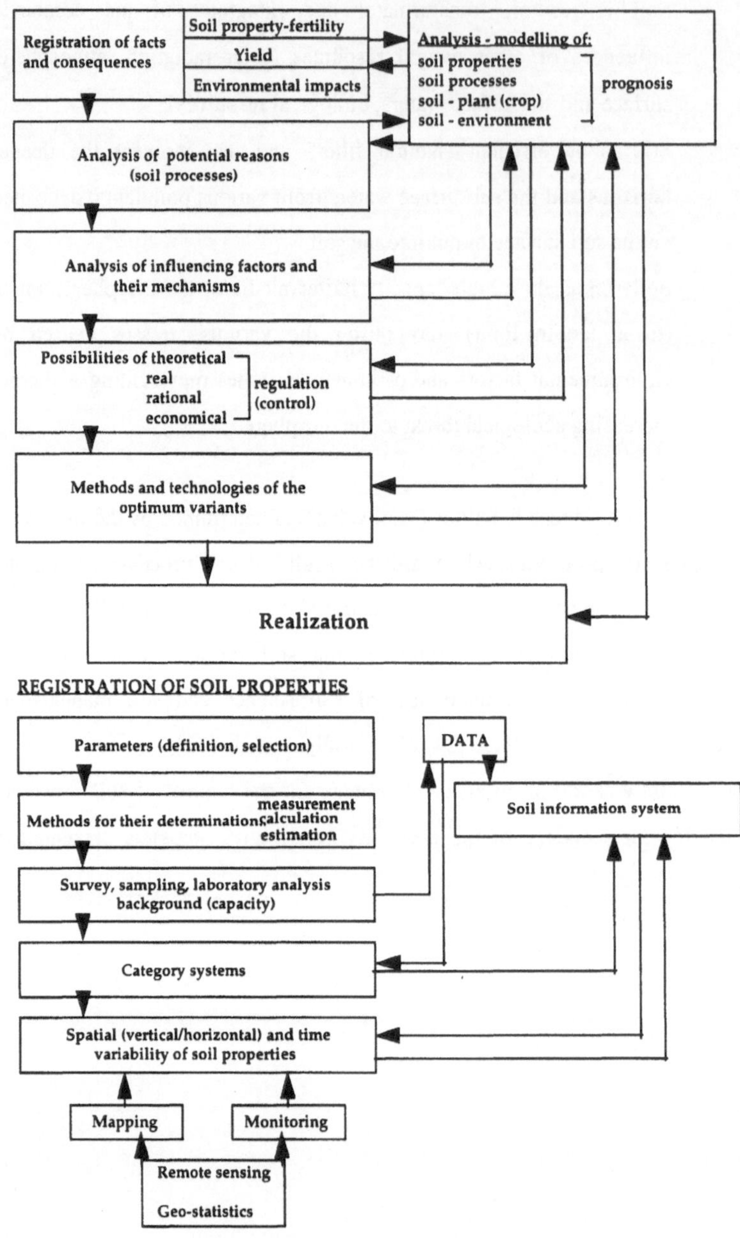

Figure 2. Control of soil processes

From the viewpoint of soil pollution, special attention has to be paid to the definition, exact description and quantification of the following *soil processes*:

- solute transport;
- (mainly) abiotic transformation, such as: dissolution - precipitation; adsorption-desorption; oxidation - reduction; acidification - alkalization; mobilization - immobilization; release - fixation;
- (mainly) biotic transformation, such as: decomposition of organic matter and humus formation; biotic mobilization - immobilization.
- solid - liquid - gaseous phase interaction and the following storage-, filter- and buffer- *capacity controlling soil properties* (CCPs): clay content and type of clay minerals; state of soil structure; organic matter content and quality; soil reaction, pH, carbonate status; salinity-alkalinity; redox conditions; absorption capacity, base saturation.

Hazards to the Environment

As we are interested in determining the long-term environmental hazards or risks, the question to be answered is: when and with what intensity (i.e. under what conditions) will soils and sediments start to leak toxic elements into the environment (i.e. the chemical time bomb effect) (Várallyay et al. [32]).

In a classification of hazards the time-aspect and the intensity of leakage of chemicals from the targets should be taken into account. It was recommended that the matrix should be constructed on which the time and intensity-aspect would be coupled to the changes of the physical, chemical and biological parameters within the targets (vulnerability) and to chemical loads into the targets.

To take into account the mentioned time- and intensity aspect one needs to look at the "hot spots" and relatively clean reference areas. Studies should be made in this respect to "hot spot" areas where (retrospective) data on chemical loads are available as well as information on the kind of soil, groundwater or sediment concerned. This is important to answer the question: "when did the polluted target start to release toxic elements into the environment" (Time aspect). Furthermore, studies will be needed of relatively clean (reference) areas of the same kind of soil, groundwater or sediment as the "hot spot" where the question: "does this kind of target have a low or high storage capacity for toxic chemicals" has to be answered (intensity aspect).

In determining the classification of environmental hazards for *a certain kind of target (soil, sediment, groundwater) and for a certain kind of pollution in a certain area a*

- First matrix should be made on the time and intensity aspects of considered chemicals leaking from the considered target into the environment. We can do this with the help of studies on "hot spot" and "clean" reference areas.
- Next a matrix on the vulnerability of the target could be superimposed on the top of this.
- On top of these two matrices a matrix on chemical loads could be superimposed for determining the environmental hazard.

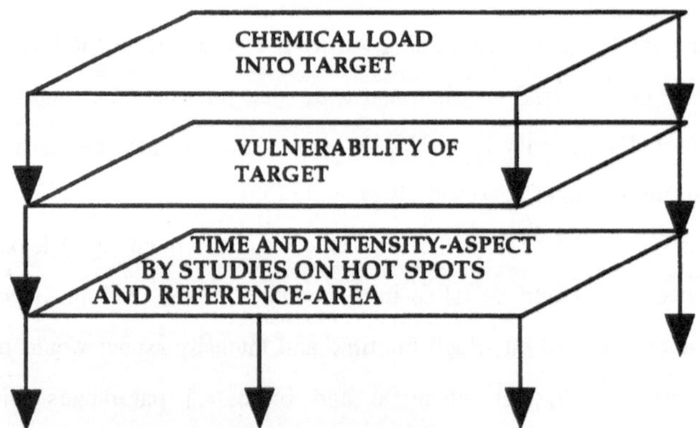

Strategy for Soil Pollution Control

The main potential possibilities of soil pollution control are schematically illustrated by Figure 3. Its main elements are:

- emission/imission reduction (preventing or reducing the quantity of pollutants deposited or transported to the soil surface or into the soil);
- prevention of the mobilization of potentially harmful chemical compounds or elements which are already present in the soil but in temporarily - immobile form;
- decrease of the susceptibility/vulnerability of soil against various pollutants (with the increase of the buffering capacity of soils) which tolerate higher critical load of pollutants, consequently reduce the "exceedance-risk" and the unfavorable ecological consequences.

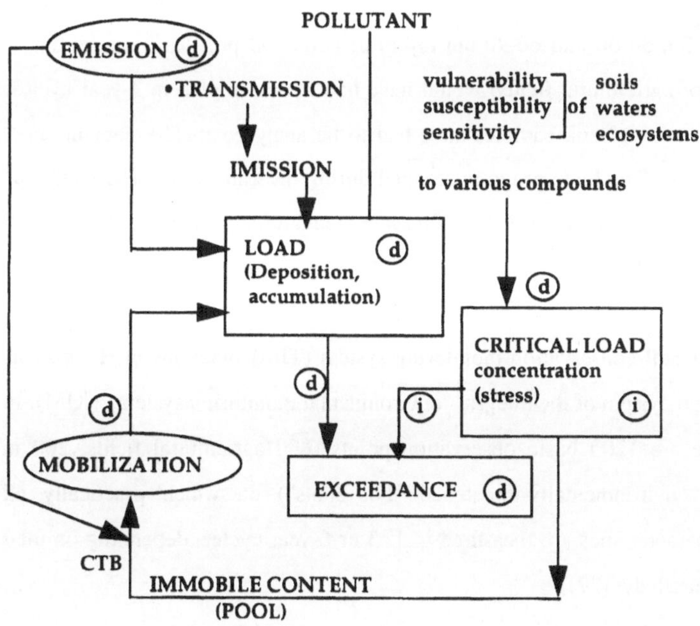

Figure 3. Strategy for pollution control (i: increase; d: decrease)

For the comprehensive assessment of the status and regime of these elements in the soil and for the evaluation of their ecological impacts and environmental hazards

- the character of the contaminants (their total, soluble, mobile, plant-, animal-, and human-available and toxic quantities);
- their sources and pathways (atmospheric wet and dry deposition; vertical movement within the soil; horizontal transport, such as surface runoff, seepage in the unsaturated zone, groundwater flow; abiotic and biotic transformation, etc.) has to be identified and quantified.

In the last years, great efforts have been made in Hungary for such assessments. It is a fact that many of these failed because of financial difficulties and even the operated programs were stopped or radically reduced. For example, the national program started in 1987-1988 for the monitoring of micronutrients and potential pollutants. According to the plan the 0-30, 30-60 and 60-90 cm layers of 6000 soil profiles (representing 5 million hectares of agricultural fields) would have had to be sampled in 3-year cycles, and 1000 "representative" soil samples have had to be analyzed for 20 elements in 5 various soil extracts. The huge program stopped during the 2nd cycle. The results of the first cycle were indicated on 1:2 000 000 scale thematic maps using a 6x6 km grid system (Várallyay [27]).

From 1992 a new soil conservation monitoring system (TIM) functions in Hungary as an independent sub-system of the integral "environmental monitoring system" (KIM). In this system there are 1200 basic observation points (800 agricultural fields, 200 in forests, 200 in environmentally-threatened "hot spots") on which practically all important soil characteristics are measured in 1, 3 or 6-year cycles, depending on their changeability (Várallyay [27]).

During the last years, thematic maps have been prepared in Hungary on the susceptibility/vulnerability of soils against various pollutants (Várallyay [24, 29], Várallyay et al. [30]). Their critical loads were calculated and efficiently used as guidelines for soil pollution control, for non-harmful waste, waste-water and sewage sludge disposal.

In the last years a comprehensive Project Proposal was elaborated and submitted to EC for funding on the "Long-term environmental risks for soils, sediments and groundwaters in the Danube Catchment Area". The framework for the first phase of this Project is shown in Figure 4. It will give a basis for the second phase which will be a complex Environmental Master Plan for the whole region (Várallyay et al. [30]).

Future Tasks

Production is economy driven. In contrast, the maintenance of the multifunctionality of soils, the quality of surface and subsurface water resources and the protection of the natural environment are not (fully) economy-dependent elements of sustainable development, but imperative tasks of the human society. Only their efficient and most economic alternatives (methods, technologies) can be selected on the basis of cost/benefit analysis.

For the efficient realization of the above-summarized elements of sustainable development:

(i) the criteria of sustainable development and production have to be defined and quantified;

(ii) The necessary economical regulations have to be elaborated (such as: tax, price, credit, subsidy regulations, etc.) guaranteeing the fulfillment of these criteria;

(iii) the defined and quantified criteria and the economy regulations have to be formulated in various legal documents (laws and related official documents);

(iv) the potential possibilities and efficient ways (methods) of sustainable production have to be elaborated, adopted, published and demonstrated, which needs the establishment of appropriate mechanisms for research, training and education, demonstration, extension and advisory service;

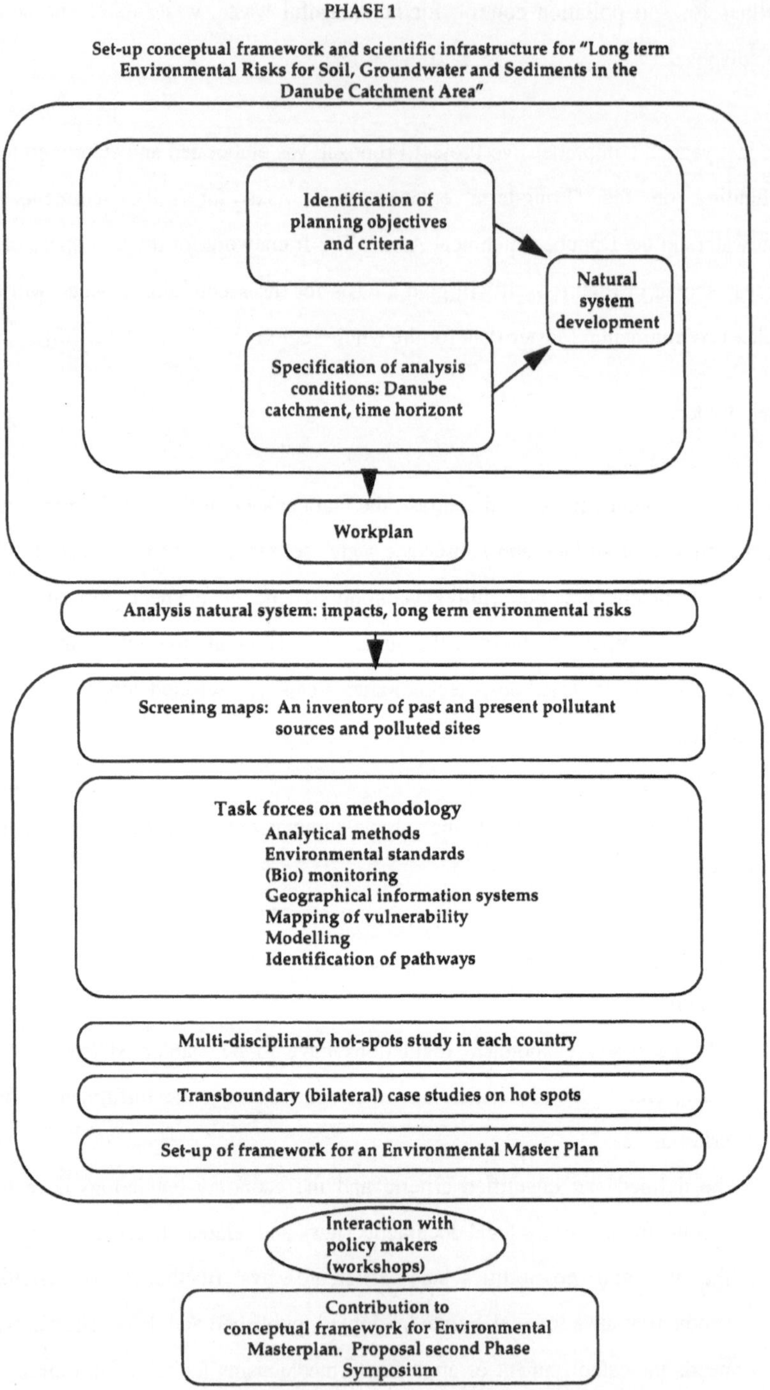

PHASE 1

Set-up conceptual framework and scientific infrastructure for "Long term Environmental Risks for Soil, Groundwater and Sediments in the Danube Catchment Area"

Identification of planning objectives and criteria

Natural system development

Specification of analysis conditions: Danube catchment, time horizont

Workplan

Analysis natural system: impacts, long term environmental risks

Screening maps: An inventory of past and present pollutant sources and polluted sites

Task forces on methodology
- **Analytical methods**
- **Environmental standards**
- **(Bio) monitoring**
- **Geographical information systems**
- **Mapping of vulnerability**
- **Modelling**
- **Identification of pathways**

Multi-disciplinary hot-spots study in each country

Transboundary (bilateral) case studies on hot spots

Set-up of framework for an Environmental Master Plan

Interaction with policy makers (workshops)

Contribution to conceptual framework for Environmental Masterplan. Proposal second Phase Symposium

Figure 4. Framework for first phase for the project "Long-term Environmental Risks for Soils, Groundwater and Sediments in the Danube Catchment Area"

(v) the necessary mechanisms for continuous control have to be built up;

(vi) environment-friendly, long-term society moral has to be developed on each level (global, continental, regional, national, sub-regional, local).

The efficient, scientifically-based strategy for soil pollution control includes simultaneously the emission/imission reduction; the increase of the buffering capacity of soils decreasing their "vulnerability"; and the prevention of CTB effects. All of the related research - education - implementation - legislation activities will be summarized in a comprehensive Environmental Master Plan, within the scope of the National Environment Protection Program (Agro-21 [2], Hungary.... [10]).

These actions are joint tasks of the state, decision makers on various levels, the owners, the users, and - to a certain extent - of each member of the society. Only their joint efforts can be efficient towards a rationally privatized, market-oriented, sustainable production harmonized with successful environment protection, ensuring a pleasant environment and a promising future.

References

1. Adriano, O. C. (1986) Trace Elements in the Terrestrial Environment. Springer Verlag. New York - Berlin - Heidelberg - Tokyo.
2. AGRO-21. (1995) Future View of the Agriculture. "AGRO-21" Brochures. Budapest. 10. 5-26.
3. Bulla, M. (1989) (Ed.) [Studies on the state of the Hungarian environment.] (H). Min. of Envir. and Reg. Policy. Budapest. 1-176.
4. Chemical Pollution: A Global Overview. 1992. Earthwatch, UNEP, Geneva.
5. Chemical Time Bombs, 1990. Report of an European Workshop, De Bilt, Utrecht, June 21-23. 1990.
6. Chemical Time Bombs, 1992. Proceedings of the European State-of-the-Art Conference on Delayed Effects of Chemicals in Soils and Sediments, Veldhoven, 2-5 September 1992, the Foundation for Ecodevelopment, Hoofddorp.

7. Downing, R. J., Hettelingh, J. P. and Smet, P. A. M. de (1993) Calculation and Mapping of Critical Loads in Europe. Status Report 1993. RIVM, Bilthoven, RIVM Report No. 259101003.

8. Greenland, D. J. and Szabolcs, I. (1993) (Eds.) Soil Resilience and Sustainable Land Use. CAB International. Wallingford.

9. Hinrichsen, D. and Enyedy, Gy. (1990) (Eds) State of the Hungarian Environment. Statist. Publ. H. Budapest.

10. Hungary: Towards strategy planning for sustainable development (1994) (National information to the UN Commission on Sustainable Development). Hungarian Commission on Sustainable Development, Budapest, 1-53.

11. Kádár, I. (1991) (Heavy metal content in soils and crops in Hungary) (H,e) KTM-MTA TAKI kiadása. Budapest. 1-104.

12. Molnár, E. (1995) Soil pollution and risk management in Eastern Europe. EERO ETC-course on "Pollution Analysis and Risk Management".

13. Molnár, E., Németh, T. and Palmai, O. (1992) Problems of heavy metal pollution in Hungary - "State of the Art". Proc. of the SETAC Conference, Prague, October 13-17, 1992.

14. The National Atlas of Hungary (1989) Akadémiai Kiadó. Budapest.

15. National Review for the Environmental Program for the Danube River Basin. Phase II. Vol. 1-2. Ministry for Environment and Regional Policy and VITUKI, Budapest, 1993.

16. Polunin, N. and Burnett, J. (1993) (Eds) Surviving with the biosphere. Proc. of 4th ICEF, Budapest, 22-27 April, 1990. University Press, Edinburgh.

17. Salomons, W. (1992) Non-linear responses of toxic chemicals in the environment: a challenge for sustainable management. Chemical Time Bombs, Proc. European Conf. Veldhoven, 2-5 September, 1992, the Foundation for Ecodevelopment, Hoofddorp, 31-43.

18. Salomons, W. and Förstner, U. (1984) Metals in the Hydrocycle. Springer Verlag, Berlin-Heidelberg - New York - Tokyo.

19. Stigliani, W. M. et al. (1991) Chemical time bombs: predicting the unpredictable. Environment. 33. 4-30.

20. Stigliani, W. M. et al. (1992) Overview of the chemical time bomb problem in Europe. Chemical Time Bombs, Proc. European Conf. Veldhoven, 2-5 September, 1992, the Foundation for Ecodevelopment, Hoofddorp, 13-29.

21. Szabolcs, I. (1974) Salt Affected Soils in Europe. Martinus Nijhoff, The Hague, Research Institute for Soil Science and Agricultural Chemistry of the Hungarian Academy of Sciences, Budapest.

22. Szabolcs, I. (1989) (Ed) Ecological Impact of Acidification. Proc. Joint Symp. "Environmental Threats to Forest and Other Natural Ecosystems, Oulu, Nov. 1-4, 1988. Budapest, 1989.

23. Várallyay, Gy. (1989) Soil degradation processes and their control in Hungary. Land Degradation and Rehabilitation. 1. 171-188.

24. Várallyay, Gy. (1991) Soil vulnerability mapping in Hungary. Proc. Int. Workshop on "Mapping of soil and terrain vulnerability to specified chemical compounds in Europe at a scale of 1:5 M, Wageningen, March, 20-23, 1991. 83-89.

25. Várallyay, Gy. (1992) Control of soil processes - a challenge for Hungarian soil science. Hungarian Agricultural Research. December, 1992. 37-44.

26. Várallyay, Gy. (1993a) Environmental aspects of soils and land use in Hungary. [Proc. 3rd Polish-Hungarian Seminar on the "New Environmental Aspects of Land Use in Poland and in Hungary" (Lublin, Nov. 7-8, 1991)] Zeszyty Problemowe Postepow Nauk Rolniczych. z. 400. 53-72.

27. Várallyay, Gy. (1993b) Soil data-bases for sustainable land use: Hungarian case study. In: Soil Resilience and Sustainable Land Use (Eds.: Greenland, D. J. & Szabolcs, I.), CAB International, 469-495.

28. Várallyay, Gy. (1994a) Precision nutrient management - impact on the environment and needs for the future. Commun. Soil Sci. and Plant Anal. 25. (7-8) 909-930.

29. Várallyay Gy. (1994b) Soil management and environmental relationships in Central and Eastern Europe. Agrokémia és Talajtan. 43. 41-66.

30. Várallyay, Gy., Salomons, W. and Csikós, I. (1993) Long-term environmental risks for soils, groundwaters and sediments in the Danube Catchment Area: the

Danube Chemical Time-Bomb Project. Soil Degradation and Rehabilitation. 4. (4) 421-432.

31. Várallyay, Gy., Buzás, I., Kádár, I. and Németh, T. (1992) New plant nutrition advisory system in Hungary. Commun. Soil Sci. Plant Anal., 23. (17-20) 2053-2073.

32. Várallyay, Gy., Rédly, M., Murányi, A. and Szabo, J. (1993) Map of the susceptibility of soils to acidification in Hungary. Agrokémia és Talajtan. 42. 35-42.

33. Welte, E. and Szabolcs, I. (1988) Agricultural waste management and environmental protection. Proc. 4th Int. Symp. of CIEC, Braunschweig, 11-14 May, 1987.

II.6

SOME ASPECTS OF GROUNDWATER AND SOIL REMEDIATION IN RUSSIA

I.S. ZEKTSER
Institute of Water Problems
Russian Academy of Sciences
Russian Federation, Moscow

It is impossible to characterize in one report all the main technical, institutional and socio-economic problems of soil and groundwater remediation that appear in Russia. These problems are extremely complex and differ greatly for various climatic, engineering and economic conditions of Russia western territory. Thus, I'll try to characterize in this report only the main problems of contamination, control and remediation of fresh groundwater as one of the main components of the environment and the main source of potable water supply. The problem of soil remediation will be considered in the report only partially.

At present fresh ground water is of great importance in supplying the population with potable water in the Russian Federation like in many other countries. There is a tendency to a constant increase of groundwater function in a common system of domestic water supply. This is due to the advantages of the ground water over the surface one as a source of water supply. The main advantage is that it is, as a rule, of a better quality, better protected from contamination and less dependent on seasonal and perennial climate fluctuations. Total volume of ground water use amounts to

E.A. McBean et al. (eds.) Remediation of Soil and Groundwater, 113–124.

13.8 cu.km in Russia according to the data of 1994 year, among them 10.4 cu.km per year are use for public and agricultural water supply, 2.9 cu.km per year for industrial purposes and 0.5 cu.km per year for irrigation. More than 60% of towns in Russia have centralized underground sources of water supply. About 30% of towns with population over 250 thousand use only groundwater for drinking water supply. Taking into account the more frequent cases of accidental surface water pollution the function of ecologically pure ground water in water supply of the population is increasing. Our principal position is to increase the reliability of water supplying systems through the obligatory use of reliably protected ground water in supplying the population with drinking water.

However, the use of ground water for water supply is limited not only by its quantity, but also by its quality. Figure 1 is an illustration of constant degradation of drinking water quality. Nowadays every eighth water sample doesn't correspond to epidemiological requirements and every fifth one doesn't conform to drinking water standard according to its chemical composition.

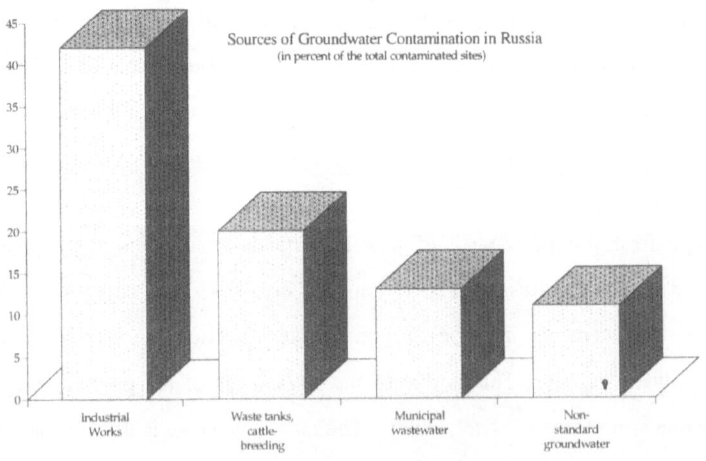

Figure 1. Sources of groundwater contamination in Russia

Groundwater quality in Russia is formed under the impact of both natural and man-induced factors. The impact of anthropogenic factors in forming the composition and quality of fresh ground water, used for water supply, has increased greatly in the last years. The idea that ground water is ecologically pure should be considered correct in some regions. At present about a thousand points of groundwater pollution were revealed in the territory of Russia, including 146 acting well fields in 87 towns and settlements. Groundwater pollution is mainly of a point character, however polluted areas amount to tens and in some cases even hundreds of square kilometres. Here, a notion of ground water contamination should be given. To our mind, contamination is such deterioration of ground water quality that makes it impossible to use them for concrete purposes.

Well fields where maximum permissible concentrations of some components are considerably exceeded are shown in schematic maps. The excesses are observed for both natural components of chemical composition and component production of which is caused by human activities such as oil products, pesticides, etc.
Industrial plants are the main source of groundwater contamination, consisting of 42% of all the contaminated sites, then comes waste accumulators and filtration fields, waste water irrigation of cattle-breeding farms, filtration from agricultural fields where pesticides, manures and fertilizers are used (20%). 14% of the sites are contaminated with waste water and public service wastes. Non-standard ground water also serves a source of contamination due to its leakage to the well fields when the regime of exploiting is disturbed (Figure 2).

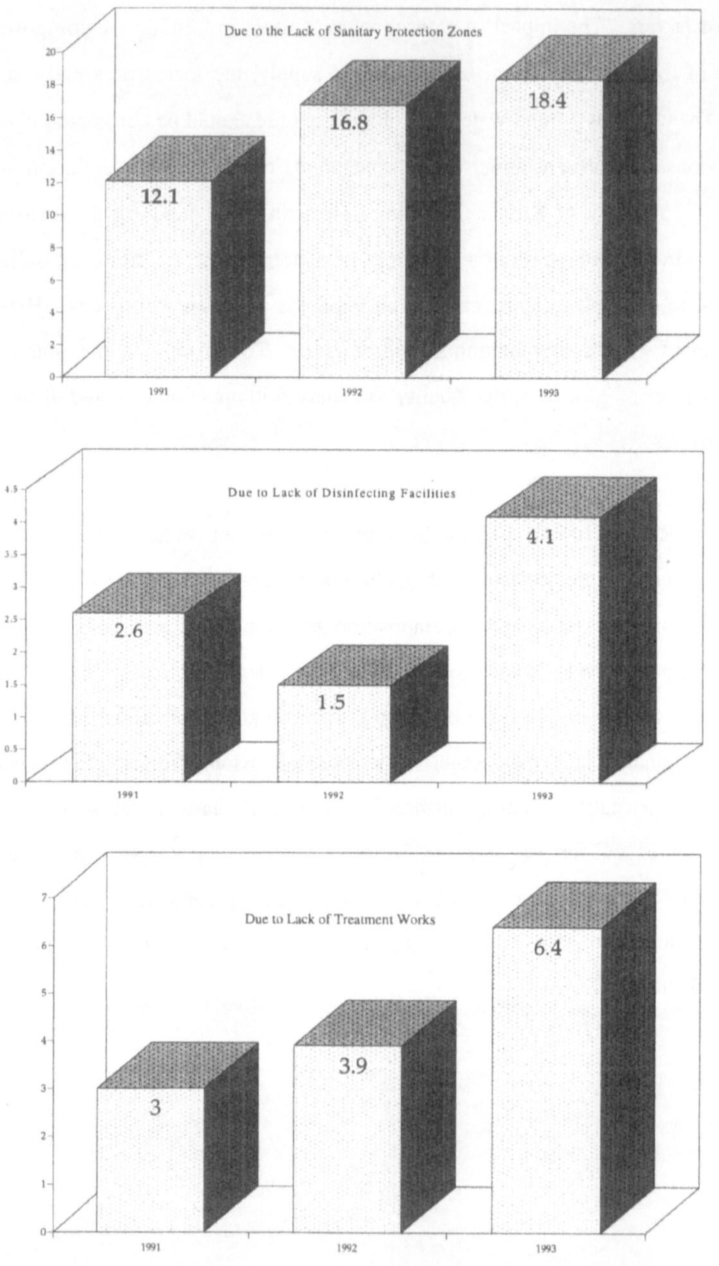

Figure 2. Sanitary standards in Russia (in percent of the total contaminated sites)

Well fields in Cherepovets (phenols, chlorobenzene, toluol), Lipetsk, Tula, Voroniezh, Tolyatti, Volgograd, Stavropol, Mozdok and Grozny (oil products), Chelyabinsk (phenols, lead, iron), Novakuznetsk (phenols, fluorine), Abakan, Angarsk (oil products) etc. are the most contaminated ones.

As it has already been mentioned, besides groundwater pollution in separate wells and well fields, regional groundwater pollution occurs. Regional changes of groundwater composition and properties are usually caused by both point and areal pollution sources. A very important aspect that should always be considered when solving the problems of groundwater protection and use is that groundwater is one of the main components of the environment. Any changes of other components of the environment (for instance, precipitation, river runoff, soils etc.) inevitably cause changes of groundwater regime, resources and quality. And vice versa any changes of groundwater (for instance groundwater level lowering under its intensive pumping out) cause changes in the environment, for example land surface subsidence, karstification etc.

Close interaction of groundwater with the environment in other components is especially well demonstrated as obvious in the last decades when the impact of man-induced factors on the environment progressed greatly. Thus, regional environment pollution results in regional groundwater pollution.

I'll give some examples. Urban impact on the groundwater quality on a regional scale is the most intensive one. It is the result of increasing mineralization of precipitation in the urban areas and "acid" rains, oil products leakages out of gas pumps, the impact of industrial and sewerage waste. Groundwater salinity in urban territories is usually 2-3 times higher than in regional ones.

The impact of "acid" rain is the second example. It is known that atmospheric emission of chemicals doubles every 10 years, that results in their concentration increase in atmospheric precipitation. Under infiltration of the atmospheric precipitation and the snow melt water, several tens of different elements get into the groundwater that change the hydrochemical regime of the groundwater, its composition and quality. The content of heavy metals in the groundwater essentially increased in the areas of infiltrating "acid" atmospheric precipitation. For instance concentration of aluminum, zinc and manganese in the snow cover in Moscow increased 15-20 times over the last years.

The most tragic example is associated with the impact of Atomic Power Stations. The Chernobyl disaster impact on the groundwater is observed at great distances, about hundred kilometres from the place of catastrophe. Considerable radionuclide accumulations were formed in the upper soil level in the Chernobyl area. Radionuclides migration through the vadoze zone resulted in the growth of their concentration several tens and even hundreds of times and even at large depth (up to 100 m), if compared with the situation before the failure.

There are a lot of similar cases. All these examples make it clear that problems of groundwater protection from contamination are closely related to a general problem of environment protection from contamination.

Groundwater quality remediation in Russia is mainly limited to making a system protecting well fields and surface water streams and reservoirs from contamination. The technique of calculations and predictions of mass transfer in aquifers when there are sites of pollution including analytical methods and methods of modelling are well enough developed. In the last years there is a change from deterministic models of predicting mass transfer to stochastic simulation of media. There are many examples of protecting well fields and rivers from pollution. Thus in the 70's a major complex of water protecting measures in the Sev.Donets river valley (the Ukraine) was designed, including vertical drainage for entrapment of polluted water and getting it into purifying facilities, and also injection wells for localizing groundwater pollution zones etc.

A good volume of work for groundwater purification is made by VODGEO Institute. In particular, a system of pumping out and purifying polluted groundwater from six valence chlorinum was made in the town of Aktyubinsk (Kazakhstan), basing the investigations carried out by this Institute. Entrapment of polluted groundwater with a system of wells is the main idea of these structures.

The problem of contamination of soil and water-bearing layers is not a new one, however its urgency has significantly increased in the last 10 years. It is due to the fact that the contamination affects a vitally important habitat of humans, are difficult to be eliminated and, the most important, effect negatively human health.

Some towns in Russia (Nizhniy Tagil, Kamensk-Uralsky, Orsk, Bratsk, Cherepovets etc.) where water resources are heavily contaminated, including groundwater, are declared to be zones of ecological disaster. There, concentrations of heavy metals, oil products, fluorine and other components are ten times higher than maximum permissible concentration (MPC).

Leakages of oil and chemicals from the territories of refineries, from oil tanks and pipe-lines are the most hazardous ones. They penetrate through the soil into the water-bearing layers causing groundwater contamination. Besides, oil products in the underground layers give off fumes, that migrate through soil and are dangerous for health and even the lives of humans.

The situation in the Volga river area can serve an example. There are the largest oil refineries in Russia. These plants were built in the 50's. At that time all the pipe-lines were laid at a depth up to 5 m. It did not help to effectively control oil and oil products leakages. At that period the largest power electric stations were built in the Volga. In particular, filling in the Saratov water reservoir and water level rise up to 5-6 m in the Volga caused flooding of the flood-plain, groundwater level rise and, as a result, groundwater and soil contamination. There were cases of observing oil products in wells, poisoning people. In the Spring of 1989 a thick layer of oil products emerged in the cellars of apartment houses at the levels close to high flood plain level in the area of the town of Novokuibyshevsk, Samara region. A hazard of fires and peoples poisoning appears. This situation can be called an ecological catastrophe. In spring 1991 there were victims. Federal Commission investigating the reasons of the tragedy has stated that a thick layer of oil products had been formed at a depth from 3 to 70 m. Only in the area of Novokuibyshevsk town the layer of oil products mainly benzene, in the underground amounted to 1.1 min.t. Getting into aquifers, the oil products migrate with the groundwater towards the river and contaminate the river water. A complex of geological studies, gas and emanation survey, seismic and electric exploration were carried out in the area that made it possible to develop a model of oil products migration and to prove the measures for purification of soil and groundwater. These measures include biological purification with micro-organisms, cleaning tanks and settling basins in the oil refineries, pumping out anthropogenic layers of oil products (with their reuse), waste water treatment, degassing etc. A plan has been worked out for nature protection in Saratov region and now a problem of financial support of this work is being solved in the Government of Russia.

Oil products pumping out from the soil and groundwater surface is an important state of eliminating oil contamination that appears as a result of leakages from oil tanks and pipe-lines. There is such a lens of oil products in Grosny town (Chechnya). This lens thickness is 3 m and is spreads over 1.5 sq.km. To study conditions of lens spreading and ways of its elimination, the Russian firm HYDEC (Hydrogeological Research and Design Committee) made a complex of drilling and test filtration work and developed a model of water and oil products filtration under local hydrogeological conditions. The volume of oil products of the delineated part of the lens amounts to 700 thousand tons. 70 thousand tons of oil products were pumped out from 1 sq.ha. in the test field from 1990 to 1994. Special variants were suggested to further intensify oil products pumping out for localization of the lens and then its elimination.

Much lesser as to its size but not less dangerous lens of kerosene was formed as a result of oil products leakage in Eysk, on the shore of the Taganrog Bay in the sea of Azov. Kerosene lens amounts to 0.7 km^2 and is at a depth of 20-22 m. The hazard is, that due to hydrogeological conditions, polluted groundwater and the lens are moving towards the Bay contaminating it. Since 1990 kerosene seepage occurs along 700-800 m of the shore line on the Taganrog bay that caused the sea water pollution. The square of kerosene spot on the sea surface is about 400 m^2. Coast-protecting measures are realized here that will help to prevent kerosene intrusion into the sea. Work for eliminating contamination is under way, and it is expected to take about 10 years.

In some cases groundwater contamination is a threat to the river water quality. The town of Uhta (Komi Republic) can serve an example. Oil products leak from oil refineries and storage areas causing surface soil pollution as well, oil products infiltrate through the aeration zone and accumulate in the groundwater surface in the aquifer. The area of contaminated groundwater spreading is about 5 km^2. Groundwater containing oil products is drained by the Uhta river and contaminates it. To make recommendations on preventing the river contamination the sources of contamination were investigated, special wells were drilled, test filtration were carried out, chemical analyses of water and soil were made,a numerical model of groundwater filtration in the river basin was developed by VODGEO Institute. As a result the source of contamination, its area and intensity were determined, and measures were worked out for blocking the river linear horizontal drainage combined with antifiltration wall, removing oil products from the groundwater surface, blocking contamination inflow into the river from polluted territories etc.

Nitrate contamination caused by fertilizer applications occupies an appreciable place in the problem of soil and groundwater pollution. Thus, groundwater contains up to 100 mg/l of nitrates, i.e. 10 times in excess of MPC in the Moscow region.

These norms amount to 250-300 kg/ha per year in the countries with intensive agriculture, such as France, Sweden, and Holland.

Nitrogen introduction in soil and groundwater is considerably less in Russia. It is due to the fact that manure applications have substantially reduced in 1990-1994 and now it doesn't exceed 50 kg/ha a year. Thus, nitrogen contamination is not regionally spread but is caused mainly by cattle farming and domestic waste.

In the recent years the problem of groundwater protection has been paid great attention to in Russia like many other countries. When studying the problem of groundwater protection, different ways are used: development of monitoring the aeration zone and aquifers, designing regional permanent analogies, proving zones of aquifer sanitary protection, experimental studies of contaminant filtration and migration.

Assessment of natural groundwater vulnerability to contamination is a hydrogeological proving the measures for groundwater protection under different natural and man-induced conditions. According to experience of some countries, such as Russia, USA, Germany,and Italy, it is possible to make regional assessments and mapping of natural vulnerability of aquifers, used for water supply and irrigation. This assessment is usually based on the analyses and processing all the available hydrogeological data and first of all data characterizing protective properties of the aeration zone.

Assessing groundwater vulnerability to contamination is made in two directions:

1) a qualitative assessment of the territory is made according to the intensity of the impact of different natural and man-induced factors on the aquifers' vulnerability, that allows us to compare different parts of the territory from the point of view of their vulnerability;
2) a quantitative assessment of the time (rate) for a certain possible contaminant penetration into the aquifer accounting for natural properties of water-bearing and overlaying rocks and migration abilities of the contaminant.

Assessment of groundwater vulnerability to contamination is made only basing the analysis and treatment of available geological and hydrogeological data and making calculations and designing models without carrying out too costly and labour consuming test filtration works.

Comprehensive data, terms and cost of the work are determined with the aims, quality and volume of the data available. Basing the experience obtained by specialists from Russia and the US in California, a total cost of assessment and mapping the groundwater vulnerability to contamination with compiling a small scale map of the territory of the state doesn't exceed 100 thous. dollars.

Assessment and mapping groundwater vulnerability to contamination is made in Russia for some artesian basins and their parts. For this purpose both techniques are used: the ones worked out in Russia and basing the technique of Prof. V.M. Goldberg and foreign ones (DRASTIC etc.).

Groundwater vulnerability maps are of great practical value. These maps make it possible:

- to develop a strategy for groundwater in the areas with different natural vulnerability;
- to prove the plans of placing and development of large industrial and agricultural projects with hazardous wastes and waste water;
- to prove the groundwater use for water supply and irrigation and places for potable water supply well fields, as well as to predict groundwater quality changes under human impact;
- to give a hydrogeological proving for different water protecting measures;
- to prove a choice of places for accumulating and storing wastes. Besides, these maps can be used by municipalities and other organizations for planning the measures for improving the ecological situation.

When speaking about contamination and soil quality remediation I want to note that control of soil secondary salinization is of great practical value in the territory of Russia and particularly in the Middle Asian Republics of the former USSR. Secondary salinization of soils is caused primarily by intensive irrigation, when the volume of water supplied to the fields considerably exceeds the norm need for the plants. Such excessive watering results in the groundwater level rise that under intensive evaporation and poor drainability brings about soil and land salinization.

Washing the soils is the most widely spread method of preventing soil secondary salinization. It is one of the most effective ways of land remediation as the excessive salts in the soil are dissolved and salt solutions are transported into deeper layers or outside the territory being washed. Washing of saline soils is combined with drainage that makes the process more intensive. The efficiency of washing the land is exemplified in the Golodnaya steppe in the Uzebek Republic. Totally there were about 5 mln.hectares of saline soils in the former USSR. Annual washing covered the territory from1 to 1.2 mln. hectares. Standard washing amounted to 16-20 thous.cubic metres per hectare and depended on the soil type, salinity level and drainage efficiency. Now washing irrigation is carried out in the Volga basin and the south of Russia at a square of 200 thousand hectares. Totally about 6 mln. hectares are irrigated and about 10 percent of this territory is subjected to secondary salinization.

It can be concluded that groundwater contamination is insufficiently studied on the whole in Russia. Steady-state monitoring is basically made only for the wells, used for centralized drinking water supply. Groundwater monitoring as a part of a complex environmental monitoring on a Federal scale is needed.

World experience shows that it is impossible to work out water protection policies under market economy and not supported with a corresponding public opinion. Scientists warned long ago about the catastrophic state of some largest water projects. However governments started with corresponding programs (e.g. the Great Lakes, the Rhine, the Thames) only when the ecological situation had become critical and every ordinary person had felt it.

The ecological situation in the most water projects in Russia is established. Experience shows that preventive water control measures cannot be formulated spontaneously. Federal purposeful activities are needed for this. The Complex Program "Volga Regeneration" that has been examined and accepted by the Government of Russia lately and will be financed by the state can be given as an example. About 40 research and planning institutes participate in the program. Regenerating the quality of natural ground water and soil, solving the most acute environmental problems, related to preserving and use of natural resources in this region are the main goals of this Program

II.7

SOME PROBLEMS OF GROUNDWATER INDUSTRIAL AND AGRICULTURAL POLLUTION IN BELARUS

Y. SEDLOUKHO
Polotsk State University
Department of Water Supply and Disposal
29 Blokhin St., Novopolotsk
BY-211440
BELARUS

Introduction

The useful supplies of fresh groundwater in Belarus are estimated as 15.7 km3/year. The process of groundwater creation is time-changing and is affected by climatic changes and human activities. For centralized water supplies basically shallow-located (50-200 m deep) aquiferous layer interconnected with overlying soil and surface waters are used.

About 4.0 million m^3/day groundwater are used in the republic for various purposes, including more than 2.5 million m^3/day for water supply.

Groundwater hydrochemical pollution in the Republic tends to aggravate, in particular in response to intensive use areas. Groundwater is mostly affected by industrial activity. Weak protection of groundwater results in contaminants penetration into

E.A. McBean et al. (eds.) Remediation of Soil and Groundwater, 125–127.

underlying strata. The main factor which may indicate the trend of groundwater pollution is the increase of mineralization e. g. since 1965 average mineralization has doubled (from 0.15 to 0.3 g/l). Soil water chemical composition is mainly affected by melioration, industrial and agricultural production and city agglomerates.

In intensive melioration areas groundwater is characterized by a considerable increase of sulfates, chlorides, calcium mineral types of nitrogen.

More than one-half of the Republic area is intensively used for agriculture. Ineffective use of mineral fertilizers, large-scale construction of gigantic cattle-breeding complexes has resulted in the intensive pollution of groundwater accompanied by an increase in chlorides content is mostly characteristic of the Republic. It had already covered more than 10% of the region's area.

Hydrochemical conditions are considerably affected by large industrial enterprises, settlements, former military objects, highways, oil and gas pipelines. Localized highly polluted groundwater sources are found in plant and municipal waste dumps and their clean-up facilities. Contamination levels in these places exceed the limiting permissible values by hundreds and thousands of times.

High contamination levels have been established at 150 -200 meters away from the source. Total mineralization in slurry storage area of PO "Belaruskaly" reaches 2-7 g/l while near salt disposal places it is about 98 g/l.

Considerable excesses of sulfates, chlorides, phosphates, nitrates, ammonia nitrogens, iron and fluoride have been found as dependent upon plant type. At chemical and refinery plants waste treatment facilities (Novopolotsk, Mozyr) groundwater contains oil products, phenolics, and other compounds exceeding limiting permissible values as much as ten times.

Persistent sources of heavy pollution are of a local character; however, substantial groundwater quality change is observed not only in the upper strata, but at over

50-70 m depths. This has necessitated the elimination of some water supply wells, as well as some groundwater wells.

In the Republic, groundwater condition is monitored by PO "Belarus geology". Special attention is paid to quality control if sources supply potable water to large industrial centers and cities. About 170 centralized water supplies and a large number of separate wells are located in their influence zones. Regular examinations of groundwater level and composition in the Republic have been started in 1947 and 500 wells have been monitored. The effect of separate industrial objects on the hydrochemical situation is monitored at 1,500 wells.

Regular issue of environmental bulletin have been commenced. A bulletin is prepared by the Academy of Sciences and Ministry of Natural Resources and Environmental Protection. Bulletin papers make it possible for the managing and planning authorities to account for environmental situation in the Republic and its regions.

The economic situation and imperfect understanding of ecological laws do not allow necessary measures to be implemented to the full extent and to prevent aggravation of hydrochemical conditions in the Republic.

II.8

GROUNDWATER RESOURCES AND POLLUTION EFFECTS IN ROMANIA

GABRIEL CONSTANTIN TOMESCU

Senior Hydrogeologist Researcher

National Institute of Meteorology and Hydrology

Bucharest, Romania

Introduction

The water resources of Romania, relatively poor and nonuniform in time and space, consist of surface waters (rivers and lakes), groundwaters and the Danube River. The available resources, in accordance with the actual level of the hydrographic basin management, are represented by 43.0 billions m^3/year, from which 17.0 billion m^3/year are from surface water, 20.0 billion m^3/year are from the Danube River and 6.0 billion m^3/year are from groundwater. The water usage corresponds to 1700 m^3/inhabitant/year.

Groundwater Resources in Romania

At the present time, the exploitable groundwater resources of the country are estimated to be 9.0 billion m^3/year, from which 4.0 billion m^3/year are shallow aquifer resources and 5.0 billion m^3/year are deep aquifer resources.

E.A. McBean et al. (eds.) Remediation of Soil and Groundwater, 129–142.

The groundwater resources represent 45% of the surface water resources but they are of superior quality, thus giving priority as potable water supplies.

There are areas with a good shallow aquifer potential, able to ensure important water discharges like the alluvial fan of the rivers: Somes, Mures, Timis, Olt, Prahova, Buzau, Putna, Moldova, Siret. The deep aquifers with a high exploitation potential are: the "Fratesti aquifer" and "Candesti aquifer" of the Romanian Plain; the Cretaceous and Jurassic aquifer of the South Dobrodja; the sandy Dacian aquifer of Oltenia, and the Pleistocene and Pannonian aquifer of the West Plain (Cris-Banat).

The areas deficient in groundwater are those of the Central Dobrodja, Transilvanian Plateau and Moldavian Plateau (The groundwater resources are presented at a regional level on Figure 1 and Table 1).

Groundwater Pollution

In Romania, the protection of the environment in general and of water resources in particular, constitutes a relatively new activity, a problem of great concern to the Ministry of Water, Forest and Environmental Protection, as well as of its specialized and research units: Central and Regional Water Authorities; County Environmental Protection Agencies; National Institute of Meteorology and Hydrology and Environmental Engineering and Environmental Engineering and Research Institute.

Analyzing the evolution of the quality of the shallow aquifers over the eight year interval (1985-1993), during which the monitoring activity has been carried on a number of aspects have been distinguished as described in the following sections.

Diffuse Pollution

In 1985, there was diffused chemical pollution of shallow aquifers with ammonium and organic substances; this pollution is a result of the intensive agricultural practices and forced industrialization.

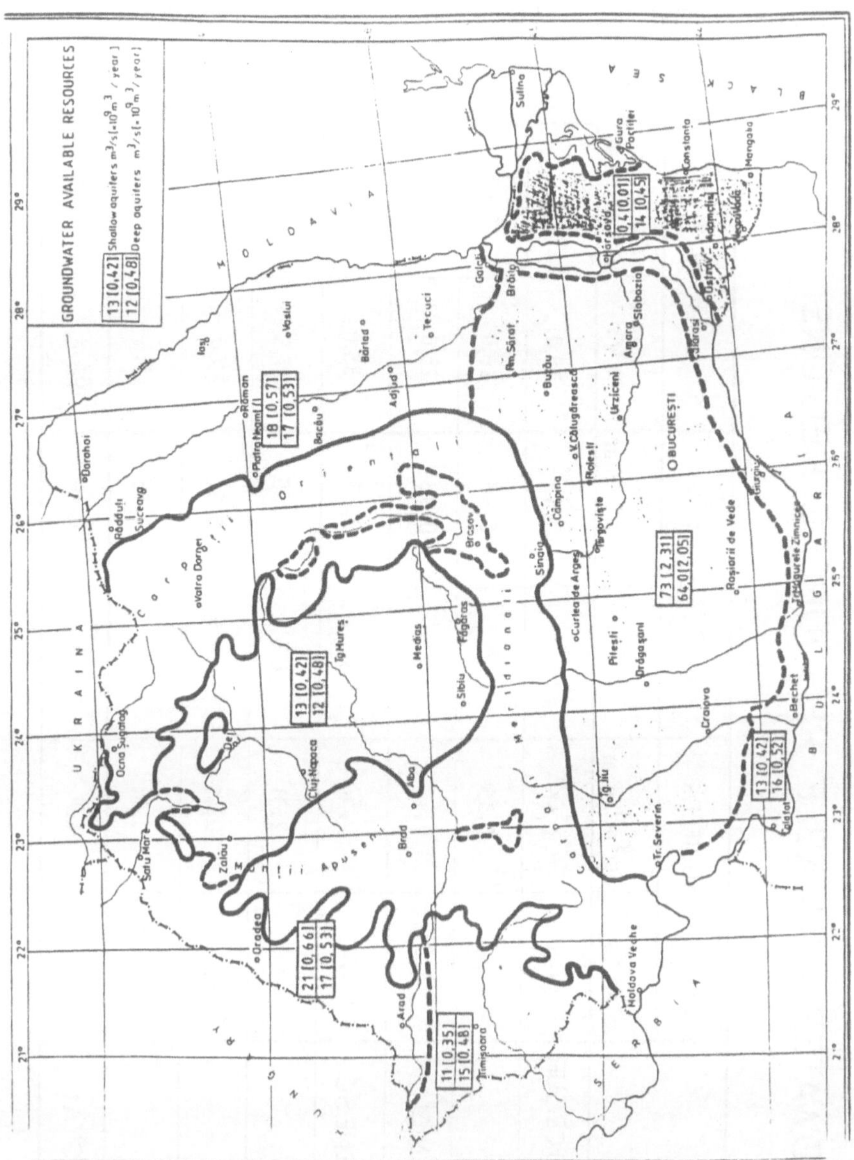

Figure 1. Regional groundwater resources

Table 1. Regional groundwater resources

GROUNDWATER AVAILABLE RESOURCES REGIONS

REGIONS	SHALLOW AQUIFERS		DEEP AQUIFERS		TOTAL	
	m^3/s	$\times 10^9 m^3/year$	m^3/s	$\times 10^9 m^3/year$	m^3/s	$\times 10^9 m^3/year$
MARAMUREȘ-SOMEȘ-CRIȘURI	21,0	0,66	17,0	0,53	38,0	1,19
BANAT	11,0	0,35	15,0	0,48	26,0	0,83
PODIȘUL TRANSILVANIEI	13,0	0,42	12,0	0,38	25,0	0,80
PODIȘUL MOLDOVENESC	18,0	0,57	17,0	0,53	35,0	1,10
CÂMPIA ROMÂNĂ	73,0	2,31	64,5	2,05	137,5	4,37
DUNĂRE-DELTĂ	13,0	0,42	16,0	0,52	29,0	0,93
DOBROGEA	0,4	0,01	14,0	0,45	14,4	0,46
TOTAL	149,4	4,74	155,5	4,94	304,9	9,68

In 1993, the pollution transformed itself from a diffused one into a general one at the level of almost all shallow aquifers, especially in the main corridors of the rivers valley (alluvial aquifers).

This general pollution appeared due to the increased number of specific pollution parameters, to which nitrates and pathogenic germs were added with significant intensities in nitrate, phosphate and pesticides concentrations.

The general pollution is situated at present in an incipient to average stage, affecting with different intensities, all withdrawals from the shallow aquifers.

The practical implications of the general-diffuse shallow aquifer pollution as well as of the intense regional pollution monitored in 1995, consists mainly of:

- the modification of all categories of shallow aquifer withdrawal sources which supply water directly for drinking and domestic purposes, as these sources show a real potential risk for the users' health, differentiated both on a long-term basis, up to a short or immediate one.
- the impossibility to use now or in the future the shallow water resources directly from the aquifer, without treatment or, alternatively, other water sources which will lead to higher costs.
- the potential danger which the shallow aquifers have acquired, by assuming multiple dimensions of pollution.

Point-Source Pollution

The most obvious impact of industrial activities on groundwater is through point-source pollution. In Romania, such impacts were aggravated mainly as follows:

- placing of industrial estates and factories on areas of high groundwater vulnerability;

- applications of technological processing involving large water quantities and significant pollutant discharges;
- the carelessness in construction and maintenance of storage, circulation and discharge systems of contaminant materials and wastes;
- lack of interest concerning the leakage and the seepage loss of non-hazardous and hazardous industrial wastes;
- inefficiency of groundwater management by monitoring and use of incentives and penalties to prevent contamination cases.

The vastness of the industrial impact on groundwater resources in Romania was pointed out in numerous comprehensive reports. However, the multitude, the diversity and the complexity have not been addressed systematically.

In order to identify at least the main types of industrial impact on groundwater resources in Romania, the causes and the effects of qualitative impact have been analyzed on the basis of subsequent data obtained through Romanian published and unpublished studies and reports and especially by the hydrogeological quality-network. The types of industrial impact on groundwater resources were specified in Figure 2 and Table 2.

Water consumption for industrial purposes in Romania represents more than 35% of the total abstraction of groundwater.

Pollution from industrial activities in Romania is spreading not only in shallow aquifers but also in deep aquifers involving a qualitative deterioration roughly estimated at $40 m^3/s$ of groundwater or about 15% of the available resources for groundwater supply.

Improvement of groundwater quality requires the achievement of some measures of conservation such as maintenance, restoration and regeneration of resources, involving interdiction of any possible sources of direct pollution and reduction of the risk of indirect pollution, removing or isolating the sources of pollution, qualitative reconditioning and remediation of resources.

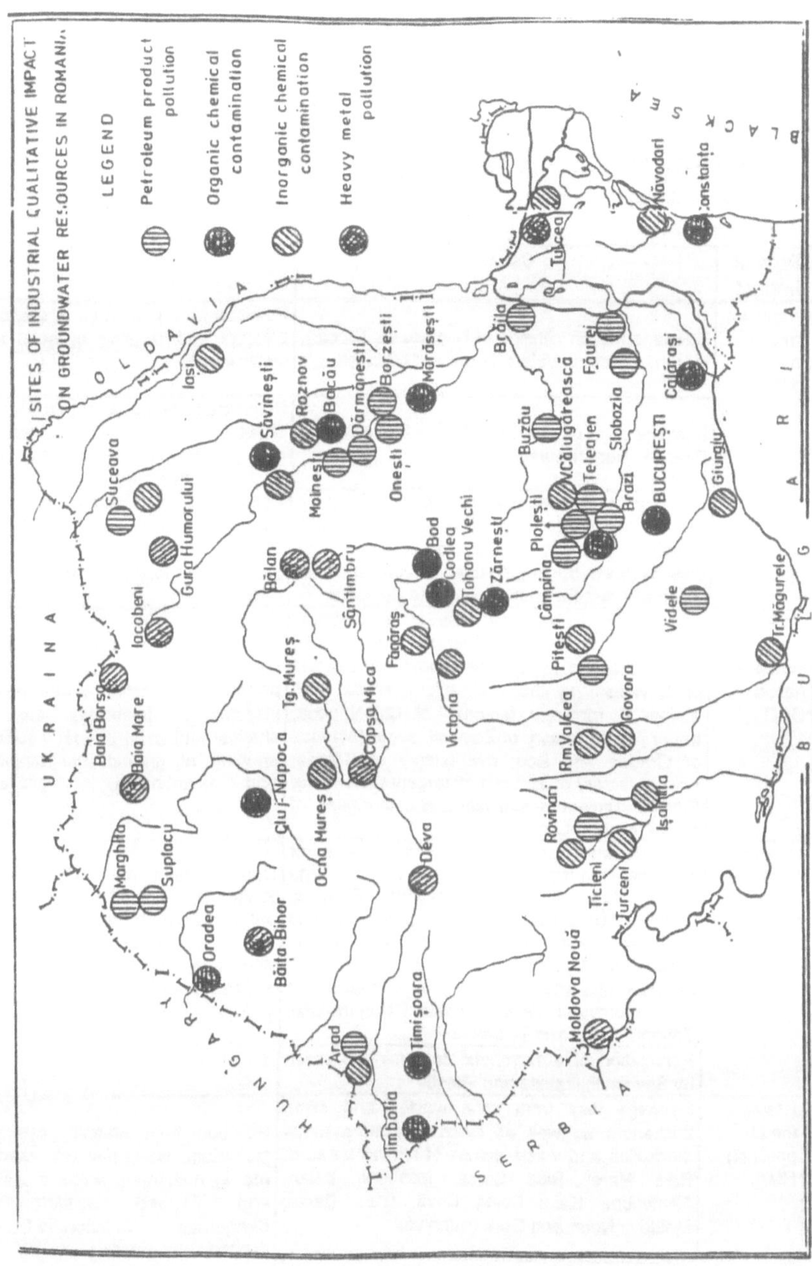

Figure 2. Sites of industrial qualitative impact on groundwater resources in Romania

Table 2. Causes and effects of industrial qualitative impact types

Table 2. Causes and effects of industrial qualitative impact types

Type of impact	Cause	Effect
Petroleum product pollution PPP	Seepage loss in oilfields of Suplacu de Barcau, Marghita, Ticleni, Videle, Braila and Moinesti	Pollution from petroleum in shallow aquifers and shutting of wells for water supply
	Leakage from oil pipelines Braila - Slobozia and Ploiesti - Reni (Faurei)	Pollution from oil and benzine in shallow aquifers and deterioration of groundwater resources for domestic water supply
	Leakage and seepage loss from petroleum refining industrial estates of Ploiesti, Brazi, Teleajen, Câmpina, Pitesti and Darmanesti, as well as from other industrial estates such as those of Arad, Borzesti, Onesti and Suceava	Pollution from gas-oil, benzine, gasoline, fuel oil etc. in alluvial shallow aquifers and shutting of groundwater tappings and domestic wells
Organic chemical contamination OCC	Leakage and seepage loss from chemical industrial estates of Calarasi and Bacau; industrial estate for synthetic fibres and threads of Savinesti; synthetic organic chemical and leather & footwear factories of Cluj-Napoca; paper & pulp factory of Zarnesti; sugar factories of Oradea and Bod; dye factory of Codlea; solvent factory of Jimbolia; detergent factories of Ploiesti, Timiscara and Marasesti; tanneries of Bucharest and Oradea.	Contamination from aniline, nitrobenzene, trichloroethylene, phenols, solvents, detergents, insecticides, pesticides etc. in shallow and even deeper aquifers, shutting of groundwater tappings and deterioration of available resources
Inorganic chemical contamination ICC	Leakage and seepage loss from fertilizer or chemical industrial estates of Arad, Târgu Mures, Fagaras, Victoria, Isalnita, Râmnicu Vâlcea, Turnu Magurele, Giurgiu, Roznov and Navodari; material stockpiles and sludge thickeners of coal power stations of Turceni, Rovinari, Iasi and Suceava; slime thickeners of Ocna Muresului, Govora, Valea Calugareasca, Tohanul Vechi and Tulcea	Contamination from nitrates, nitrites, ammonia, chlorides, sulphates, sulphides, cyanides, caustic soda etc. in alluvial or fissured shallow and even deeper aquifers and severe deterioration of groundwater quality
	Percolation of atmospheric contaminants close by Savinesti, Isalnita and Pitesti	
Heavy metal pollution HMP	Seepage loss from mine wastes and slime thickeners as well as leachate from material stockpiles and waste-dumps of mining areas of Baia Mare, Baia Borsa, Iacobeni, Balan, Sàntimbru, Baita Bihor, Deva, Gura Barza, Moldova Noua and Gura Humorului	Pollution from arsenic, cadmium, chromium, lead, mercury, uranium etc. in recharging areas of alluvial and fissured aquifers and dangerous deterioration of groundwater quality
	Percolation of atmospheric pollutants close by beneficiation and metalworking factories of Baia Mare, Copsa Mica and Tulcea	

Some Example of Groundwater Pollution in Romania

1. In the Oradea-Felix area, the vulnerability of the thermal aquifers, as well as the ones being sources of potable water is created by the circulation of the fluids along the technical faults, which put in contact the Neozoic detritical series with the Mesosoic carbonates ones.

For their protection, it has been proposed to establish a hydrogeological protection area (100km^2), and to prohibit the storing and the injection of residual waters in the Oradea area, which has a lot of industrial units.

2. The water supply of the users from Codlea-Brasov area is critical and has to face large difficulties due to pollution of ground and surface water sources. Organic pollution is due to the industrial activity in the area, the main source being the "Colorom-factory" specializing in dyes and varnishes which has polluted the aquifers up to 100m depth with aniline (a harmful pollutant of the blood with cancerous effects) and nitrophenol (particularly the p-nitrophenol has mutagen effects and it is framed as the second in toxicity).

3. Videle - Poieni-area, with oil-bearing activity, occurring on a surface of about 10.000 hectares, in Teleorman county, from which 350 hectares are polluted with oil products and salt water, resulting from the processing of oil secondary recuperation (salt-water injections).

4. The nitrates, ammonium and nitrites coming from the chemical fertilizer plant (Arad) generated the pollution in a shallow aquifer of the alluvial fan at Mures, endangering the potable groundwater system-supply situated downstream of this industrial unit.

5. Another example which deals with complex aspects concerning the pollution processes due to industrial, urban and agricultural activities and their impact on sources of potable groundwater is concerned with the Timisesti-Moldova area.

Two drains and a water catching system by wells provide 1.75 m^3/s of drinkable water, for inhabitants of Iassy municipality.

In 1980, the quality of groundwater tapped for supplying water to the population, was drinkable water (according to the chemical and bacteriological tests and analyses performed between 1967-1980.)

Since 1982, the chemical and bacteriological tests and analysis have proven widespread pollution such that, in 1992, the physico-chemical tests and analyses: nitrates - 114 mg/dm^3; NH4 = 0.85 mg/dm^3; oxidizability of 20 mg/dm^3 KMnO4, thus being the proof of the groundwater chemical pollution.

After the specific investigations, the following sources of pollution were detected:

- urban wastewater, inadequately purified (from Targu Neamt town);
- wastewaters coming from the food, furniture and building materials industries, intermittently discharged into the Targu Neamt Water Treatment Plant;
- the precipitation of the region that washes the localities, industrial and town residual deposits, farming fields where fertilizers, pesticides and fungicides have been used.

The vulnerability of the shallow aquifer being very high, it has been suggested that the impact of the groundwater polluting processes upon the drinkable water supply should be reduced by:

- re-evaluation of the pollution protection areas;
- prohibition of the storing of animal residues in the ditches;
- prohibition of the discharge into the Moldova River of industrial discharges;
- restriction of the quantities of chemical substances used by farmers in the area upstream of the tapping system;
- chemical and biological analyses and tests performed according to an inspection schedule;

- prevention of the illegal and uncontrolled storing of pollutants by regular on-spot inspections carried out by experts of the County Ecological Agency and Water Authority.

Widespread groundwater contamination has arisen by the leaching of agrochemicals from diffuse agricultural sources (nitrate, ammonium, pesticide); from urban and industrial activities (landfills, wastewater, leakage and seepage loss from petroleum refining estates or oil fields), in which the point-sources coalesce and produce industrial or urban diffuse pollution.

Industrial impact of point-sources on groundwater resources in Romania is directly or indirectly brought about by:

- mining industry (coal, ore, salt, alluvial gravel);
- oil and gas extractive industry;
- petroleum industry;
- energy and electricity generating industry (water power, fuel power, nuclear power);
- benefication and metalworking industry;
- chemical industry;
- leather and footwear industry;
- paper and pulp industry;
- food industry;
- other (alcohol, cement, ceramic, electrotechnical, glass, medicine, rubber, textile, wine) industries.

The Groundwater Quality Monitoring System

In Romania, the groundwater quality control rested on national regulations, under incidence of low priority. The enforcement units are The Water Authority and Environmental Protection Agency.

To provide groundwater quality protection, the National System of Groundwater Quality Monitoring was set up in 1984.

One of the main goals which this system has had, was to identify those aquifers storing important water resources, to know the qualitative substance of these resources, to identify the potential polluting sources of the aquifers and to draw up immediate but also long term strategies for the protection of these water resources.

Within this system, over 2500 hydrogeological observation wells were established for a systematic investigation of the water quality (2-4 chemical analyses per year).
The results obtained following these extensive monitoring activities, have led to the yearly preparation of the "Groundwater Quality Year Book".

The analyses that are being carried out periodically on the significance of the information obtained have contributed to the structuring of this monitoring system into two subsystems:

- groundwater background monitoring (national);
- groundwater impact monitoring (local).

These two subsystems are interdependent. For each aquifer, the observation well within the background subsystem makes evident the qualitative and quantitative dynamics of the aquifer in correlation with the major sources contributing to the aquifer supply.

The wells within the impact monitoring subsystem are located in the area of the main groundwater's polluting sources. These wells indicate the intensity and type of pollution in view of eliminating the source and protection measures. From the structured point of view, these wells enable the possibility to carry out pumping to dewater systems to perform possible inhibiting substance injections.

For the shallow aquifers in which centralized systems for large communities of people, depending on the vulnerability of the groundwater protection methodologies have been worked out. For these areas, analyses performed with the cost-benefit equation emphasize the timeliness of establishing these areas which will vary in surface and manner of exploitation of the land.

The data gathered has lead to extremely valuable information, which on the one hand has contributed to understanding the process of groundwater's quality in formation and evolution and to substantiate the strategies that must be adopted in the near future in order to provide an adequate protection of the groundwater resources .

To date in Romania, no significant remediation actions have been undertaken in groundwater pollution, in most of the cases in advance. The cost of remediation action has been supported partially by the Water Authority, but, principally, by the polluter himself.

Soil Pollution

Soil pollution in Romania affects an area of more than 7.0 million ha.

The main pollution sources are:

- spreading of chemical fertilizers (nitrates, ammonium, phosphates); and
- the excessive use of the organic chlorides - pesticides for agriculture and silviculture;

Conclusions

Groundwater is an important part of Romanian water resources, but widespread contamination of groundwater from diffuse agricultural and point-source discharges have drastically reduced available groundwater resources.

Unless measures of prevention, protection and remediation are taken, the groundwater resources will be reduced 30% with the most affected being the shallow water which will be reduced by 40-45%. The most unfavorable situations coincide with the greatest groundwater resources areas and with high vulnerability to pollution (alluvial aquifers and karstic areas).

II.9

POLLUTION OF SOIL AND GROUNDWATER IN SLOVAKIA

PAVOL POSPIŠIL

Racianska 95/28

Bratislova, 83102

SLOVAKIA

Introduction

The pollution of soil and groundwater in parts of Slovakia must be differentiated between industrial activities, transport (mainly railways), agriculture and from population (urbanization).

Pollution is the result of unsuitable legislation or almost no legislation on environmental problems from the past. The owners over the last four decades (state organizations and cooperative agriculture) were neither pressed to take measures against creating pollution, nor act for its remediation. It was always possible to find political and/or economic reasons for exceptions from valid laws and to find the way, how not to react to increase ecological problems. For this reason many contaminated sites are of very old origin and also of great extent. The remediation will be very complex and will create a great economic problem.

E.A. McBean et al. (eds.) Remediation of Soil and Groundwater, 143–149.

Kinds of Pollution

On the basis of origin, the pollution sources are based as follows:

-those created by industrial activity;

-by agriculture; and

-by urban activity.

In this case the reason is usually low environmental education of management and the political situation. As an example, we can show the construction of oil refinery on the brim of the hydrogeological structure which is well known for the largest accumulation of groundwater in the former Czechoslovakia and perhaps of European importance. Despite the opinion of one of our specialists in groundwater and his warning against underrating the groundwater pollution, his opinion was not taken into consideration. The basic principle of manipulation with oil products was violated and the investor (at the time the State) saved money on construction, sewers, and so on.

A different type of pollution arose in the industrial zone of the town Zilina, where the chemical industry has existed for more than 80 years. It was explained in the last decades without taking appropriate measures to prevent pollution.

According to present statistics, more than 7,200 landfills exist on the territory of the Slovak Republic. It must be noted, that before 1990 there was no law on garbage treatment. As a results, the majority (about 1270) of uncontrolled landfills of urban garbage (from 2,950 of the total number) pollute the surface water and more than 800 of them the groundwater. To be successful in the remediation of this situation, we need: to finish legislation, to appoint the responsibility, to find the model of financing and to appoint the share of the public.

Regarding the range of pollution, we distinguish the areal ones (mainly from agriculture) and the point ones (small landfills, small industrial activities - the dry cleaners, filling stations and underground storage tanks).

The pollution of groundwater by nitrates creates a serious problem. It comes from agricultural activities during the last two to three decades. The official agricultural policy was focused on production without taking into consideration the natural conditions and the financial expenses. The self-sufficiency in agricultural products was secured for political goals. To reach these goals, enormous quantities of fertilizers were used which resulted in pollution by nitrates, sulfates, phosphorus and by other components of fertilizers.

Fuel filling stations as sources of small scale pollution prevail. The reasons are usually underground tanks not adequately isolated, and the incorrect design of pipes.

In the case of landfills, the reason for pollution is the way they were created in the past. Not very long ago, so called "wild landfills" were created by inhabitants mainly in the country. No exploration for natural (geological) conditions was done and no measures to prevent the seepage and spreading of leachate outside the landfills were taken.

A special category is pollution of both the soil and the groundwater at the Site where the Soviet Army acted on greater scale. To realize the possible range of pollution from this source it is enough to know that at the Site of an airbase more than one meter thick layer of oil products was found on the groundwater table during drilling. One can call such case, a man-made deposit of oil products.

Present State of Remediation

Groundwater

The offices of the former system realized in some cases the necessity o solve the problems of groundwater pollution especially in these cases, where it was not possible to make them secret. Well known is the case of refinery of Slovnaft near Bratislava. The well field with output of water for 1000,000 people had to be abandoned and a new one constructed. In this case, a hydraulic barrier was used to brake off the path of oil from the place of pollution to the well field. From that time (in early seventies) pumping of water from wells has been done and several hundred million crowns were spent to cover the expenses for groundwater remediation. The result is that pollution is now localized and the water sources in the important structure are relatively safe.

In general, the methodology of pollution is as follows: exploration of the kind and spreading of pollution (usually by boreholes), hydrogeological evaluation of natural conditions, sampling and assessment of all results, and project design for remediation is settled and submitted.

The procedure of remediation is usually a long lasting one.

When we deal with pollution by oil, usually the hydraulic method is used. The pumping of groundwater from the bottom of the aquifer is done from boreholes. This creates the drawdown on the groundwater table. Then the oil products from the places of the greatest drawdown are pumped from special tanks which are sunk below the water level to the well in order to gather the oil from the groundwater table. Because the water pumped to create the cone of depression is usually polluted as well, it must be treated in towers by aeration and sometimes by special filters.

The fees for the outflow of pumped water to the river are high. Therefore, the infiltration of this water back into the aquifer is done instead. It needs suitable natural conditions and special adjustment of the infiltration wells.

In cases where pollution has not been produced during remediation, after the layer of oil from the water table was pumped off, the remediation continues by sparging and venting.

When groundwater is polluted by another chemical other than oil, the usual way to purify the groundwater is to pump the water and then treat above the surface. The purification technology depends on the kind of pollution and must be chosen very carefully.

Soil

Several methods for remediation of soils are used in Slovakia . It depends on the kind of pollution to determine which one is chosen. Because the pollution from oil products is the most frequent we have the most experience in this kind of treatment.

In the last years biological methods prevail. The excavation and transport of the polluted soil to a safe place has been done and remediation by biological technology took place until recently. Today, if local and natural conditions are suitable, the best way of soil purification is in situ or on site bioremediation. We prefer local microorganisms in oil degradation to special ones cultivated in laboratory conditions. Their development in local conditions is secured by adding water, nutrients and oxygen. First the soil must be homogenized and then it is possible to add nutrients.

The duration of remediation is different from case to case. It depends on the degree of pollution and the kind of pollution. It usually takes one-half a year (spring and autumn) to get rid of the pollution. If the pollution is very high it might take two seasons to get the soil clean. The soil can be then used for different purposes including agricultural production.

Our results and experience are quite good as we can see from the next table:

TABLE 1

Oil Content Before Remediation (mg/kg)	*Oil Content After Remediation (mg/kg)*	*Time (month)*	*Efficiency (%)*
4800	360	6	92.5
5700	800	5	86
66800	11100	7	83.4
31200	4230	7	86.4
20500	3295	4	84

(Ref. Vúrup)

As for expense, they range from 400 to 900 Sk. It depends on the pollution level and local conditions.

Problems

Since 1990 governments and society have realized that the condition of the environmental in our country is bad and in some parts of Slovakia alarming. The remediation in all directions is slow but it is possible to find progress in all spheres. We can find the cause for this slow progress in areas as follows: legislative problems, financial problems, and low level of environmental education of people.

The legislation problem is complex. The people in high positions are very sensitive to political pressure and the pressure of different lobbies which have an interest to solve the problems for their profit. From many drawbacks we should mention the law on responsibility for garbage of different types and also the fact that the villages and towns have not enough money because of inconvenient allocation of taxes.

A part of the financial problem comes from the present stage of the whole economy. Another problem will arise in the future if the new owners will not be educated enough to understand the problems of environment and the problems of mutual connections between the economy and the environment. Clear and unambiguous laws will be very helpful in environmental policy.

Let us look at the problem from the other side. People who work in environmental non governmental groups are very often zealots without appropriate knowledge and information in the sphere where they are trying to do "their best". They sometimes create more mess than benefit. As examples we can mention the Gabcikovo Project case from not long ago and Tichy Potok Project at present.

Conclusions

In my opinion the inevitable condition to prepare better and faster solutions for the future is good legislation which would be compatible with the present stage of our transformation and with the laws of European Union Countries as well. It is a very difficult task which should be solved as quickly as possible.

Another important condition is to finish our transformation in a proper way and simultaneously solve our educational problem in the sphere of the environment.

References

1. Eszényiová A.: Ponukový list na dekontamináciu zeminy zneccistenej rop.látkami VÚRUP Bratislava (Rukopis.)

II.10

IDENTIFICATION AND EVALUATION OF ECOLOGICAL LOSSES CAUSED BY PRESENCE OF THE RED ARMY IN POLAND

MAREK NAWALANY

Institute of Environmental Engineering Systems,

Warsaw University of Technology

Nowowiejska 20, 00-653 Warsaw, Poland

Introduction

The Red Army has been stationed in Poland from the time of the Second World War until as recently as 17 September 1993. The army occupied some 59 military objects covering in total ca. 70,000 ha. A Special Report has been prepared by the Polish Ministry of Environmental Protection, Natural Resources and Forestry (MOŒZNiL) on types of ecological losses caused by the presence of the Red Army, their extent, potential harm that can be posed by the environmental pollution and also the scope of the protection and remediation actions that need to be undertaken to avoid further deterioration of the situation. This paper is exclusively based on the Special Report. The institution responsible in Poland for monitoring environment - State Inspectorate for Environmental Protection, has made an assessment that 35 post-Red Army objects pose

E.A. McBean et al. (eds.) Remediation of Soil and Groundwater, 151–163.

or may pose a danger for the environment. On the basis of economical analysis the most urgent 21 cases (objects) with their total area of ca. 60,000 ha has been investigated. The data gathered is at present the basis for making decisions as to which objects must be dealt with as the first priority. The Report consists of two parts : General Part and Detailed Part. In this paper only the General Part is presented.

Scope and Methodology of Analysis

The major identified pollution generated by the Red Army (R.A.) relates to soil and groundwater pollution by all kinds of petroleum products. The *scope* of the analysis (measurements/observations) made consisted of :

i) *pollution of soil and groundwater by petroleum products and other chemicals*

The research has been subdivided into three parts : pollution of soil petroleum products, pollution of soil by other chemicals (mainly heavy metals) and pollution of groundwater by petroleum products and other chemicals.

ii) *pollution of surface waters*

iii) *pollution and devastation of surface of the soil*

The research has been devoted to biological cover of soil and to waste disposal sites.

iv) *pollution of military terrain by poison gases*

v) *radioactive pollution*

vi) *forestry devastation.*

The research has been conducted using :

- *aerial photos*
- *hydrogeological reconnaissance*
- *physical examination and chemical analysis of soil and water samples*
- *dosimetric measurements of radioactivity of soil and water samples.*

Characteristics of Military Objects Investigated

In most cases military objects were old German military objects which have been captured by the R.A. in 1945 and later modified according to needs of the R.A.. Types of the military objects left by the R.A. and investigated by the PIOŒ were as follows :

- main airfields
- auxiliary airfields
- all kinds of stores, magasines and fuel distribution stations
- naval basis
- military complexes with the expercises ranges
- munitions magasines
- garrisons located in towns.

Figure 1 shows location of the major military objects of the R.A. in Poland.

Figure 1. Location of the R.A. military objects investigated by PIOS

State of Pollution and Devastation

1. General data

Investigations made have proven the following general facts :

i) pollution of soil and groundwater has been found the most dangerous - especially hydrocarbons mixed with all kinds of chemicals. In most cases they form a film floating over the surface of water. Ten percent of the total area of groundwater beneath of military objects of the R.A. has been found unusable for drinking purposes. Also surface waters are endangered due to their interaction (drainage capacity) with groundwater.

ii) another dangerous pollution of soil and groundwater is caused by dumps, landfills and waste disposal sites under which toxic substances have been found.

iii) presence of toxic substances stored in magasines, closets and the like. Chemical contents of these substances is in many cases unknown or not easily identifiable.

iv) There were no contamination by combat gases nor radioactive substances has been found.

2. Contamination of soil by petroleum products

In all objects investigated there were stores of liquid fuels. They represented different capacities of fuel storages, from 200 to 102,000 m^3 of fuel. The total capacity of them was ca. 460,000 m^3. Search for petroleum contamination of soil has been done by drilling. Some of the boreholes have been turned into observation piezometers. The Drager-Stitz probes have been used to evaluate the (horizontal) extent of the soil pollution. To evaluate the vertical extent small diameter drillings have been made. To evaluate a thickness of the petroleum films formed over the groundwater table large diameter boreholes have been drilled. They were also used for taking soil samples. Contamination of soil by petroleum products has been encountered in all locations sampled. The major sources of this kind of contamination were :

- fuel reloading stations
- fuel pipelines
- fuel distribution stations
- fuel containers
- garages
- maintenance workshops
- cleaning stations
- places for heating the aircraft engines.

Petroleum product contamination has been found over the total area of 406 ha. The total volume of these products was ca. 18,4 mln m^3 . Both, the aeration zone of subsoil - 13,2 mln m^3 and the saturation zone - 5,2 mln m^3 , have been contaminated. Groundwater flow cause spreading of contaminants outside the direct contamination zone. In 15 cases thickness of the petroleum product film ranged between 1 cm and 5 m. Total area of the film exceeds 90 ha. The estimated volume of the floating fuel contamination is ca. 95,000 m^3 whereas the total volume of these products in subsoil is about 155,000 m^3.

3. Contamination of soil by other chemicals

Locations of sampling points for evaluation of contamination of soil by other chemicals have been chosen in the vicinity of :

- petroleum contamination sites
- exercise ranges for chemical military units
- sites of waste combustion

- storage for chemicals
- storages for electrical batteries
- purification stations
- communal objects
- terrain in which a biological cover has been changed.

Also, the local communities have delivered a considerable number of specific information about the (undetected) excavations made by the R.A.. The level of contamination by chemicals has always been related to the background values (concentrations) measured at places in which the influence of above sources and agricultural activity have been excluded.

The major conclusion from these investigations was that in most of the objects, concentrations of heavy metals have exceeded the background values many times, e.g.

- chromium 3,6 to 93 times
- lead 2,2 to 61 times
- mercury 1,7 to 38 times
- cadmium 1,3 to 18 times
- cooper 1,8 to 15 times
- cobalt 1,1 to 14 times
- arsenic 1,3 to 10 times
- zinc 1,9 to 5 times.

In many places phenols have been detected in concentrations between 0,0085 and 626 mg/kg of dry mass of soil. Many other chemical like cresols, nitrogen, chlorinated HC, aromatic HC etc. Total volume of soil contaminated with chemicals was ca.1 mln m^3 with the total area contaminated of 15,300 ha.

4. Contamination of groundwater.

The major sources of groundwater pollution were :

- contaminated soils
- exercise ranges for chemical military units
- stores for electric batteries
- stores of chemicals
- contaminated surface waters
- waste disposal sites.

In the course of investigations, the following parameters deciding about the spread of groundwater contamination have been measured :

- thickness of the water table aquifer
- directions and velocities of groundwater flow
- extent of pollution plumes
- type and degree of contamination
- state of groundwater intakes.

The major conclusion obtained from the investigation was that the first (water table) aquifers have been contaminated in the vicinity of 20 objects of the R.A. In most cases they were contaminated with petroleum products and heavy metals. Contents of petroleum products in groundwater ranged from trace concentrations to 55,500 mg/dm^3. Heavy metals contents depended on type of the object investigated. The most common heavy metals and the corresponding exceedence of allowable values were : copper (1,9 - 600 times), nickel (1,2 - 85 times), lead (1,5 - 310 times), mercury (2 - 41 times), cadmium (1,6 - 96 times) and chromium (1 - 600 times). Also a number of other chemicals has been detected with their exceedence level ranging from 1,5 - 227 (aluminium).

The total volume of contaminated groundwater is ca. 144,7 mln m^3 and its area ca. 6500 ha.

5. Contamination of surface water

Surface waters are the part of environment that are particularly vulnerable to all kinds of external factors. It is due to the high solubility of inorganic chemicals in water, specific reactions between organic matter with constituencies of natural waters and, above all, because of the easy migration of contaminants in surface waters. Surface waters (like rivers, lakes, creeks, canals etc.) were present in 14 objects investigated. Their total area was 495 ha. The following indicators of water quality have been taken into account :

- COD
- total iron content
- total content of suspended matter
- concentration of heavy metals (lead, cooper, mercury, cadmium and zinc).

Contamination with heavy metals has been found the most dangerous one. In some places concentration of heavy metals were exceeding by 2 -15 allowable concentration for the III-class water.

6. Contamination and devastation of soil surface

The basis for assessing the devastation of the soil surface were the aerial photos (scale 1:10,000) which allowed search for major changes in physiography of the soil and plant cover distraction. Especially the terrain of former exercises ranges of total area of 15,300 ha has been found devastated. Second type of considerable soil changes are 68 waste disposal sites covering in total an area of 98 ha. The latter are a potential sources of groundwater contamination. In their vicinity a considerable exceedence of heavy metals concentrations has been found. For instance, in extreme cases in relation to background samples, lead has been exceeded 187 times, copper 140 times, zinc 120 times, chromium 92 times, cadmium 80 times, mercury 40 times, cobalt 8 times and arsenic 3 times. It was also concluded that all remediation, removal or processing of waste stored at these waste disposal sites need to be made exceptionally carefully for there is a danger of encountering undetected/nonreported explosives there.

7. Contamination with poison gases and radioactivity

In the vicinity nor in stores poison gases nor radioactivity have been detected.

8. Losses in forestry

The total area of forests within the borders of the R.A. military objects is estimated 38,100 ha. All the forests have been found devastated. In particular the following detrimental changes have been found during the investigations made by PIOŒ :

- losses in wood productivity of forests
- lowering of the quality of wood produced
- illegal deforestation and forest fires
- mechanical devastation of young forests
- distraction of forest plants
- distraction of forest soil.

The forests losses were the heaviest in the areas of exercising ranges.

Financial Evaluation of Losses

1. Methodology of evaluation of ecological losses

The basic assumption is that the total ecological losses caused by the presence and operations of the R.A. in Poland are equal to the total costs of cleaning the contaminated and devastated elements of the environment to extent that is considered noncontaminated. However it will never be the same degree of cleanliness expected for all area to be remedied. It will depend on the future destination of the particular part of environment at particular places. Different future applications will define the actual costs. These costs will depend in general on :

- degree of contamination or devastation
- local conditions (e.g. local geology)
- assumed remediation technology.

Table 1. Unit Costs

No.	Type of remediation	Unit	Unit costs [10^6 zl]
1	Remediation of subsoil contaminated by petroleum		
	products	m^3	1,20
	- permeable soils	m^3	4,00
	- semipermeable soils		
2	Intakes of petroleum floating on groundwater table		
	or at the upper part of impermeable strata	m^3	0,30
3	Remediation of soil contaminated by chemical		
	- harmful substances	m^3	0,08
	- ammonia and medicals	m^3	1,20
	- toxic substances	m^3	10,00
4	Cleaning of groundwater	m^3	0,10
5	Cleaning of surface water	m^3	0,07
6	Cleaning of surface water reservoirs from wastes	ha	150,00
7	Remediation of devastated terrain	ha	14,00
8	Remediation of waste disposal sites		
	- with area less than 2 ha	ha	4 564,00
	- with area grater than 2 ha	ha	3 564,00
9	Hydraulic isolation		
	- down to 4.0 m	m	4,00
	- down to 7.0 m	m	8,00
10	Losses in forests	ha	55,00

Table 2. Total costs

No	Type of contamination		Area [ha]	Volume [10^6 m^3]	Total costs [10^9 z^{3}]	Costs of urgent works [10^9 z^{3}]	Fraction of area [%]	Fraction of costs [%]
1	Soil and groundwater pollution	soil	427,7	19,4	34 765	1 993	0,71	66,58
		groundwater	6 500,0	145,0	14 765	-	10,85	28,28
2	Pollution of surface waters		17,5	0,4	20	5	0,03	0,04
3	Land devastation	devastated land	15 330,0	-	215	-	25,60	0,41
		waste disposal sites	98,0	2,1	356	356	0,16	0,68
4	Loses in forests		38 100,0	-	2096	-	63,62	4,01
	Total				52 21	2 354		

Tables 1 and 2 illustrate the unit costs and total costs of cleaning particular elements of environment within the area used by the R.A. in Poland :

Conclusions

As the result of the investigations made the following most urgent remediation actions are needed :

1. Removing of the floating petroleum product films from the soil and groundwater
2. Removal or isolation of waste disposal sites
3. Remediation of soils contaminated with chemicals
4. Cleaning of surface waters
5. Construction of hydraulic isolation to protect groundwater intakes

The following classifications of the future destination/application of the former R.A. objects has been made :

I-st group : national parks, infiltration areas for groundwater and protection zones of groundwater intakes

II-nd group :agricultural areas, forests, urban areas, recreation areas

III-rd group :airfields, exercises ranges, stores, industrial terrain, roads.

The total losses are estimated to be 52,2 trillion z³. The urgent remediation will cost 2,3 trillion z³.

Bibliography

1. PIOS „Identification and evaluation of ecological losses caused by presence of the Red Army in Poland (final report), Warsaw, June 1994

II.11

WATER RESOURCE PROBLEMS OF MOLDOVA

A. O. GAVRILITSA
Scientific Research Institute of Water Problems and Melioration in the Republic of Moldova

Introduction

The Republic of Moldova is situated in the center of Europe, with a territorial area of 33,700 sq. km. From north to south it is 350 km and from east to west up to 150 km. Moldova occupies a large area by Dniester-Prut interfluve and narrows along the left beach of the Dniester. In the territory of Moldova lives 4,341,000 people, The density of population is high with 128.8 people/km^2. Nature has created the land with a cheerful, beautiful landscape with the very fertile soil - chernozem (black soil), which occupies more than 75 percent of the land area. Most of Moldova is located in a zone of sufficient precipitation, the central and southern regions are prone to droughts. The quantity of precipitation is in the range of 427 to 591 mm.

The water resources of Moldova are limited. The main rivers of Moldova are the Dniester and the Prut which originate from Karpats. In the territory of Moldova the flow is 3085 constant and temporary watercourses, from this only 7 rivers extend more than 100 km, 247 more than 10 km. From the small rivers the most significant river is the Prut with a length of 286 km and the southern-eastern territory of Moldova (the

E.A. McBean et al. (eds.) Remediation of Soil and Groundwater, 165–170.

region Jurjulesht) a short distance (less than 1 km), the river Dunai flows by the left beach.

Resources and Regimes of Surface Water

The Dniester River by inflow on the Dniester has an average annual flow of 10.7 cu km. The Dunai River has an average annual flow of 203 km^3 and represents the second river with an influence on Europe. The Dniester and the Prut during average and lower flow periods consists, 50% flow for use in Moldova, including with sanitary (ecological) resources. The river Dunai possesses superfluous flow storage which can be utilized with the intent of giving sufficient water to the south region of the Republic. The main characteristics of Moldova rivers in summary are given in Table 1.

TABLE 1. Main river characteristics in Moldova

	Name of Rivers	*Length (km)*	*Catchment Area (sq km)*	*Annual Flow (mill cu m)*	*Discharge (cu m/sec)*
			TRANSITE RIVERS		
1.	Dniester	1352(630)	72100(19070)	10700	339
2.	Prut	967(695)	27500(7900)	2906	92
3.	Dunai	28502(1)	817000(3)	203000	6430
			MAIN SMALL RIVERS OF THE REPUBLIC		
4.	Reut	286	7760	313	9,9
5.	Ikel	101	814	20,5	0,7
6.	Buik	155	2150	91,3	2,9
7.	Botna	152	1540	33,6	1,1
8.	Cogilnik	243(125)	3910(1030)	59,1	1,9
9.	Yalpug	142(135)	3180(3165)	91,3	2,9
10.	Kahul	39	605	9,2	0,3
11.	Chugur	97	724	21,8	0,7

For the territory of Moldova, the volume is normally 1.34 km^3 of natural flow. Interval annual river flow distribution seasonally on average for annual water content in percent from the norm are as follows: Dniester contains annual flow in spring 44.4%, summer 25.5, autumn 16.0, winter 14.1; Prut correspondingly 39.1, 33.5, 12.3 and 15.2; small rivers (Example of Reut) respectively 50.0, 23.0, 14.0 and 13.0. In the summer, small rivers are shallow, sometimes fully dry. The formation of floods in these rivers can be at any time in a year. However, in the summer, most intensive floods happen by rainfall in the catchment area.

Chemical contents of the river water in Moldova are variable. The Dniester-Prut interfluve character SO_4^{2-}, Ca^{2+}, Na^+ water with mineralization 1000-2000 mg/L. In the north and in Kodras, the River contains hydrocarbonate-magnesium-nitrate water with the mineralization 500-1000 mg/L, and in the eastern regions hydrocarbonate-nitrate water with mineralization 500-1000 mg/L. In the river valley of Dniester is the same water but only with mineralization of 200-500 mg/L.

During the last 10 years the river network of Moldova including the Dniester and Prut has endured root changes from its natural condition in the past. It is connected with hydrotechnical structures for many aims. Huge water reservoirs have been developed (Dubosser and Novodniestrovsk on the Dniester and Costesht-Stiinka on the Prut). These structures not only change the flow regime but also the ecological condition of the rivers.

Flow regulation of the small rivers by construction of cascade ponds and water reservoirs (3000 units) have given new condition (ecological) and requirement for management of local water resources. Management of the river regime on the new condition demands scientific elaboration in all the sides regarding optimum utilization of water resources.

Underground Water Resources

Moldova presents hydrogeological structure of artesian slide and represents part of the Black-sea artesian basin. The basic direction of the underground flow presently is oriented from north to south in the side of Black Sea. The natural resources of underground water of Moldova comprise 1.0-1.2 cu km per year. About 25% from them are useful for drinking water and supply without extra water preparation.

The total number of wells exploited with the aim of water supply are more than 4,000. Entrails of Moldova are rich in a valuable mineral waters which are utilized as drinking ("Varnichna", "Kishinevski", "Balski" and others). Some are considered medicinal. Most underground water is impacted by nitrate pollution. It has been recorded that 48% of the wells and 35% of the springs exceed permissible limit concentration by 2-9 times. Because of the specific hydrochemical condition, the territory has caused enrichment in underground water with toxic microelements and compounds: fluoride, selenium, stronium, ammonium nitrate and others. These high concentrations have various organic substances, hydrogen sulfate and metals.

About 50% of the rural population utilized water with harmful substances at non-permissible concentration. In 75% of the Republic the concentration of fluoride exceeds by 2-10 times.

Lake and Water-Reservoir

In the territory of Moldova, there are 57 lakes with a total area of water of 62,2 km^2. These are predominantly small lakes with the areas up to 0.2 km^3. Not big lakes and pond type lakes are located mostly on the floodplain of the Dniester and Prut.

Most large floodplain lakes are located on the ending of the Prut (Belj-6.26, Dragale-2.65, Rotunda-2.08, Phontan-1.16 km2). In the valley of the Dniester, parts of pond type lakes have been liquidated in the melioration operation.

Except natural lakes in Moldova there are 3,000 ponds and water storage reservoirs, with the total volume of more than 1.8 cu. km and area of water about 333 sq. km. The most significant water storage reservoir is located on the river Prut in section Costeshti-Stiinka and on River Dniester at city Dubossari. The total water storage reservoirs are counted 82 in Moldova. From them 75 are from 1 to 5 mil. m^3 volume. From the total number of artificial ponds more than 60% are small water bodies with volumes up to 100,000 m^3m about 30% from 101,000 to 500,000 m^3, approximately 5% from 50,000 to 1,000,000 m^3. A summary of the given water storage reservoirs are given below on Table 2 as following.

TABLE 2. Chacteristics of some water storage reservoirs

Water Storage Reservoir	*River*	*Year Construction*	*Full Volume mil m^3*	*Area of Water Plot km^2*
Navo-Dniestrovsk*	Dniester	1981	3000,	142,0
Costeshti-Stiinka	Prut	1976	1085,0	92,0
Dubossarsk	Dniester	1954	485,0	67,5
Cuchurgan	Cuchurgan	1964	88,0	27,3
Yaloveni	Ishnovech	1978	21,7	4,4
Gidigich	Buic	1963	40,0	8,0
Ulma	Botna	1961	2,1	0,7
Costeshti	Botna	1962	3,3	1,8
Rezeni	Botna	1963	3,4	1,9
Komrat	Yalpug	1957	4,0	1,7
Congaz	Yalpug	1961	9,9	4,9
Old Badrajii	Racovech	1989	4,9	1,0
Kniazevk	Sarata	1967	2,8	1,0
Minjir	Lapushna	1982	12,2	2,6
Chiaga	Chiaga	1960	4,1	2,7
New Sarata	Sarata	1967	2,2,	1,5
Caplan	Caplan	1983	8,3	1,5

*Flow regulation has carried out with calculation of requirement in Moldova

Quality of Water Resources

The main causes of progressive pollution of the water resources is unsatisfactory wastewater discharges. In the basins Dniester and Prut throw off flow 3 million m^3 volume in a year but clears 195 mil. m^3, i.e. only 6%.

From 1960 to 1990 the rivers Dniester and Prut became very polluted. In the past, the river Prut was very clear in Europe, However, situated in its basin is an enormous animal husbandaries complex, industrial and agricultural factories without waste treatment, construction which brings in the fields high pollution and too deteriorated sanitary-epidemiological condition. The sanitary-epidemiological condition of Dniester is little better, yet has been demonstrated permissible limit concentration of phenol, pesticide fluoride and other pollutions. The main sources of surface water pollution letting out impure or not completely clear flow, wastewater from animal husbandaries complex, washing fertilizer and chemicals from the agriculture fields.

Significantly, all these deteriorate the quality and contribute to increased levels of diseases related to stomach infection. Utilization is with the aim of drinking water, containing elements and compound-polluter of nature and antropogenic origin, bring to appearance may numbers of diseases by insufficient fluoride, dental decay and diseases of the blood.

II.12

GROUNDWATER CONTAMINATION IN TURKEY

M.T. GÖNÜLLÜ

Assoc. Prof.

Yildiz Technical University, Civil Eng. Faculty, Env. Eng. Dept.,

Yildiz, Istanbul 80750 Turkey

Introduction

Approximately 95% of the world's usable fresh water is underground. In almost all countries on the globe, it also supplies irrigation needs and drinking water needs in rural regions, at much higher rates than urban needs. Some countries having sufficient surface water sources for the purpose of drinking water have not recognized groundwater pollution problems. However, numerous countries depending on ground water, due to limited useable surface water sources have found out the importance of ground water contamination. For example, in North America and the Netherlands, the rate of usage of groundwater as drinking water is around 50% and 80%, respectively. It has been estimated that about 1% of ground water supply in US is contaminated (1).

In Turkey, having a substantial amount of surface water potential, groundwater is used only in rural areas. In highly populated big cities (e.g. Istanbul has over 10 Million), due to deficits in surface water reserves by the arid weather in the last decade and being relatively polluted of surface water sources, in recent years, new well fields have been opened. However, the share of ground water usage in Istanbul is very little.

E.A. McBean et al. (eds.) Remediation of Soil and Groundwater, 171–182.

Turkey's water budget is presented below(2):

Water Sources:

Annual mean precipitation height	642 mm
Annual mean precipitation amount	501 km^3

Surface Waters:

Annual flow	186.1 km^3
Rate of annual flow/annual precipitation	0.37
Consumable annual water amount	95.0 km^3
Realized annual consumption	25.2 km^3

Groundwater:

Existing annual water reserve	10.0 km^3
Annual pumping capacity	6.6 km^3
Realized annual consumption	5.4 km^3

As we see from this budget; realized ground water usage rate in Turkey is only 17%.

In this paper, ground water contamination potential and its related legal aspects in Turkey are addressed.

GroundWater Contamination Sources in Turkey

In general, ground water contamination sources may be grouped into four categories:

1. Waste and wastewater disposal activities: landfills, abandoned dump sites, municipal and industrial wastewater treatment plants, lagoons, septic tanks, wastewater land treatment operations, artificial recharge of ground water with contaminated water, deep well disposal systems.

2. Industrial and commercial facilities: leaks and spills occurring during transport, storage and utilization activities of chemical substances.

3. Agricultural facilities: the application of fertilizers and pesticides to the land.

4. Mining facilities and geological formations: Oil wells, several mining activities and ground water contaminations from natural geological formations.

Substantial levels of contamination from the first source occur only in heavily residential areas. In 1990, 44% of the total population was in the west, and 26% of that was at Central Anatolia(3). On the other hand, the population of province and district centers were 66% of the sum. The 3 largest provinces are in western region of Turkey; Istanbul, Izmir and Bursa. Densely-populated regions, at the same time, have been more industrialized. By considering the distribution of the manufacturing industry, it is seen that the share of the western region of the Country is higher than 50%(4). Accordingly, Turkey's western region is subjected to much higher pollution potential. Other regions in which agricultural facilities are common will be exposed to unsuitable usage of fertilizers and pesticides to land, and the quality of ground water will be affected from these chemicals. In the Country, minerals such as sulfur, phosphate, antimony, chrome, iron, copper, boron, magnesite, petroleum, mercury etc. are also produced(5). Also, mining facilities can cause local ground water contamination.

By examination of 1991 Environmental Statistics for Turkey(6), the following findings have been obtained:

- total number of municipalities: 2027
- the number of municipalities where garbage is collected: 1979
- average amount of garbage collected in summer(ton/day): 49 535.70
- average amount of garbage collected in winter(ton/day): 65 329.10
- distribution of municipalities by removal method of garbage is as shown in Table 1.

TABLE 1. Removal method of garbage in the municipalities.

	Number
Metropolitan municipality dump	49
Municipality dump	1334
Another municipality dump	64
Sanitary landfill	-
Burning in an open area	159
Incineration plant	-
Sea disposal	6
Lake disposal	2
River disposal	300
Compost plant	6
Burial	34
Land fill	132
Public and private dumps	3
Dumped to agricultural field	7
Burning at municipality open dumps	25

- Total number of dumps in 1991: 1415
- Municipalities undertaking geologic, hydrologic and meteorologic studies, and criteria for locating dump sites in 1991 are given in Table 2. The number of municipalities which carried out research is only 69(5.13%), in 1991.
- In Table 3, municipalities by dump site problems are presented. The number of municipalities without problems (as declared) is 323.
- Table 4 shows municipalities by the route of leachate from the dumps, 1991.
- Total number of closed dumps by 1991: 667
- Municipalities by problem associated with closed dumps are obtained as Table 5. number of municipalities having problems with dump is 292.

All these findings show that ground water pollution problem resulting from solid waste disposal in Turkey is in a considerable in size.

TABLE 2. Municipalities undertaking geologic, hydrologic and meteorologic studies, and criteria for locating dump sites

Criteria which are taken into consideration	Number
Transportation distance	32
Capacity	21
Situation of arrival	1
Final usage of area	32
Topography	17
Local and environmental factors	45
Hydrologic and geologic factors	17

TABLE 3. Dump site problems in the municipalities.

The problem caused	Number
Water pollution	121
Odor	839
Flies	872
Fire	121
Explosion	22
Animal without a protector	387
Used as a meadow	104
Unsightly to the environment	587
Soil pollution	1

TABLE 4. Route of leachates from the dumps in the municipalities

The route of leachate	Number
Sea disposal	21
Lake disposal	14
River disposal	158
Absorbed by soil	1256
Forms a puddle	33
Merges into the city sewerage	3
Pumped back onto the dump again	1
Given to the treatment plant of dump site	-

TABLE 5. Closed dump problems in the municipalities

Problems	Number
Subsidence of earth	9
Water pollution	62
Smell	194
Flies	185
Fire	22
Explosion	2
Animal without a protector	52
Used as a meadow	30
View pollution	168
Soil pollution	3

Studies Made on Ground Water Pollution in Turkey

Unfortunately, there have been very few monitoring studies made on ground water contamination in Turkey. In the literature, especially, works on hot spring waters are numerous. Also some works have been done on health problems resulting from natural minerals in well water; for instance, fluoride problem of Isparta City's ground water(7).

Some earlier works, in order to determine leachate qualities from dumps have been made. In Table 6, the range of leachate results from some operating and abandoned dumps for the years 1985 and 1986 has been given(8). Still today this sort of work is quite new. Recently, the results from Marmaris and Izmit uncontrolled dumps are presented in Table 7 and 8 (9,10).

Some data on the pollution of Istanbul's ground water has been given in a work(11). That work finds that about 300 km^2 area through different depths in Istanbul is polluted chemically and biologically. In fact, some biological experimental observations on well waters from especially densely-populated districts in Istanbul confirm this finding. In addition, in that work, it was found that polluted depths under Istanbul range from 10 - 150 m.

By means of well water examinations made around Izmir, Bursa and Adana Cities, some pollutants from domestic (microorganisms), industrial(cyanide, chloride etc.) and agricultural (nitrate, pesticides) facilities have been determined(12).

TABLE 6. Leachate characteristics from operating and abandoned dumps in Istanbul.

	Kemerburgaz (today in operation)	Ümraniye (abandoned in 1994)	Merdivenköy (abandoned in 1980)
pH	7.35-8.20	7.30-8.25	8.35
BOD (5 days)	250-655	215-5760	55
COD	1020-4800	960-12100	1135
T.Solids	936440-22680	7210-18200	12450
TKN-N	152-1445	235-1485	470
Chloride	1080-7095	980-3670	2310
Hardness(as	1680-3750	1090-7600	1040
$CaCO_3$)	605-3660	425-2300	1080
Na	140-2880	275-1900	1130
K	254-760	225-1105	193
Ca	2.6-16.9	3.2-56	11
Fe	0.09-0.79	0.15-1.51	1.6
Zn			

All values in mg/L, except pH

TABLE 7. Leachate from Marmaris uncontrolled dump (composite sample).

Cl	mg/L	2450
SO4	mg/L	240
NH3-N	mg/L	560
pH	-	8.19
COD	mg/L	13200
CN	mg/L	< 0.05
Phenol	mg/L	< 0.05
Oil and	mg/L	44
grease	mg/L	0.4
Cr	mg/L	5.2
Fe	mg/L	1.3
Ni	mg/L	0.5
Zn	mg/L	0.4
Cu	mg/L	0.4
Pb	mg/L	195
Mg	mg/L	0.04
Hg		

TABLE 8. Leachates and groundwater from Izmit's uncontrolled dump.

		leachates	ground water
Cl	mg/L	7740	-
SO4	mg/L	316	-
NH3-N	mg/L	120-1680	< 5
COD	mg/L	4000-	96
Phenol	mg/L	24400	< 0.05
Oil and grease	mg/L	0.5-2.4	6.5
Cr	mg/L	56-90	-
Fe	mg/L	2.87	13.5
Ni	mg/L	3.9-126.4	1.8
Ca	mg/L	4.7-17.7	140
Mg	mg/L	120-361	61
Hg	mg/L	49-243	< 0.01
		0.09-0.12	

Legal Situation and Evaluation of GroundWater Problems

"The Act About Ground Waters"(13) is quite an old act (1960) and a related regulation (1961), Although these legal regulations have had some rules related to pollution they have not been used to solve pollution problems. In this situation, the existence of surface water being in sufficient amount has taken a substantial role. However, the amount of clean surface water reserves is being decreased year by year. Although "The Regulation of Water Pollution Control" (1988), includes the regulation of pollution of all the water resources, usually it was used to solve only surface water pollution problems (15). After "The Regulation of the Control of Solid Wastes" (1991) became effective, ground water pollution problems have begun to take concern (16). After 1991, municipalities have begun to seriously consider rehabilitation of uncontrolled solid waste dumps which is an important source of ground water contamination. In the last four years, it has been observed that many dumps were closed and rehabilitated. "The Regulation of the Control of Dangerous (Hazardous) Wastes" which is the newest regulation (1995) seems to bring a much higher acceleration on the matter(17).

References

1. Environmental Engineering Research Council of ASCE (1991) Ground-Water protection and reclamation, *Journal of Environmental Engineering* **116**, 654-662

2. Erdogan, E.(1991) Ground and surface water potentials of Isparta City, *Symposium on Preservation of Fresh Water Resources and Environmental Problems in the Lakes Region, SW of Turkey, 3-5 June 1991 Isparta, Proceedings*, VIII-XV.

3. State Institute of Statistics Prime Ministry Republic of Turkey (1991) *1990 General Population Census.*

4. State Institute of Statistics Prime Ministry Republic of Turkey (1993) *1992 General Census of Industry and Business Establishments-First Stage Results of the National Total.*

5. State Institute of Statistics Prime Ministry Republic of Turkey (1992) *1991 Mining Statistics.*

6. State Institute of Statistics Prime Ministry Republic of Turkey (1992) *1991 Environmental Statistics- Municipal Solid Waste Statistics.*

7. Özgür, N., Pekdeger, A., and Bilgin, A. (1991) Origin of the high fluorine contents in aqueous systems of the Gölcük Area/Isparta, W-Taurides, *Symposium on Preservation of Fresh Water Resources and Environmental Problems in the Lakes Region, SW of Turkey, 3-5 June 1991 Isparta, Proceedings*, 233-250.

8. Gönüllü, M.T. (1989) Municipal-Industrial solid wastes and the leachate problem, *Environment '89, Proceedings of the 5th Environmental Science and Technology Conference*, Adana, 934-941

9. Gönüllü, M.T. etal. (1995) *The report of measurements and analysis of decomposition gases and leachate in Marmaris Uncontrolled Dumping Area,*

Yildiz Technical University.

10. Gönüllü, M.T. etal. (1995) *The report of measurements and analysis of decomposition gases, leachate and ground water samples in boreholls in the Izmit Dump*, Yildiz Technical University.

11. Ercan, A. (1993) The steadiness and pollution of Istanbul's ground water, *Akabe Business World* Bulletin, No. 1, p.8.

12. Foundation of Turkey's Environmental Problems (1991) *Turkey's Environmental Problems '91*, Ankara, 174-178.

13. Turkish Official Gazette (1960) The Act About Ground Waters, 12,23,1960, No. 10688.

14. Turkish Official Gazette (1961) The Regulation of Ground Water, 07,20,1961, No. 10875

15. Turkish Official Gazette (1988) The Regulation of Water Pollution Control, 09,04,1988, No. 19919

16. Turkish Official Gazette (1991) The Regulation of the Control of Solid Wastes, 03,14,1991, No. 20814

17. Turkish Official Gazette (1995) The Regulation of the Control of Dangerous(Hazardous) Wastes, 08,27,1995, No. 22387

Part III:

Remediation Technologies

III.1

AUDITING APPROACHES TO THE IDENTIFICATION OF PROBLEMS

J. BALEK

ENEX Tabor

Czech Republic

What is an Environmental Audit?

According to the Merriam-Webster Dictionary an audit is "a formal examination and verification of financial accounts". Since the advent of increased environmental concern in the early 80's, a term "environmental auditing" has been introduced with the aim of extending the financial examination to a broader scope of the environmental issues and perform an evaluation of the environmentally sound performance of already-existing facilities. As it will be shown, sometimes the financial aspects also form a significant part of the environmental audit. In a broader sense, the environmental audit covers a variety of methodologies and various approaches related to one or another part of the environment.

Very often the internal environmental audit should serve to assess a company's environmental status and improve, for one reason or another, the effectiveness of its environmental policy. In some controversial cases a preparation of the external environmental audit can be imposed by the inspection authority such as representing the government, on the enterprise or technology with rather dubious environmental policy.

E.A. McBean et al. (eds.) Remediation of Soil and Groundwater, 185–196.

While the actual content of the audit can vary from place to place, the general objectives remain more or less the same. A typical audit should:

- ensure the cost-effective systems of environmental protection are in use;
- ensure that the standard of the environmental protection is sufficient to meet current and future regulatory demands;
- ensure that the environmental protection standard is sufficient to promote good relations at local, national and international level;
- minimize actual and/or potential liability;
- promote environmental awareness; and
- make recommendations on environmentally sound management and technology.

Historical Context

One of the very first official approaches toward auditing was the Environmental Auditing Policy Statement published by the Environmental Protection Agency (EPA) in the USA on November 8, 1985. The statement was then considered to first of all become effective as an interim guidance. EPA solicited written comments until early January, 1986. From the comments submitted then by various institutions,a difficulty was recognized in identifying the purpose and content of the audit. Some companies felt that identifiable auditing programs (the incentives most frequently mentioned in this context) were fraught with legal and policy obstacles. Based on that observation EPA, stated that "because audit quality depends to a large degree in genuine management commitment to the program and its objectives, auditing should remain a voluntary program."

An attention was then also paid to the content of the audit. It was recognized very soon that the environmental audits are only a part of a broader successful environmental program and therefore should not be expected to cover every environmental issue. Therefore, it was recommended by the commentaries to EPA to limit the audit documents to specific questions.

Following such a development, the EPA has introduced a general policy in the environmental auditing as a systematic, documented, periodic and objective review of the facility operations and practices related to meeting the environmental requirements.

In Canada the promotion of environmental audits falls under the Canadian Environmental Protection Act dated May 1988. Environment Canada recognized the power and effectiveness of environmental audits as a management tool for companies and government agencies with the intention to promote their use by industry or others. Nevertheless, it was decided that inspectors would not request environmental audit reports during routine inspections in order to verify compliance with the Act. Actually, an access to the environmental audit reports may be required only when inspectors have reasonable ground to believe that:

- an offense has been committed;
- the audit's findings will be relevant to the particular violation, necessary to its investigation or as evidence; and
- information being sought cannot be obtained from other sources.

A rather liberal approach in the field of auditing was followed by the International Chamber of Commerce (ICC). The ICC expressed a strong belief in maximizing the use of self-regulation by the business community in the spirit of responsible case. Such a belief is based on two presumptions considered by the ICC as essential: First, if properly applied, self-regulation is frequently more effective than reliance on legislation and official regulations. Secondly, excessive proliferation of regulations is counter-productive. According to the ICC, an effective protection of the environment can be best achieved through the appropriate combination of the legislative regulations and of policies and programs (such as auditing) established voluntarily by the industry.

ICC concluded that the primary and obvious advantage of the environmental auditing is to help safeguard the environment and to assist with any substantive compliance with local, regional and national laws and regulations and with company policy and standards. As another related advantage was considered a reduced exposure to litigation and regulatory risk. ICC specified following benefits of the audits:

- facilitating comparison and interchange of information between operation or plants;
- increasing employee awareness;
- identifying potential cost-savings including those from waste minimization;
- evaluating and performing training programs;
- evaluating the effectiveness of emergency response arrangements;
- assuring updated environmental database;
- assuring the decision-making in relation to plant and technology modifications;
- enabling management to give a credit for good environmental performance;
- helping to establish good relations with respective authorities; and
- facilitating to obtain an insurance coverage for the environmental liability.

Major international industrial enterprises have developed their own approaches suitable, first of all, for the needs of respective companies and their branches. In other words, each of the large companies have their own current method of planning, conducting and using the environmental audit. In 1989 the Industry and Environmental Office under United Nations Environmental Program (UNEP) invited 12 senior-level industrial experts to a workshop on environmental auditing. Some of the major global companies such as Shell, BP and Ciba were invited. After the presentation of their own approaches in the field of the environmental auditing, the experts tried to define and summarize benefits of the environmental auditing in generally acceptable terms.

The representatives concluded that environmental auditing is a tool consisting of a systematic, documented, periodic and objective evaluation of how well the organization, management and equipment are performing. The aim was to help safeguard the environment by:

- facilitating management control of environmental practices; and

- assessing compliance with company and public policies, which includes meeting regulatory requirements.

Remarkably, some of the large companies tried to avoid the term "environmental audit" and replace it with such terms as the "environmental surveillance", "environmental review", "environmental quality control" or "environmental assessment".

This indicates that some of the environmental audits can be considered as another form of the Environmental Impact Assessment (EIA) performed in one alternative for already existing enterprise or technology. Therefore, a person looking for the results of environmental auditing should not be surprised when they are found in the files under different topics and key words.

Another example of the environmental auditing is from the Czech Republic. Since 1990 the privatization process has been occurring in the country. Initially, the value of state-owned companies and their performance had to be evaluated with the aim of setting a price for which the state property would be sold to the private sector. After several months of such practice it was discovered that in many cases the assessed value of the enterprise had to be significantly reduced as a result of the need to remedy serious environmental damages to the soil, groundwater, surface water and air which had occurred on the factory premises and in their vicinity. Therefore, a legal notice signed in May 1992 declared that an evaluation of the commitments related to the damaged environment had to be a non-separable part of any major privatization project. In other words, a special form of the audit was requested for all major privatization projects. The notice specified rigidly the audit's content. The following components were given as obligatory:

- description of the technology, its historical development, processed materials including their transport, produced waste, risks and accidents;
- assessment of the treatment of the waste water and comparison with legally accepted standards, also possible impact of the dangerous components in waste water;

- assessment of the air pollution based on actual and permitted emission limits; also the emission limits planned to be reached at the peak stage of the factory development. Based on the difference between the actual and permitted limits, appropriate technological and organizational measures should be recommended;
- assessment of the waste management and conclusion whether the existing measures comply with the Waste Management Act. Particularly an evaluation of the waste management produced by heating and burning systems was required;
- evaluation of other environmental components such as the pressure on the landscape and the treatment and damage to the agricultural and forest lands;
- specification of fees and levies which the enterprise was obliged to pay for any pollution of the environment through water, air and waste management and for any destruction of the vegetation, soil, landscape and protected areas;
- financial assessment of all damages to the environment and cost of the remedial works; and ;
- constraints the auditor faced when performing his duties.

Based on the audit's result the cost of the privatized enterprise was reduced and the new owner could apply to a special Fund for a reimbursement of money to be spent on the remediation of those damages to soil and water developed prior to privatization. The remediation of air pollution remained a responsibility of the new owner.

This so-called "Financial participation of the government in the environmental remedial" is under the jurisdiction of the "Fund of the National Properties of the Czech Republic" and any financial support from that fund requires prior consent of the Czech government.

Next to the obligatory audits, some factories ask for special internal audits which serve as a source of environmental information when the factory intends to improve its management, establish a joint venture with a foreign company and introduce new technology. For such an audit, an environmental auditor is recruited; he/she prepares an internal audit more or less in accordance with the request of factory management. The certificate is issued by the Czech Ministry of Environment to the applicant only after the technical, environment, medical and legal knowledge has been officially examined.

Based on the author's experience this example of auditing is sometimes rather difficult, because some factories try to hide significant features and factors related to the environment, health and pollution. This is because the certified auditor is considered another governmental inspector and not a professional working in the interest of the factory. Also the scoping and screening of the problems to be audited is not widely practiced, indicating that almost always the auditor has to cover all aspects as required for a standard complete EIA.

No doubt exists that the audits performed in other eastern states have to satisfy different demands depending on their own legal system; alternatively in some states the auditing has not been commonly practiced so far. Nevertheless, it can be expected that after these sites will join EC, certain standards will be set up in the future for internationally recognized audits. Of course, with increased environmental awareness some of the enterprises will develop their own auditing rules modified in accordance with specific needs and the stage of their own development.

Basic Steps in the Auditing Process

Based on the previous experience we can specify steps which at present should be generally recommended when performing the environmental auditing (Figure 1).

The complex program consists of pre-audit activities, post audit activities (both marked in Figure 1 by normal frames) and activities at the Site (marked by shadowed frames).

The program should start with the selection of the enterprise or its part to be audited and the choice of the priorities to be assigned to the audit.

As a next step, the auditor discusses the audit program with the enterprise representative, asks for the supporting data to be prepared in advance, administers a questionnaire, defines the scope, determines applicable requirements, priority topics, protocol and resource needs.

The site visit is the most important part of the exercise. During the visit to the site the team reviews background data and files, opens the in situ meeting, tours the facility, reviews the audit plan and confirms the understanding of internal controls.

When assessing internal controls, the auditors identify their strengths and weaknesses, adapt the audit plan and resources allocation and define testing and verification strategies.

When gathering the evidence the auditors identify testing and verification strategies, collect data, and ensure that protocol steps have been taken. The auditors review their findings and observations and ensure that the findings are factual.

When evaluating the findings, a complete list is prepared, working papers, documents are assembled, findings are summarized and a short report for the closing meeting with locals is prepared.

Preliminary findings are presented at the closing meeting and discussed with the personnel. As a part of the post-audit activities the closing report is finalized, corrected and distributed as a draft report for further comments.

After receiving the comments and corrections the draft report is finalized, signed and distributed.

The above steps may be followed by the preparation of the action plan based on the results of the audit.

As another option, a so-called follow-up on the action plan is prepared. This is being based on further negotiation with the enterprise representatives.

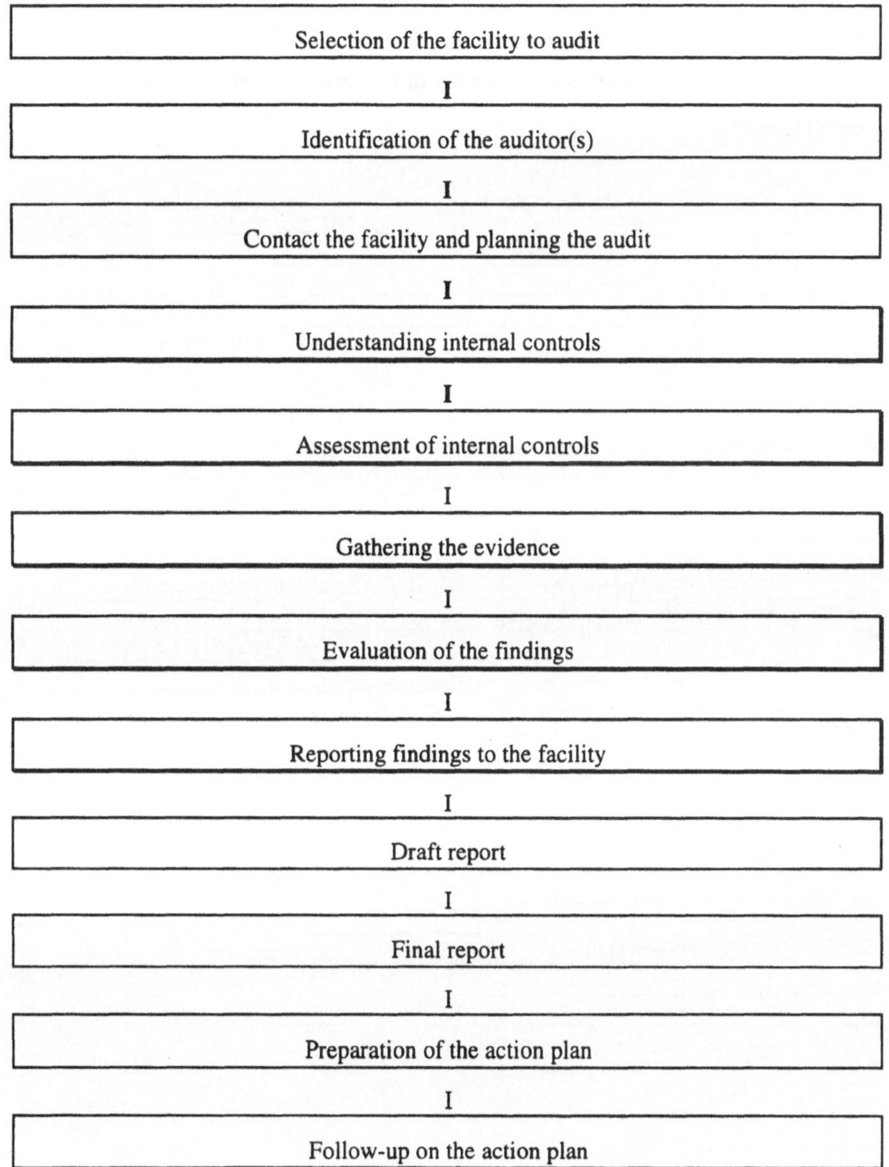

Figure 1.

The Content of Audit

There is no uniform approach in this respect. Some audits may be scoped so that they are focused on several important items, while others cover all possible aspects. If not prescribed by law, the auditor has to scope the topics accordingly to the specific needs. The following is an example of the content of the audit focused on the environmental health aspects:

Environmental review elements	
Policy and responsibilities	Land impact
Communication	Solid waste
Training	Hazardous waste
Risk assessment	Transportation of hazard. elements
Monitoring	Fuel and chemical storage
EIA and research	Water supply
Air emission control	Security
Water effluent control	Miscellaneous
Industrial hygiene elements	
Policy	Equipment
Regulatory compliance	Health hazard evaluation
Staffing	Record-keeping
Facilities	Health hazard control
Industrial hygiene	Training
Health hazard recognition	
Occupational health elements	
General	Company health department
Medical surveillance	Meetings/awareness
Workplace contamination	Equipment
Emergency preparedness	
Policy and responsibilities	Training
Risk evaluation	Community response
Emergency plan	Legal aspects
Emergency equipment	Security services
Internal communications	Plant maintenance

The Environmental Auditing Program

While the previously discussed audits were performed more or less as a non-repeatable exercise, some large companies achieved a stage when the separate audits were replaced by a more systematic program. As an example, the auditing program of Noranada, Inc., one of the major wood processing companies based in Canada is useful. The Company has considered the environmental auditing as a systematic objective method of verifying that environmental, health, industrial hygiene, safety and emergency preparedness standards, regulations, procedure and corporate guidelines are being followed. In support of these objectives Noranda has established specific goals which are to:

- audit all operations at least once every four years;
- correct all deficiencies and findings in a timely and cost-effective manner;
- reduce liability and risks to a minimum by correcting previous practices, by improving engineering design, by process modification and by chemical substitution;
- improve awareness and understanding of environmental regulations, standards, guidelines and codes of practice;
- transfer technology and develop an improved awareness of good environmental practices
- improve the efficiency and the cost-effectiveness of the environmental program.

Conclusion

There is no doubt that the environmental audits are beneficial both to the environment and to the enterprise. Systematic internal auditing at regular time intervals is a good preventative action against external, legally-imposed audits, examinations and other legal steps, which may be taken against the company by state authorities when the facility performance does not meet requested standards. This does not mean that any voluntary environmental auditing generates the necessity to respond to government reporting requirements and compliance inspections.

The environmental auditing can be considered as a universal approach, applicable world-wide. This means that the know-how transfer and methodological support can be provided relatively easily to the companies in need of such support. The basic condition is that the external auditors have free access to any kind of available information and data they may need in order to accomplish their work successfully.

Last, but not least, it should be mentioned that although environmental auditing essentially remains an internal management tool, the enterprises in their own interest shall respond to public information requests on the environmental issues by using much of the audit's content as possible.

III.2

AN OVERVIEW OF SOIL REMEDIES POTENTIALLY SUITABLE FOR USE IN CENTRAL AND EASTERN EUROPE

BRUCE C. CLEGG
Conestoga-Rovers & Associates
8615 West Bryn Mawr Avenue
Chicago, Illinois 60631, USA

Introduction

Over the past fifteen to twenty years, a wide variety of technologies have been developed for remediating contaminated soils. Abundant operational data for a large number of settings now exist to effectively assess the efficacy and implementation costs for many of the fully developed technologies. Using this knowledge and our current understanding of the various physical, social, and economic differences that exist between Central and Eastern Europe and many western countries, it is clear that not all the remedial alternatives appropriate for use in the West may be good candidate technologies for Central and Eastern Europe.

Specific constraints that serve to narrow the range of applicable remediation technologies include the absence of funds necessary to address each environmental problem to a risk-free or nominal-risk condition. Thus, the most 'bang-for-the-buck' may be considered a necessary, albeit not always desirable, threshold criterion for selecting appropriate technologies. Secondly, adverse climatic conditions may further

E.A. McBean et al. (eds.) Remediation of Soil and Groundwater, 197–204.

narrow the number of suitable remedial options. Prolonged low temperatures or extremes in ambient temperature may be directly problematic and can also lead to ancillary problems such as relatively thick frost layers. A third important consideration is the absence of a large remedial contractor infrastructure with expertise in some of the more esoteric remediation technologies.

The following discussion has been developed with a view to summarizing only those soil remedies that may be most suitable for use in Central and Eastern Europe, given some of the important considerations noted above.

Technology Overview

A wide variety of appropriate soil remedies exist for use both in-situ and ex-situ. Of these, biological and physical methods predominate.

The biological methods considered further include direct degradation by a naturally occurring biomass, bioventing, and landfarming. Physical methods reviewed include soil vapor extraction, solidification/stabilization and incineration via fuel blending.

In-Situ Soil Treatment

1. Biological Treatment

In-situ biological treatment is a process whereby oxygen and nutrients are added to the soils to enhance the breakdown of contaminants by non-native microbes or naturally occurring microorganisms. This remedy is typically implemented by adding nutrient-enhanced water to the system through infiltration basins at the ground surface or through recharge wells. Once added, the water is circulated through the soils to be remediated. In addition to acting as a transport media for nutrients, added water can also work to dissolve adsorbed contaminants. The extracted water can be treated on site, if required, to remove dissolved chemical species prior to reinjection. This treatment technology can provide a substantial reduction in organic contaminant mass in soils without the high added cost of soil excavation.

Several factors influence the effectiveness of an in-situ biological treatment process. These factors include:

- available oxygen concentration;
- appropriate levels of macronutrients and micronutrients;
- redox potential;
- soil pH;
- degree of water saturation;
- soil temperature;
- chemical species to be treated and their concentration;
- hydraulic conductivity of soils; and
- competition, predators and the presence of toxins.

In-situ biodegradation is often used in conjunction with a groundwater pumping and reinjection system to circulate nutrients and oxygen through a contaminated zone. Under favorable conditions, soil microorganisms are known to degrade many organic species. Microorganisms are capable of completely degrading many organic compounds into water and carbon dioxide in the presence of sufficient oxygen and nutrients such as nitrogen and phosphorous, at near neutral pH and under optimal soil temperatures. Anaerobic degradation of organics is possible although the rates of degradation are often too slow to constitute an active remediation.

Biological treatment techniques are generally not suitable for soil contaminated with organo-chlorine pesticide compounds and metals present in concentrations sufficient to inhibit microbial growth. However, this technology is well suited to soil contaminated by petroleum hydrocarbons and especially the more volatile aromatic component (i.e. benzene, toluene, ethylbenzene and xylenes) present in most petroleum products and waste residues. Bench scale and/or pilot-scale tests are often required to ascertain the effectiveness of biological treatment at any particular site.

2. Soil Vapor Extraction

Soil vapor extraction (or soil vacuum extraction) (SVE) is a technique used to remove volatile organic compounds (VOCs) and certain semi-volatile organic compounds (SVOCs) from the vadose or unsaturated zone. SVE is typically an in-situ process that makes use of vapor extraction wells or trenches installed in the contaminated zone. The extraction wells/trenches can be used alone or in conjunction with air injection wells that either utilize atmospheric air or actively use forced air injection. However, regardless of the mode of air introduction the influent air strips volatile compounds from the soil and carries them to the vapor extraction well/trench.

The vacuum extraction process removes chemical vapors trapped in soil pore spaces, but also affects, to a limited extent, residual liquid contaminants and dissolved contaminants from groundwater. Water in the collected air stream is condensed and separated from the air stream and is transferred to a water treatment system or is discharged directly to an acceptable receiving body. The air stream is then treated, if required, prior to reinjection or exhausting to the atmosphere.

Several factors impact the effectiveness of in-situ vapor extraction at any particular site. These factors include:

- physicochemical character of the chemical species to be treated;
- soil temperature;
- soil air conductivity;
- moisture content;
- contaminant concentration;
- geological conditions; and
- soil sorption capacity.

The applicability and ultimate efficacy of the SVE process is very site specific. The technology is best suited for use in permeable, well drained soils with low organic carbon content. Since SVE works only in the vadose zone, it is sometimes plausible to lower the groundwater level to increase the volume of the unsaturated zone. One method of depressing the water table elevation is by placing an impermeable cap over the treatment site to minimize surface water infiltration. An impermeable cap can also serve to increase the area of influence by preventing short circuiting of airflow directly to the surface. Factors such as stratigraphy and soil heterogeneities influence the flow of air as well as the location of contaminants.

Once the area to be treated has been defined, the extraction wells/trenches can be optimally located such that airflow within the area is maximized while airflow through unimpaired areas is minimized. The vapor extraction wells usually consist of screened pipe placed in a permeable packing. The top few feet of the well is grouted to prevent a short-circuited airflow to the surface. Vacuum pumps or blowers reduce gas pressure in the extraction wells and induce subsurface airflow to the wells.

As the air travels through the soil, it passes through a series of pores providing the least resistance to flow. Air that passes through pores containing vapor and liquids will strip the volatile contaminants from the soil. Many species existing in a condensed phase may vaporize and this process will continue until the condensed-phase organics are removed from the higher permeability soil.

The airflow draws chemical vapors and entrained water from the extraction wells to a vapor-liquid separator. In this unit, the liquid is separated and contained for treatment or discharge and the vapor is exhausted to the atmosphere or advanced to a vapor treatment unit. Monitoring probes can be installed to measure the soil vapor concentrations in situ. Moreover, sampling ports can be installed at many stages after extraction from the well.

Collected vapors are typically treated using carbon adsorption, thermal destruction or condensation. Carbon adsorption is the most common vapor treatment method and can be used to accommodate a wide range of VOC concentrations and airflow rates.

3. Bioventing

Bioventing is an effective technology for promoting biological degradation of volatile and non-volatile hydrocarbons in contaminated soil. However, the presence of organo-chlorine pesticide compounds and various methods could potentially inhibit the effectiveness of bioventing. This system, engineered to increase the rate of microbial biodegradation in the unsaturated zone using forced air as the oxygen source, is a potentially cost-effective alternative to many conventional systems.

By using air as an oxygen source, a more complete recovery of contaminants can be achieved due to the higher diffusivity of gases over liquids. At many sites, geological heterogeneities create a problem with waterborne oxygen sources because fluid pumped through a geological formation is channeled into the more permeable pathways. In a gaseous oxygen delivery system, diffusion can take place at a rate several orders of magnitude higher than for aqueous systems.

The technology relies on air flow through contaminated soils, at rates that are optimal for aerobic biodegradation. However, the addition of nutrients and moisture may be desirable to increase biodegradation rates. Gas monitoring points can be installed to sample short vertical sections of the soil.

Ex-Situ Soil Treatment

1. On-Site Biological Treatment

This technology uses biodegradation techniques to degrade contaminants in the soil or permit them to volatilize into the air. The basic concept involves providing a favorable environment to enhance microbial metabolism of organic contaminants resulting in the breakdown and detoxification of those contaminants.

This biological treatment technology involves aeration and biological degradation of the soils by tilling, typically on an engineered treatment pad. Tilling is often conducted on a regular basis to aerate the soil. Tilling also promotes volatilization of the contaminants to the surrounding air. Moreover, additives can be used to enhance the biodegradation process. This process usually continues until acceptable contaminant levels are reached.

The implementation of a biological treatment remedy utilizes common construction techniques; however, depending on the volume and physical nature of the material requiring treatment as well as climatic conditions, the remedy may require a long treatment duration. Biological treatment commonly requires from three months up to two years for completion per lift of soil. The length of treatment time can be estimated by treatability studies. Remediation of each batch is typically confirmed by sampling and analysis.

2. Solidification/Stabilization

The process of solidification/stabilization results in immobilization of the contaminants in the soils by binding them in a soil or concrete-like, leach-resistant matrix. Contaminated soils are collected, screened to remove oversized material, and introduced into a batch or continuous mixer. The waste material is mixed with water and various admixtures, (typically cement, kiln dust, or flyash and/or proprietary additives or agents). The material is thoroughly mixed until homogeneous in color. The treated waste is compacted and allowed to cure into a solidified mass (commonly referred to as a monolith) with significant unconfined compressive strength, high stability, and a rigid texture similar to that of concrete.

3. Thermal Destruction

An often overlooked resource for treatment of hazardous organic waste is the availability of cement kilns and fossil-fuel fired power generation facilities. Fuel blending may be a vary favorable remedial alternative for wastes with low moisture content, optimal particle size and, although not essential, some residual energy content. Coal tar wastes and hydrocarbon-saturated soils may be excellent candidate wastes for this form of remedial technology.

The advantages of fuel/waste blending are obvious in that a treatment system infrastructure may already exist through available kilns and power plants and it is often an inexpensive treatment technique for wastes with some existing residual energy content.

Disadvantages to fuel blending include an ongoing concern for toxic air emissions and disposition of ash that may be rendered hazardous due to non-combustibles (e.g. heavy metal) present in the original wastes. In addition, waste feed rates may be low due to waste characteristics and unique operational constraints at individual facilities.

III.3

UNCERTAINTIES OF COSTS OF REMEDIATION OF SOIL POLLUTION

K. VERSCHUEREN & J. CAPKA

IMd-Micon

Kempenlandstraat, 1, 5260 Vught, The Netherlands

Heidemij

Utrechtseweg 68, 6800 Arnhem, The Netherlands

Introduction

Restoring a polluted site to its original state can be very costly. Soil remediation costs have been estimated to be as high as 50 billion dollars for a small country such as the Netherlands. A soil clean up approach which leads to these enormous costs will have a negative impact on development opportunities in the industrial and agricultural sector and therefore it is important to find out which factors create the largest influence on the costs of remediation. In this paper an approach to deal with uncertainties in estimating soil pollution clean up costs is presented.

Pollution Investigation

1. Pollution Inventory

"Hard data", based on field information, is too narrow a basis from which final conclusions on the number of size of polluted areas can be drawn. Thus, besides the

E.A. McBean et al. (eds.) Remediation of Soil and Groundwater, 205–225.

existing field information, a "suspect spots" approach is typically followed. This approach incorporates "soft data" and has been used successfully by Heidemij in a number of projects. The approach is characterized by the following phases:

1. inventory of (past) activities that may (have) lead to soil pollution;
2. assumptions about type of spots and their layout, e.g. average surface area and depth of pollutant penetration, depending on the hydrogeology, type of soil and pollutant;
3. assumptions about methods of cleanup and cost estimates (cost/efficiency curve);
4. sensitivity analyses of the cost estimates, with respect to variability and uncertainties in the assumptions.

After carrying out an inventory of potentially polluted spots, some pollution sources should be considered as line-sources as opposed to point-sources. This could be the case for old, leaking sewers and leaking gas pipes (gas condensate). Line-sources can be presented as a number of point sources.

2. Distribution of the Pollutants

A number of polluted areas can be identified as the result of accidents and spills of dangerous products. From these polluted spots, the pollutants may leach into the groundwater and be transported in the direction of the groundwater flow. Pollutants which are readily soluble in water such as phenols and ammonium sulphate will be transported with the velocity of the groundwater. The main area of concern with regard to these pollutants is the contamination of the groundwater. On the other hand, pollutants which have a very low solubility (e.g. heavy fuel oil), will have only a small impact on the groundwater. The dissolved fraction of these pollutants will move much more slowly than the groundwater. More than 99% of the volume of these pollutants will be adsorbed in the zone above the level of the groundwater.

Industrial sites typically show a number of polluted spots of various dimensions. Based on the experience of hundreds of soil investigations at industrial sites, it has been concluded that the majority of the polluted spots have dimensions with diameters ranging between 5 and 25 metres.

In those cases where quite soluble organic products have been spilled, such as gasoline or chlorinated solvents (e.g. trichloroethylene), a plume of contaminated groundwater can be detected at distances normally not exceeding 100 m from the source. However, there are a number of well-documented cases at which the dissolved pollutant has traveled over distances up to 2,000 m from the source. These cases are rare and only occur in regions with an unconfined aquifer and very high groundwater velocities. Pollution of soil and groundwater at any site can conveniently be described in terms of a number of polluted spots, as mentioned above, in which the bulk of the contaminants are concentrated, in the unsaturated zone (the zone above the groundwater level) in combination with pollution of the groundwater as a result of the many individual sources.

3. Evolution of Legislation - The International Experience

Learning from past experience

Confronted in 1980 with a case of illegal dumping of dangerous wastes in the community of Lekkerkerk, which caused several complaints about smell and health problems in this small community, the Dutch Government developed an ambitious soil and groundwater cleanup plan. This plan aimed at restoring the so-called "multifunctionality" of all polluted sites within 20 years.

Soil and groundwater quality standards (the so-called A-, B- and C-trigger values) were developed as a tool in the decision-making process.

In the mid eighties, a commission of experts concluded that insufficient environmental data were available to define acceptable pollution levels and advised that the so-called A-values should be used as cleanup values. The A-values refer to the natural background concentrations in protected areas. This resulted in very costly cleanup operations.

After 10 years of experience with these A-, B- and C-values, most advisors involved in soil remediation studies, as well in the Netherlands as in other countries agreed that the A-values are too stringent if used as cleanup value for the following:

- higher values are environmentally acceptable, especially for organic compounds;
- the high cleanup costs are prohibitive in many cases;
- a trigger value expressed as a concentration of a pollutant in the soil or in the groundwater is not always the best parameter for environmental risk assessment. The amount of pollutant present in a certain area should also be considered.

The general trend which followed internationally in cleanup approaches is the following:

- more emphasis on the study of the fate and effects of pollutants in the soil whereby degradation processes are taken into account;
- a cost/benefit analysis of the various cleanup scenarios (the 20/80% approach).

A more towards a more practical, less costly, approach is evident. In several countries, alternative lists of so-called "trigger values" have been developed. They are all different because they all represent a trade-off between the desired environmental quality goals for soil and groundwater and the very high cleanup costs if stringent soil and groundwater quality standards are imposed.

Uncertainties of assessment of cleanup costs

1. Operational Costs Versus Realestate Value

Soil and groundwater cleanup costs cannot be regarded as investment costs. They are either part of the operational costs or, if the pollution has not been removed, the costs will feature on the company's financial balance as a future liability which will reduce the value of the property.

At present it is not possible to make a distinction between the costs which have to be incurred in the coming 10 to 20 years and the reduction of the value of the property. For this reason, cleanup costs are usually being calculated as though all the polluted areas would have to be cleaned up in the following 20 years.

This approach always result in an overestimation of the costs because of the following reasons:

- a number of polluted spots will not have to be treated, because the pollution will disappear as a result of natural biodegradation processes;
- a number of spots will be removed as part of the modernization process of the enterprise, resulting in less costs than calculated;
- the cleanup activities will be spread out over 20 years. Cheaper cleanup methods will become available in the future.

Consequently, the calculated cleanup costs are the some of the costs of the expected cleanup activities and the reduction of the real estate value caused by the residual pollution.

2. How Many Polluted Areas Will Be Identified and What Are Their Dimensions?

The second uncertainty results from a fragmented knowledge of the presence of the pollutants. Only a small fraction of the polluted area is usually being investigated through boreholes and tests. The number of polluted spots and their dimensions is estimated, based on the number of potential sources of soil and groundwater pollution. The effect of this kind of uncertainty can be displayed using the so-called "Monte Carlo" simulation approach. This will yield a range between which the real cleanup costs will be situated with a 90% of certainty. The cost calculations can be made for different removal percentages of the volume of pollutants.

Figure 1 illustrates the "old" approach. This approach is based on drawing the contour lines for the so-called A- and C-trigger values, with the assumption that all pollutants exceeding the A-level (background) should be removed.

An approach which is followed to establish a cost efficiency curve is presented for the same site in Figure 2. In this figure the polluting activities and sampling and analysis.

3. How Much of the Pollution Should Be Removed?

The uncertainty which has, however, the greatest effect on the cleanup costs is caused by the crucial discussion on the cleanup values. Should the pollution be removed entirely up to the background concentrations or is a certain residual pollution level acceptable?

The impact of this choice can best be illustrated by the cost efficiency curve on Figure 9. The curve shows how the total cleanup costs will increase with increasing percentage of pollutant removal. This curve is commonly called the 20/80 curve, because it shows that cleanup costs for an 80% removal of the volume of the pollutant are about 20% of the costs of the total removal of the pollutants. Beyond a certain point the costs increase exponentially. This is caused by the fact that the overall concentrations of the pollutants will decrease exponentially with the distance from the source. This results in a rapid increase in the volume of soil and or groundwater to be treated when trying to remove all pollutants.

In a previous paragraph, reference was made to a cleanup approach based on a total removal of the volume of the pollutants. In situations where the pollutants offer a lesser risk - for example because of the nature of the pollutants - a lower removal percentage can be acceptable. The difference between 80% and 95% removal efficiency is approximately a factor of 2, (point 2 on the curve). From 95% to 99% removal efficiency, the costs will double again (point 3 on the curve).

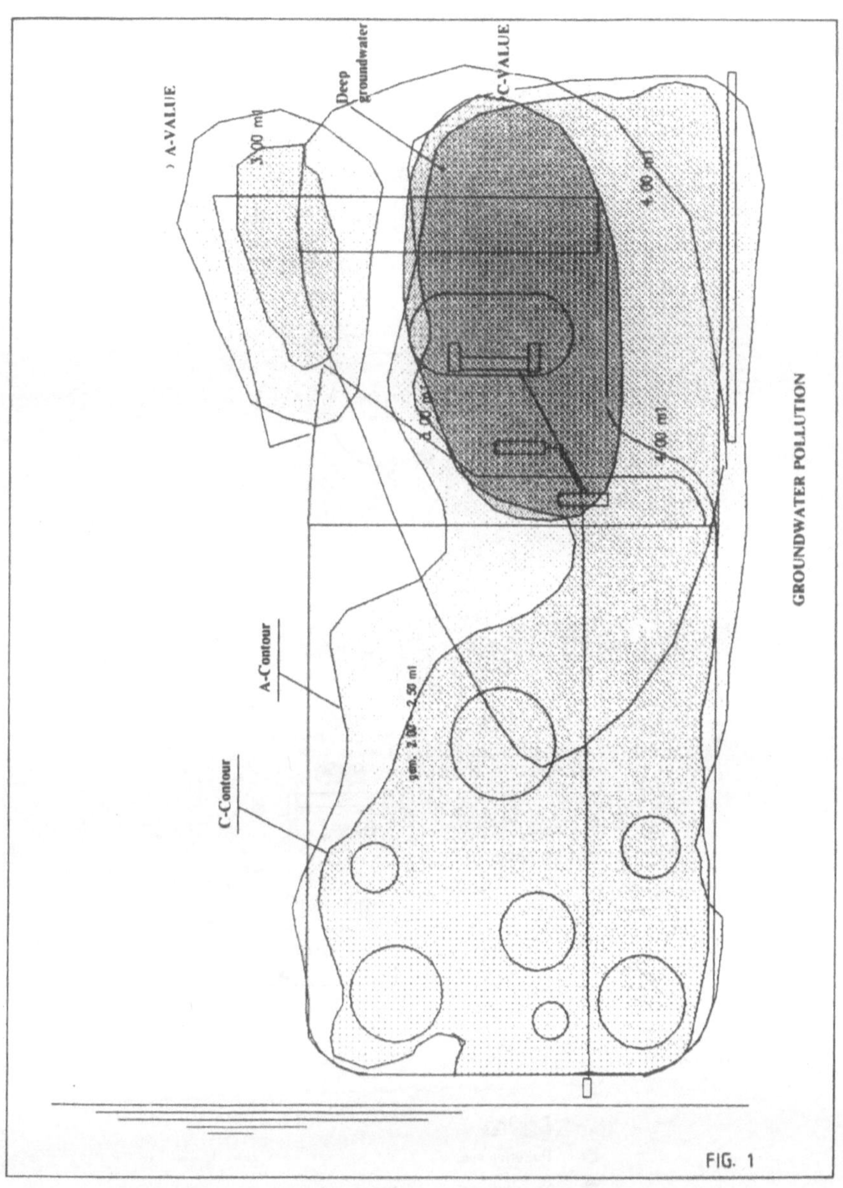
A-VALUE
3.00 ml
Deep groundwater
C-VALUE
4.00 ml
4.00 ml
A-Contour
C-Contour
gem. 2.00 - 2.50 ml
GROUNDWATER POLLUTION

FIG. 1

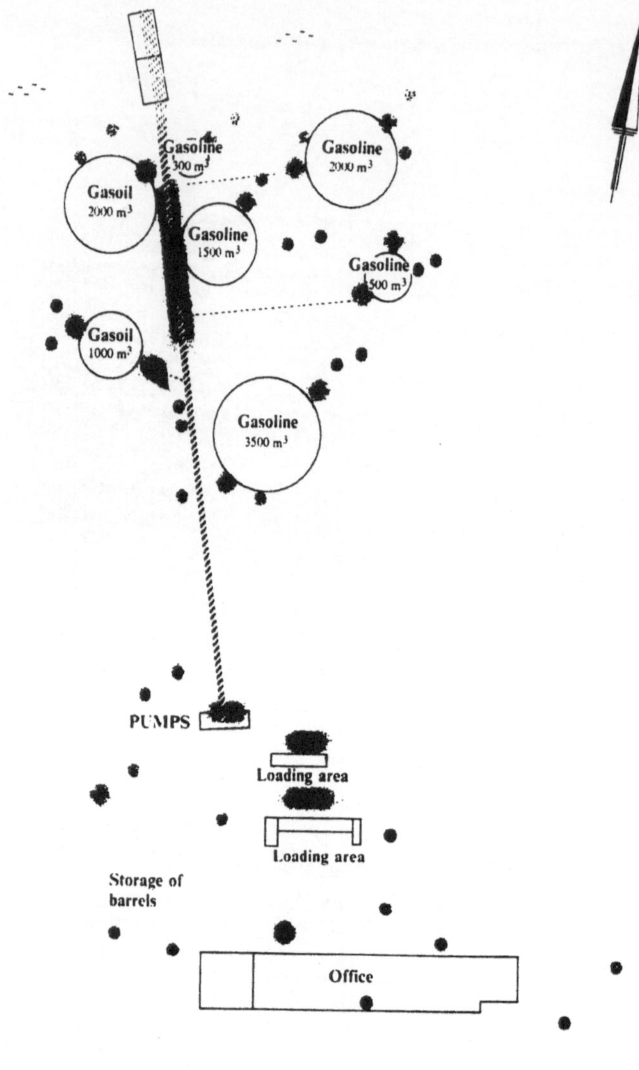

Gasoline
300 m³
Gasoline
2000 m³
Gasoil
2000 m³
Gasoline
1500 m³
Gasoline
500 m³
Gasoil
1000 m³
Gasoline
3500 m³
PUMPS
Loading area
Loading area
Storage of barrels
Office
Legend
Pollution spot
Non-polluted area (< 50 mg oil /kg soil)

FIG. 2

IMd Micon

It is, therefore, of the utmost importance to interpret the presented cleanup costs in this framework of uncertainties. Cleanup activities should be based on the removal of the risks to the environment and public health. Cleanup approaches based on concentration values only will be prone to inevitable exaggeration of the costs.

4. Evolution of the International Scene

In several industrialized countries such as The Netherlands, Germany, Denmark, Canada, and USA, a very clear trend can be observed towards a rational approach to soil and groundwater clean-up. This approach is based on the experience that a removal of 100% of the pollutants is very costly and in many cases not necessary to obtain an acceptable situation.

The clean-up approach is based increasingly on risk assessment and a lot of common sense. The cost-efficiency curve in combination with a risk assessment based on the fate and effects of pollutants in the soil are vital tools in the negotiations with the local and state authorities. The outcome of these negotiations show a clear trend towards a pragmatic approach and removal efficiencies of 80 to 95% in many cases.

The A, B and C values from the Czech guidelines have been derived from the Dutch A, B and C values developed in the mid-eighties. Slight differences between the two lists can be detected. The Czech and Dutch A, B and C values have only an advisory status. They are generally considered too stringent if used as absolute clean-up values. They are tools in the decision making process.

5. Mitigation Measures

The soil pollution clean-up approach should be "source" oriented (prevention of further pollution from sources) rather than "target" oriented (prevention of migration across site boundaries).

Through negotiations with the authorities, agreement should be sought on this "source" oriented approach. In the case of the "source" oriented approach, the investigations and remedial actions can be focused on the pollution-spots. In the case of the "target" oriented approach, mitigation measures will be designed to control the migration of pollutants across site boundaries. This implies the need for a large number of interception wells near the site boundaries and the "pump and treat" treatment of large volumes of groundwater.

For many cases a "source" approach will be considered and will therefore be taken as the starting point for making cost estimates. In order to make these estimates, a distinction is made between two groups of pollutants:

- type A: Liquid pollutants which will float on water, and which usually form a layer on the groundwater. For these type of pollutants it is assumed that the remedial action will consist of the removal of a strongly polluted unsaturated zone and only a small volume of the groundwater;
- type B: Very soluble pollutants, such a metal salts. For this type of pollutants it is assumed that the remedial action will consist of the removal of a moderately polluted unsaturated zone but a larger volume of groundwater.

A simple model has been developed to describe the distribution of pollutants in the "spots". Figures 3 to 7 present a typical pollution spot for the contamination of an oil storage plant in the Netherlands. The contaminants consist of a mixture of gasoil and gasoline.

The piechart in Figure 8 presents the percentage of the pollutant load in the various zones. Based on these data a cost-efficiency curve can be constructed. Such a curve is presented in Figure 9.

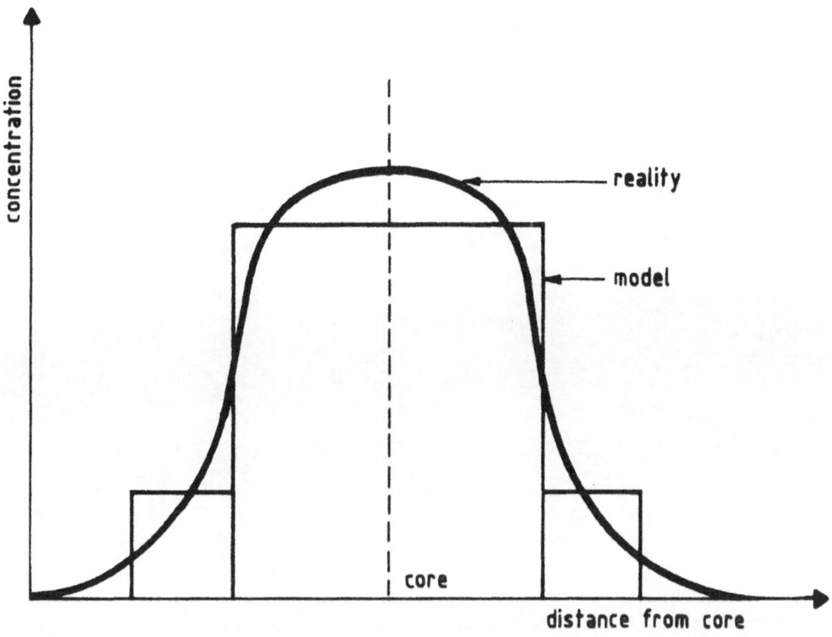

FIG. 3

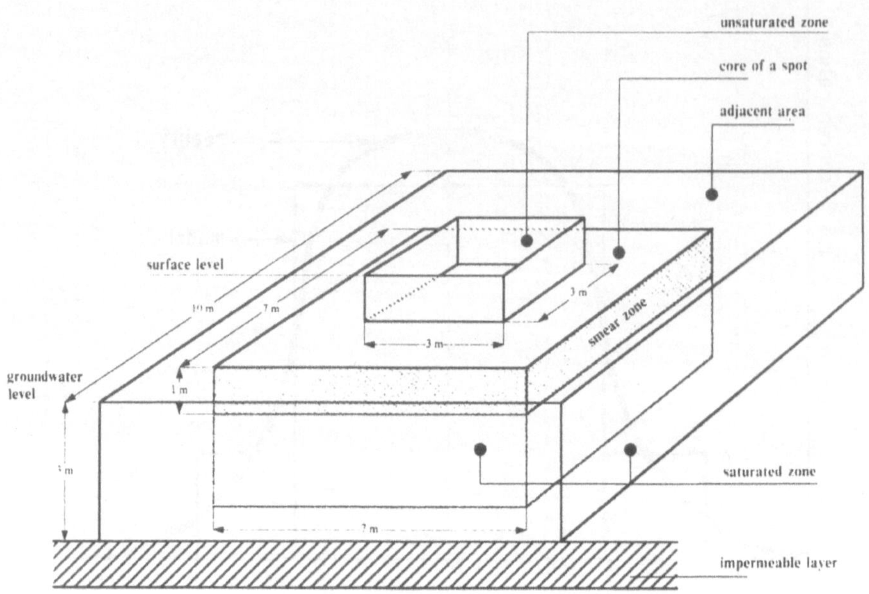

DIMENSIONS OF A CONTAMINATED SPOT

FIG. 4

OIL CONTENT

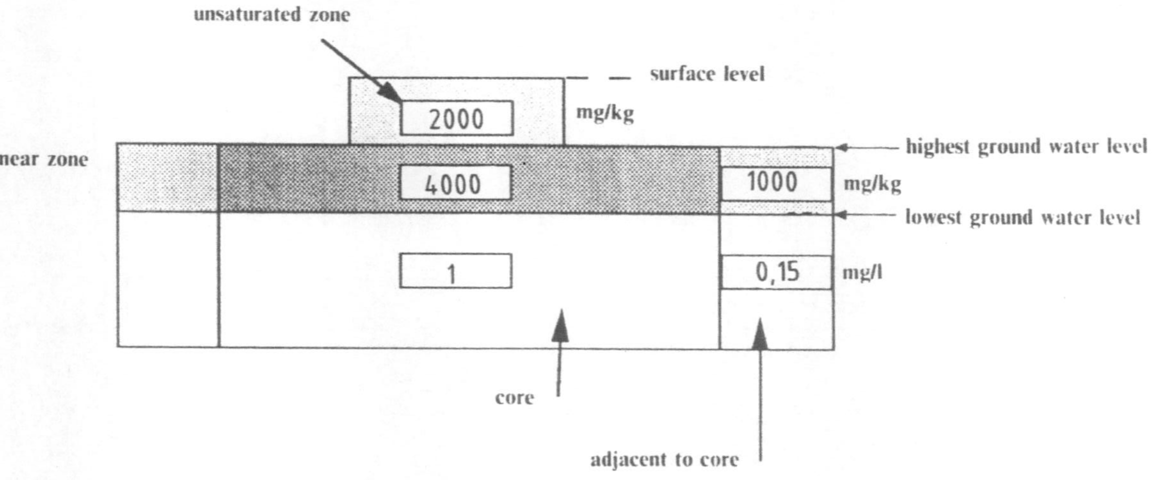

FIG. 5

BENZENE, TOLUENE, ETHYLBENZENE, XYLENE (BTEX) CONTENT

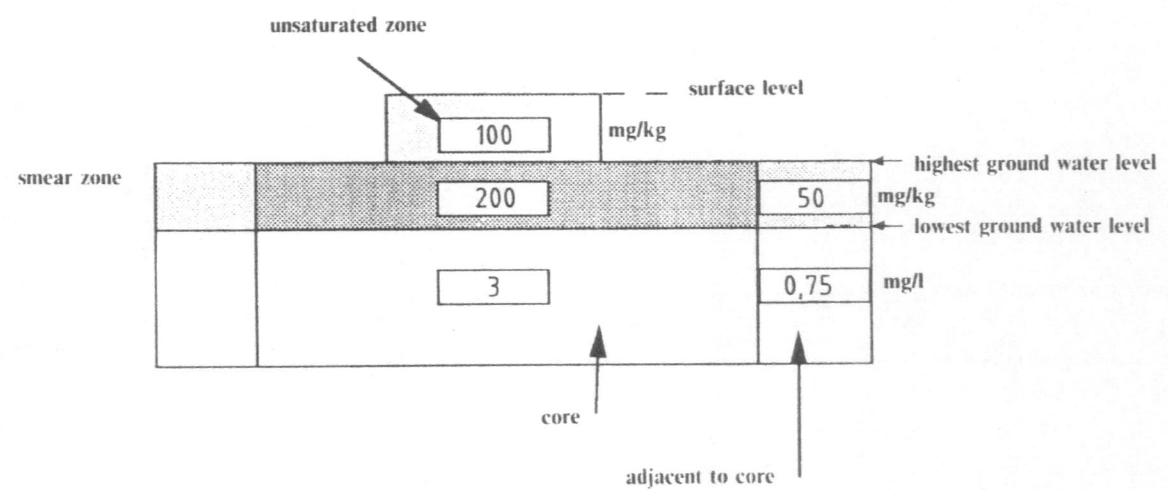

FIG. 6

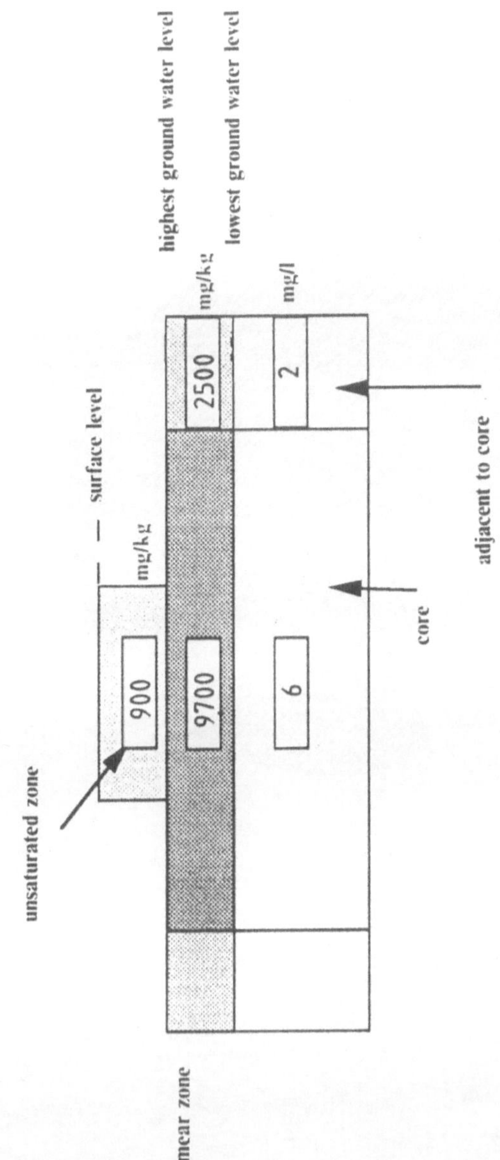
VOLUME OIL + AROMATICS
unsaturated zone
900
mg/kg
surface level
9700
6
core
adjacent to core
highest ground water level
2500
mg/kg
lowest ground water level
2
mg/l
smear zone

FIG. 7

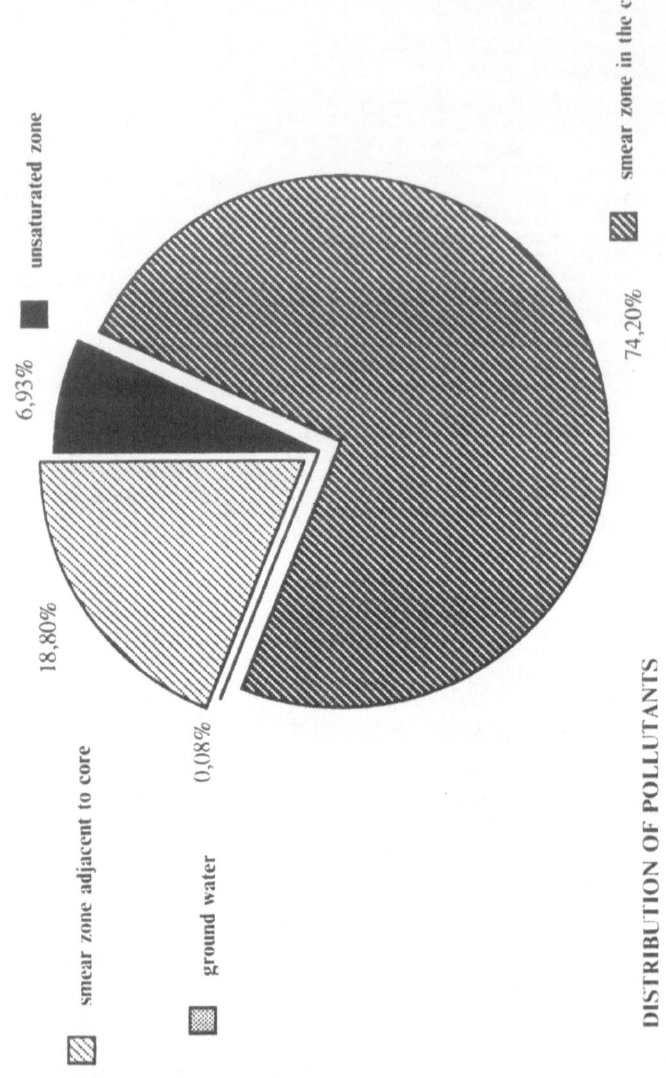

DISTRIBUTION OF POLLUTANTS

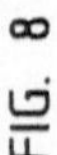

FIG. 8

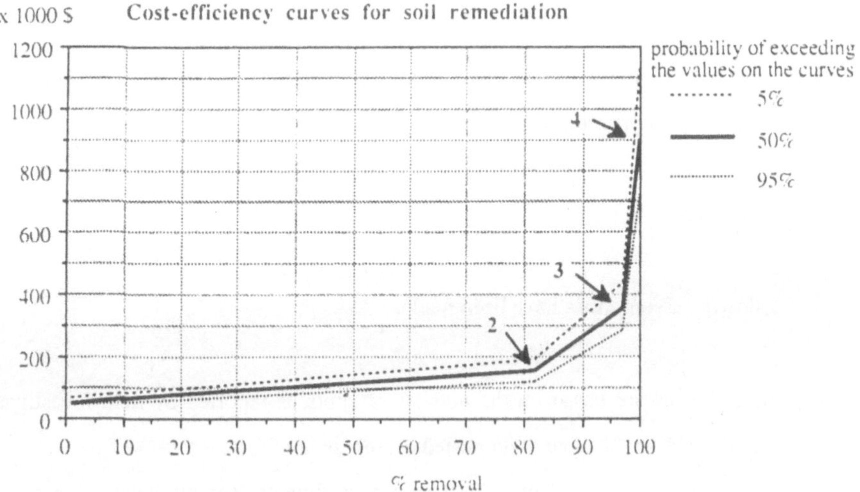
x 1000 $
Cost-efficiency curves for soil remediation
1200
1000
800
600
400
200
0
0
10
20
30
40
50
60
70
80
90
100
% removal
probability of exceeding the values on the curves
5%
50%
95%
2
3
4

FIG. 9

Case Study: Industrial Site, Czech Republic

As parts of an environmental audit, soil remediation costs have been estimated according to the approach presented.

1. Assumptions

The following assumptions have been made:

- the estimates are based on the number of spots as reported by field investigation plus the additional spots as identified during the audit;
- the estimates are based on two groups of pollutants (soluble versus non-soluble) with investigation and remedial action approach as described in the previous paragraph;
- the estimates are based on a uniform type of spot (depth, surface, volume of groundwater involved);
- the estimates are based on unit price ranges for soil (excavating, dumping/incinerating, new soil, transport) and groundwater (well installation, pump and treat, discharge) as presented in Table 1; and
- an average of approximately 95% of the pollution is removed.

Table 1.

Activity	*Unit*	*Total Cost (CZK)*
Excavation	m^3	200 - 300
new soil	m^3	180 - 350
dumping		
category: IIa - IIb	t	200
category: IIb - IIIa	t	800
category: IIIb	t	2,800
incineration	t	5,000 - 20,000*
transport	t/km	2.50 - 3.50
pumping well (8 m deep)	1 pc	12,000 - 14,000
water treatment	m^3	6
gravity separation of oil	day	500
	m^3	16
heavy metals	day	500
	m^3	14 - 22
biological treatment	m^3	14 - 22
stripping + sorption	m^3	32 - 40
ozonation	m^3	28 - 35
wet oxidation	1 year	150,000 - 350,000
groundwater monitoring(1) analyses (depends on type of contaminants)	1 pc	1,500 - 7,500

* maximum cost - 20,000CXK for incineration of PCBs

(1) depends on the number of wells and range of analysis

2. Uncertainty Analysis

Since the input for the cost estimates is largely based on estimates and assumptions, the results will be presented with a financial "risk profile".

This risk profile consists of a range of the final cost for 95% removal of the pollutants.

- all variables (depth, surface, unit prices) which carry an uncertainty are presented as a range. These ranges are the input for the uncertainty/sensitivity model;
- costs are calculated about 10,000 times with a random choice (within the ranges) for all the variable. This mathematical approach is known as the Monte Carlo simulation.

For the site under consideration, a total of 192 equivalent spots have been counted. These equivalent spots have the following dimensions:

- surface area:20 - 80 m^2, with an average of 50 m^2
- depth:5 ± 5 m, with an average of 5 m below the surface

The line sources represent an estimated 100 equivalent spots. The total clean-up costs for 95% removal amount to 2,2 M US$ with an uncertainty range of US$ 1.5 M (10% probability) to US$ 3 M (90% probability).

The 10% possibility means that there is only a 10% chance that the real cost will be lower than this value. The 90% possibility means that there is a 90% chance that the real cost will be lower than this value.

Conclusion

It should be emphasized that the above-mentioned costs are based on the assumption that all of the identified polluted spots will have to be cleaned within the coming 20 years. This, however, will not be the case. Furthermore, the cost estimates are based on a 95 removal percentage. If a lower removal percentage is acceptable, lower costs will result. Finally, further costs can be limited if low-priced soil treatment facilities were to become available in the area.

The cost-efficiency curve is an essential element in the decision making process together with a risk assessment of residual pollutant concentrations and pollutant loads.

Based on a limited risk assessment it appears that the lowest cost/benefit ratio can be obtained by preventive measures and clean-up actions near the source (source approach). Preventing pollution of migrating across the site boundaries ("target" approach) is generally a much more expensive approach.

III.4

MONITORING REQUIREMENTS FOR HEALTH AND SAFETY DURING REMEDIATION

D.H. HAYCOCK & S. BESZEDITS
Conestoga-Rovers & Associates Limited
651 Colby Drive
Waterloo, Ontario
Canada N2V 1C2

There are thousands of sites contaminated with hazardous wastes scattered throughout North America. A large number and variety of substances may be present at any given location. While toxic chemicals are the most prevalent contaminants, unique pollutants may be encountered at some sites. For example, military sites may be contaminated not only with aviation fuel, fire fighting chemicals, and deicing fluids, but also with ordinance and explosive wastes and low-level radioactive materials [1].

Remediation of hazardous waste sites is a complex and challenging task that dictates a critical assessment of environmental conditions, real and potential risks, regulatory cleanup requirements, remedial options, costs, and numerous other site- specific factors. Moreover, to carry out a remediation program effectively and within time and budget constraints demands the skills and coordinated efforts of a multidisciplinary team of personnel.

E.A. McBean et al. (eds.) Remediation of Soil and Groundwater, 227–250.

Personnel engaged in the remediation of hazardous waste sites face a myriad of health and safety concerns. In addition to chemical exposure, these concerns include fire and explosion, oxygen deficiency, ionizing radiation, biological hazards, electrical hazards, heat stress and cold exposure, and noise. Consequently, ensuring the health and safety of personnel constitutes a most important component of any remediation undertaking.

Today, a multitude of laws and regulations are in effect in both Canada and the United States which specifically address activities at hazardous waste sites in order to protect the health and safety of workers [2,3,4,5]. Similar laws prevail in most European countries [6,7]. Most of the laws and regulations in North America have been enacted recently; in the not-too-distant past laws and regulations pertaining to workers involved with hazardous wastes were either nonexistent or superficial. Until fairly recently as a matter of fact, there were few laws addressing health and safety issues for employees in any industry.

Due to the lead the United States has taken in cleaning up contaminated sites and instituting health and safety laws in the workplace, it is instructive to review the evolution and current status of worker-protection legislation in the United States.

Prior to 1960, federal laws pertaining to worker protection were scant in number. Most of the occupational safety and health laws were enacted by the individual states. These regulations varied widely, and enforcement proceedings against violators were lax. Unsafe and unhealthy conditions prevailing at many workplaces prompted a proliferation of federal laws during the 1960s. Eventually, Congress passed the Occupational Safety and Health Act (OSHAct) in 1970 [4,5,8]. In effect since April 28, 1971, this nationwide act was passed to ensure safe and healthful conditions in the workplace. It requires employers to institute steps to protect employees from recognized workplace hazards that are likely to cause illness or injury.

The OSHAct also led to the establishment of three organizations: the Occupational Safety and Health Administration (OSHA), the Occupational Safety and Health Review Committee (OSHRC), and the National Institute for Occupational Safety and Health (NIOSH). The most important of these three are OSHA and NIOSH. Created within the Department of Labor, OSHA is primarily responsible for developing and implementing workplace safety and health standards. Located within the Department of Health and Human Services, NIOSH conducts research on occupational hazards, and makes recommendations to OSHA regarding the formulation of OSHA standards based on research results. The agency is also actively involved in education, training, and disseminating information related to occupational safety and health. At the present time, in addition to OSHA and NIOSH, several other federal agencies also have specifically defined responsibilities for assisting in the protection of workers from adverse health effects.

The 1960s and 1970s were also decades in which environmental issues rose to the forefront. In 1976 Congress enacted the Resource Conservation and Recovery Act (RCRA) which mandated the management of hazardous wastes [4,5]. To prevent future long-term environmental damage from unsound hazardous waste dumps, the Comprehensive Environmental Response, Compensation and Liability Act (CERCLA), commonly referred to as Superfund, came into force in 1980. While RCRA deals primarily with active industrial sites, CERCLA deals with inactive ones. Both of these acts require that the Environmental Protection Agency (EPA) develop and enforce specific standards to attain the goals set forth within them. According to a recent estimate, it will cost about 100 to 300 billion US dollars to comply with the aims of RCRA and CERCLA.

Enforcement of these EPA standards stimulated a great expansion in the environmental industry. Demand for workers versed in handling toxic chemicals or hazardous wastes outstripped availability, resulting in the hiring of individuals with little or no experience. To provide the necessary protection for the health and safety of workers at hazardous waste sites, Congress in 1986 promulgated the Superfund Amendments and Reauthorization Act (SARA). Under SARA, OSHA was instructed to develop standard procedures and strategies designed to ensure the health and safety of all employees engaged in hazardous waste activities.

In response to these directives, OSHA standard 29 CFR (Code of Federal Regulations) 1910.120 was promulgated [4,5]. Effective as of March 6, 1990, it is regarded as the most significant occupational health and safety legislation since the OSHAct of 1970. The standard is a comprehensive document, covering virtually all aspects of hazardous waste site remediation. It requires that employers develop and implement a safety and health program for employees involved in hazardous waste operations and stipulates medical surveillance, training, planning, and several other important activities. Key provisions and ramifications of 29 CFR 1910.120 are discussed in the ensuing sections.

Site Specific Safety and Health Plan

Before any activity can commence on a hazardous waste site, an overall safety and health program for all personnel involved must be developed and implemented. The plan should be flexible enough, however, to allow appropriate revisions as new information about the site emerges. It usually consists of three broad segments: organizational structure, work plan, and site-specific safety and health plan [4,9].

The organizational structure identifies the personnel needed for the overall operations, delineates the chain of command, and specifies the responsibilities of each employee.

The work plan defines the objectives of the site operations as well as the logistics and resources required to attain the stated goals.

The site specific safety and health plan, which establishes policies and procedures to safeguard personnel from the potential hazards posed by the site, is a most important document. It must adequately address the specific safety and health hazards of each phase of the operations on the site and clearly stipulate worker-protection measures required for safe work.

A proper and acceptable site-specific safety and health plan must explicitly address a multitude of provisions. Some of the main elements to be covered include [4,9,10]:

- Key personnel responsible for health and safety
- Risks or hazards associated with each operation carried out
- Adequate training of personnel
- Personal protective clothing and equipment
- Medical surveillance
- Site control measures
- Decontamination procedures for personnel and equipment
- Air monitoring
- Contingency plan for safe and effective response in emergencies.

Detailed examples of typical site safety and health plans may be found in a number of publications [4,11].

Developing and implementing the plan is the responsibility of the site safety and health supervisor, the individual in charge of all matters relating to health and safety. Other duties of the supervisor are to conduct frequent inspections of site conditions, facilities, equipment, and activities to determine whether the plan is adequate and being complied with. Furthermore, before initial entry and then as deemed necessary, the supervisor holds safety meetings.

The plan must be posted on site and be readily available to all concerned parties, including employees, employee representatives, contractors, subcontractors, and officials of OSHA and other appropriate governmental agencies.

Site Characterization

The aim of site characterization is to generate information about site hazards and to assist in selecting worker protection methods. There are three broad stages to site characterization [4,9,12]:

- Offsite characterization
- Onsite survey
- Ongoing monitoring and hazard assessment program.

Offsite characterization consists of historical research and a perimeter reconnaissance. Historical research involves interviewing people having knowledge about the site (e.g. former owners and employees, and government officials), and examining old files and maps. A thorough historical research can yield a wealth of information about activities that occurred at the site and their duration, the nature of the wastes on the site, and the physical and chemical properties of hazardous materials that may be encountered. For example, historical research can readily ascertain whether contamination by fuels and firefighting chemicals is likely to be present at a military site. However, it is often difficult to obtain accurate information about certain aspects of military sites due to the secretive nature of past activities and the traditional reluctance of senior military officials to divulge information. Historical research can also provide valuable data on climatic, geologic, topographic, hydrologic, and environmental conditions which can be organized on a preliminary site map.

Perimeter reconnaissance is conducted following the completion of historical research. In reconnoitering around the site, it is particularly important to carefully observe and note buildings; tanks, drums, or other containers; signs and placards; size and location of impoundments; biologic indicators; unusual odors; surface water/liquids and their color; and wind direction. Monitoring the ambient air at the site perimeter, and collecting and analyzing soil, surface and groundwater samples around the perimeter constitutes the final phases of perimeter reconnaissance. With the data gathered, the preliminary site map can now be refined. Since the site map is essential to conduct remedial work, every effort should be made to ensure that the information on the map is both current and accurate. Hence, it must be continually updated to reflect changing conditions.

Onsite survey verifies and supplements the information obtained during the offsite characterization. The initial entry to the site must be well planned and carefully executed. Extreme vigilance and conservative actions are a must. While the composition of the entry team depends on the characteristics of the site, it should always consist of at least four persons; two of whom will enter the site and two who will remain outside in a supportive role [4]. Selection of the protective equipment donned by member of the team is based on the information from the historical research and perimeter reconnaissance. Air monitoring must be conducted during initial entry when site information is insufficient to rule out safety and health hazards. It is especially crucial to identify conditions that may pose inhalation or skin absorption risks and can be immediately dangerous to life and health (IDLH).

If IDLH or other threatening situations are not present, or if proper precautions can be taken, the entry team can continue the survey, noting terrain features; the types and condition of containers, impoundments, and other storage systems; the nature of the waste materials; natural wind barriers and potential pathways of dispersion; indicators of potential exposure to hazardous substances; as well as safety hazards such as confined spaces, damaged buildings, and cluttered or irregular surfaces. Collecting water, soil, waste, and other samples completes the onsite survey program.

Once the presence and concentration of specific chemicals or classes of chemicals has been established, it is imperative to determine the hazards associated with them; a task that can be accomplished through recognized procedures [9]. Information on the chemical, physical and toxicological properties of each chemical known or suspected to occur on site should be recorded on a Hazardous Substances Information Form and made known to all employees. Finally, after the site has been deemed safe for the commencement of operations, an air monitoring program for ongoing hazard assessment must be implemented.

Site Control

Site control minimizes the potential contamination of site workers and the surrounding community. Site control measures employed to reduce exposure to chemical, physical, biologic, and safety hazards include the delineation of work zones, use of the buddy system when warranted, enforcement of safe work practices, implementation of decontamination procedures, and establishment of communication networks [4,9].

Zoning is of particular significance for it will keep contaminants within specified areas and hence reduce the potential for spreading contamination. A hazardous waste site should be divided into as many different zones as needed to meet operational and safety objectives. The three most frequently used zones are [4,9]:

- The exclusion or hot zone. This is the area where the actual remediation work occurs and where the highest level of personal protective equipment is needed.

- The contamination reduction zone (CRZ). This is a buffer zone which surrounds the hot zone and is the site where decontamination takes place.

- The support zone. This is the uncontaminated area where the command post as well as the administrative, transportation, and communication facilities are located. Personnel should not be exposed to hazardous conditions here.

Ideally, these zones should be arranged concentrically. The outer boundary of the exclusion zone, known as the "hotline", should be clearly marked or enclosed by physical barriers such as fences, earth-berms, chains, or ropes. To prevent crosscontamination from contaminated areas to clean ones, movement of personnel and equipment among the zones and onto the site itself should be minimized and restricted to designated access control points.

Rendering assistance to one's partner when necessary is the primary purpose of the buddy system. However, since this arrangement alone may be insufficient, the command post supervisor should maintain visual or communications contact with workers in the exclusion zone at all times.

Two sets of communications should be established: internal communication among personnel on site, and external communication between onsite and offsite personnel.

To maintain a strong safety awareness and enforce safe procedures at a site, a list of standard operating procedures (SOPs) should be developed which state practices that must always be followed and those that must never occur in the contaminated areas of the site. If the hazards are substantially different, separate SOPs should be formulated for the hot zone and the contamination reduction zone. The list should be displayed conspicuously and given to everyone entering the site.

Safety and Health Training

Proper training is one of the key aspects of ensuring the health and safety of personnel involved in hazardous waste site activities. No worker should engage in any form of field activity without sufficient training. The objectives of training programs are [4,9,13]:

- To install an awareness in workers of the potential hazards they face.

- To provide the knowledge and skills necessary to perform the assignments with minimal risk to personal health and safety.

- To make workers aware of the purpose and limitations of safety equipment.

- To teach workers how to cope with emergency situations.

The level of training provided should conform with the worker's job functions and responsibilities, should include classroom instructions as well as practical experience, and should be up-to-date. Contents of the training course should cover topics such as health and safety regulations, specific hazards and their risks, hazard recognition, proper use of engineered controls, use and decontamination of personal protective equipment, and a knowledge of emergency procedures. Those who have successfully completed the prescribed training are given a written certificate.

According to OSHA regulations, general site workers who are engaged in activities which have a high exposure potential are required to complete a minimum of 40 hours of off-site instructions, a minimum of 3 days of actual field experience under the direct supervision of a trained, experienced supervisor, and 8 hours of annual refresher course [4,10].

Employees who work in areas that have been monitored and fully characterized indicating that exposures are under permissible levels and employees who are on site only occasionally and are unlikely to face high exposures require a minimum of 24 hours of classroom instruction, a minimum of one day on the job training, and 8 hours of annual refresher training.

Managers and supervisors must complete the same (or equivalent) training as required for the workers they oversee, a minimum of 8 hours of specialized off-site supervisory training, and 8 hours of annual refresher training.

Workers who may be exposed to unique or special hazards must receive additional appropriate training. Employees assigned to respond to emergencies must receive training on proper response procedures. Trainers must complete a training program for teaching a specific subject, or have credentials and instructional experience for that subject.

A record of training should be maintained in each employee's file to confirm that the individual assigned to a task has been given sufficient training for that task.

Medical Program

Since workers involved in remedial activities may be exposed to a number of adverse and injurious conditions, a sound medical program is the cornerstone of an effective health and safety program. A medical program should be developed for each site based on the specific needs, location, and potential exposure of employees at the site. It should be designed by an experienced occupational health physician in cooperation with the site safety supervisor. It is important to bear in mind that devising a medical surveillance program is not an easy task because employees at hazardous waste sites are potentially exposed to innumerable toxic substances, often in situations where identification or quantification of these exposures is impossible.

Components of a site medical program include: surveillance, consisting of pre-employment screening, periodic medical examination, and termination examination; emergency and non-emergency treatment; recordkeeping; and program review [4,14,15].

The main objectives of a pre-employment medical examination are to ascertain whether the prospective employee is capable of performing his assigned tasks, especially while wearing personal protective equipment, and to establish baseline profile information for comparison with future medical data. The pre-employment screening should therefore include a medical history and a physical examination. The medical history should elicit information on past illnesses and chronic diseases, while the physical examination should focus on the pulmonary and cardiovascular systems.

Frequency and content of a periodic medical examination will depend on the nature of the work and exposures; however, it should take place at least annually [14]. In terms of scope, the basic periodic medical examination is the same as the pre-employment screening, modified according to current conditions such as changes in the employee's symptoms, site hazards, or exposures.

Termination examination is given when employment is ended or when the employee is transferred to a new site. It ensures the completion of a comprehensive medical profile.

A wide range of actual and potential hazards must be considered in planning medical treatment for any possible emergencies. Key employees at the site should have some formal first aid training, especially in dealing with explosion and burn injuries, with heat stress, and with acute chemical toxicity. Standard operating procedures must be developed for emergency medical treatment of those injured workers who are also contaminated. Appropriate first aid equipment must be available on the site, and arrangements for evacuating injured or ill personnel, including transportation to a nearby hospital, must be established. Hospital emergency wards should be informed of any special or unusual risks posed by the injured or ill worker needing immediate care.

Arrangements must be made for non-emergency medical care or consultation for employees who are experiencing lingering health effects resulting from an exposure to hazardous substances. Moreover, off-site medical personnel should also investigate and treat non-job related illnesses that may put the worker at risk because of task requirements.

Proper recordkeeping is essential due to the nature of the work and risks and because of the time interval between exposures and the appearance of their chronic effects. Current OSHA regulations require that an employer must maintain and preserve medical records on exposed workers for 30 years after they leave employment [4,14,15]. The results of medical testing and full medical records must be readily available to employees, their union representatives, and OSHA officials.

Periodic evaluation is necessary to ensure the effectiveness of a medical program. The assessment should be conducted at least yearly and cover the following [9,14]:

- Ascertaining that each accident or illness was promptly investigated to determine its cause and that necessary changes in health and safety procedures were made when warranted.

- Evaluating the efficacy of specific medical testing in the context of potential site exposure.

- Adding or deleting specific medical tests as suggested by current industrial hygiene and environmental health data.

- Determining if any additional testing is required.

- Reviewing emergency and non-emergency medical protocols.

Air Monitoring

Since airborne contaminants constitute a significant threat to human health, identifying and quantifying these contaminants at hazardous waste sites is of vital importance. Reliable measurements of airborne contaminants are necessary not only to assess the health risks posed to workers and the general public but also to select appropriate protective equipment, to differentiate between areas where protection is needed and areas where it is not needed, to determine when operations must be ceased for safety or health reasons, and to determine medical surveillance requirements [4,9,12].

There are two basic approaches available for identifying and/or quantifying airborne contaminants [9]:

- The onsite use of direct reading instruments.
- Laboratory analysis of air samples obtained by gas sampling bag, filter, or some other means.

Air monitoring priorities on the site should be based on the information generated during initial site characterization. Depending on the site conditions and project goals, four categories of site monitoring may be required: monitoring for IDLH and other dangerous conditions, general onsite monitoring, perimeter monitoring, and periodic monitoring.

Monitoring for IDLH and other dangerous conditions determines the presence or absence of flammable or explosive atmospheres, oxygen-deficient environments, and highly toxic levels of airborne contaminants. The purpose of general onsite monitoring is to identify the major classes of airborne contaminants and their concentrations. Perimeter monitoring, i.e. monitoring at the perimeter of the site, helps to evaluate and confirm the integrity of the site's clean areas. Periodic monitoring is needed to account for changes in atmospheric conditions as a result of work on the site.

In addition to the general environment, individuals working on the site should also be continually monitored for dangerous gases, vapors, and particulates. One means of accomplishing this is by having persons working in a contaminated area wearing sampling devices. The samples obtained represent the actual inhalation exposure of those who are not wearing respiratory protection and the potential exposure of those who are equipped with respirators.

Personal Protective Equipment

Personal protective equipment (PPE) shields or isolates individuals from the chemical, physical and biological hazards that they may encounter. Careful choice and proper use of adequate PPE should protect the respiratory system, hearing, and all other body parts.

PPE may be divided into two categories: chemical protective clothing, including accessories (e.g. safety helmet, utility knife, radio, safety belt, ear plugs and muff), and respirators [4,16]. In selecting and using PPE, one should seek to balance the needs of protection, productivity, comfort, and cost. Selection of the equipment should be made by someone who is familiar with both the equipment and the conditions under which the PPE will be employed. However, since no single combination of PPE is capable of extending protection against all hazards, PPE should be utilized in conjunction with other protective measures.

Chemical protective clothing (CPC) can be defined as any article providing skin and/or body protection, e.g. fully encapsulating and non-encapsulating suits, aprons, sleeve protectors, and gloves. CPC is available in a wide variety of materials [17,18]. As a barrier to chemicals, the performance of CPC is determined not only by the material but also the design and quality of the clothing construction. Ideally, the chosen material should resist permeation, degradation, and penetration. The CPC selected should offer the widest range of protection available against the specific chemicals known or expected to be on the site. The most preferable materials are those which offer the longest breakthrough times and lowest permeation rates.

Another important consideration in choosing CPC is its heat transfer characteristics. Since most chemical protective clothing is virtually impenetrable to moisture, evaporative cooling is limited. To ensure comfort, clothing with the lowest thermal insulation value should be selected in hot environments or for high work rates.

Besides permeation, degradation, penetration, and heat transfer, factors which need to be scrutinized during clothing selection include durability, flexibility, ease of decontamination, and compatibility with other equipment.

Respiratory protection is of paramount importance since inhalation is one of the chief routes of exposure to hazardous gases, vapors, or particulates. Respirators are usually classified as air purifying or atmosphere supplying [4,9,19]. Air purifying respirators employ filters, neutralizing agents, and/or sorbent materials to purify the ambient atmosphere for breathing while atmosphere supplying systems protect the wearer by providing an alternative source of respirable air. Air purifying units afford effective protection in atmospheres which contain relatively low concentrations of known contaminants and are not excessively depleted of oxygen. On the other hand, atmospheric supplying respirators must be utilized when the working atmosphere contains high levels of pollutants, extremely toxic contaminants, or deficient oxygen levels. Because the use of respiratory protection places an additional physiological burden on the wearer, a medical check to determine fitness to wear such protection should be given to all potential users.

Individual components of respiratory protective equipment, chemical protective clothing, and various accessories may be assembled into a personal protective ensemble.

As the amount of information about the site increases, personnel should be able to upgrade or downgrade their level of protection with the approval of the site safety supervisor.

To properly don and doff an ensemble, set procedures should be followed. Since these operations are difficult to perform alone, assistance should be provided. Correct doffing is especially important in order to prevent contaminant migration. The person assisting during doffing must be suitably attired.

Needless to say, all PPE must be inspected frequently for damages which should be promptly repaired if possible, completely decontaminated after use, and stored correctly.

Wearing PPE puts the hazardous waste worker at considerable risk of developing heat stress. Because heat stress is one of the most common illnesses at hazardous waste sites, regular monitoring and other preventive precautions are essential. Physiological monitoring includes measuring the heart rate, oral temperature, and body water loss. Preventive measures to mitigate heat stress include adjusting work schedules, resting in cool areas, drinking large amounts of fluids during periods of heavy sweating, and maintaining physical fitness.

Decontamination

Decontamination is the process of removing or neutralizing contaminants that have accumulated on personnel and equipment. Because decontamination is crucial to health and safety, all personnel, clothing, and equipment leaving the contaminated areas of a site must undergo decontamination. Contaminants can be present not only on the surface of PPE but can also penetrate into the material. While surface contaminants are relatively easy to detect and remove, the opposite holds true for contaminants that have penetrated a material. Since prevention is one of the best methods to mitigate the extent of contamination and hence facilitate decontamination, it is important to establish standard operating procedures during work that minimizes contact with wastes.

Decontamination processes either physically remove contaminants, inactivate contaminants by chemical detoxification or disinfection/sterilization, or employ a combination of both physical and chemical means to eliminate contaminants [4,9]. Sterilizing techniques are impractical for personal protective clothing and equipment; therefore disposable PPE should be used when dealing with infectious agents.

Physical means involving dislodging/displacement, rinsing, wiping off, and evaporation can readily remove the bulk of gross contamination from PPE. Loose contaminants, adhering contaminants, and volatile liquids are particularly susceptible to physical removal from PPE. Physical removal is usually followed by a wash/rinse process using cleaning solutions. In choosing the appropriate wash and rinse solutions for decontamination, several factors have to be taken into account such as the solubility behavior of the contaminants, effectiveness of the solutions and methods, compatibility of the solutions with the contaminants and item to be decontaminated, and the hazards associated with certain cleaning solutions. The most common solvents employed are water, dilute acids, dilute bases, and organic solvents.

Decontamination facilities should be located in the contamination reduction zone. The facilities may be temporary or permanent; type and duration of site activity anticipated usually determines the choice.

The decontamination process for personnel and their clothes and equipment involves a series of steps which are carried out in a specified sequence. Decontamination and removal of protective garments and other equipment starts with the most contaminated items, followed by the less contaminated items, and ending with the least contaminated ones. Each process is carried out at a different station, separated physically to prevent crosscontamination between them. Flow design, size, and number of stations at a site generally depend on the number and frequency of workers undergoing decontamination. Before entering the clean support area, workers undergo a vigorous wash, using soap, water, and scrubbing to ensure decontamination of their bodies.

Measuring the effectiveness of decontamination for PPE is rather difficult in most cases. Visual inspection, wipe sampling, and analysis of cleaning solutions are some of the basic tests used to monitor effectiveness [9].

Ensuring a viable decontamination program at a site entails a number of other considerations. For example, provisions must be made not only for routine decontamination but also for emergency situations. Decontamination workers must also be properly decontaminated before entering the clean zone. Clothing and equipment which cannot be decontaminated or which are discarded must be properly disposed of as are waste liquids and any other residue generated during decontamination activities.

Emergency Response

Due to the nature of the activities at a hazardous waste site, emergencies are an ever-present threat. Common causes of emergencies may be work-related or waste-related and include fire, leaks, explosions, chemical exposure, medical problems, and mundane accidents such as slips and falls. An emergency may be minor or major. Whether it is one or the other, it calls for prompt, decisive, and accurate response.

In order to handle emergencies, planning is essential. Therefore, a contingency plan, a document which sets forth policies and procedures to site emergencies, should be developed. It should be integrated with the plans of appropriate government agencies, be rehearsed regularly, and be reviewed periodically. The plan should include [4,9,12]:

- The names and functions of onsite emergency personnel.

- A warning signal or system which can be detected throughout the site.
- A detailed site map.
- Locations and availability of standard emergency equipment.
- A list of telephone numbers for outside emergency services such as hospitals, fire, ambulance, and police.
- Evacuation routes and procedures.

Training of personnel responding to emergencies is absolutely essential. Therefore, all personnel involved in emergencies must attend training classes. The curriculum should be realistic and practical, focus directly on site-specific anticipated situations, and be brief and often repeated . Those having emergency response duties should have their records confirm that they had been sufficiently trained and that the training received is up-to-date.

Since onsite personnel may not have the manpower or resources to cope with an emergency, contingency should be made for a multidisciplinary support task force to provide consultation or assistance to the on-site response personnel. Other offsite groups whose help may be required in an emergency response include hospitals and local fire and police departments. The scope of assistance required from outside personnel is dictated by the nature and extent of the emergency.

Before a site can return to normal operations all emergency resources should be replenished. Until personnel and equipment are ready to handle another emergency situation, the site cannot be regarded as safe.

After the emergency has been resolved, the cause or causes should be thoroughly documented. Documentation serves to help avert recurrences, as evidence for future legal actions, for assessment of liability by insurance companies, and for review by government agencies. Consequently, it should be objective, accurate, and complete. The document should include a chronological history of the event, all known facts about it, names and titles of personnel responding, actions taken to resolve the situation, and a tabulation of all injuries, exposures or illnesses incurred during or as a result of the emergency.

Transportation of Hazardous Wastes

Trucking hazardous wastes from a remediation site requires the same safety and health concerns as any other site activity in order to avoid accidents and minimize contamination. Loading of bulk waste or containers should be carried out by using plastic ground covers and truck bed liners to prevent contamination of the surrounding ground area and the truck itself. Any part of the truck which has come into contact with contaminated materials or contamination on the ground must be decontaminated as the vehicle passes through the contamination reduction zone. All wash and rinse solutions from this operation must be properly disposed of.

Both the cleanup contractor and the transporter must adhere to the shipping regulations set forth by several federal agencies. While several agencies are involved, the Department of Transportation (DOT) is the principal regulatory agency for the interstate transport of hazardous wastes [4]. DOT regulations require that hazard class labels be affixed to all containers greater than 110 gallons (420 liters). Color, print type and size as well as placement of these labels on the containers are governed by a set of explicit and strict rules. All vehicles must bear a clearly visible placard on each end and each side.

DOT regulations also stipulate that workers engaged in loading and decontaminating trucks and labelling drums and other containers must have had specialized training in these tasks. Truckers must wear appropriate protective clothing.

Even though a variety of shipping papers may accompany transported hazardous wastes, the only legally authorized document is the original generator-prepared hazardous waste manifest. This document, properly filled out, must accompany the waste at all times.

Concluding Remarks

A hazardous waste site is a hostile terrain, posing innumerable threats to personnel engaged in remedial actions. To ensure safe working conditions and avoid accidents, it is imperative that all employees be properly trained, have the right protective equipment, and follow prescribed procedures. Above all, however, employees should exercise common sense. In most instances, common sense is the best guide in performing a task properly and safely.

References

1. Pastorick, J.P. (1994) Ordnance, explosive waste, and unexploded ordnance, in W.F. Martin and S.P. Levine (eds.), *Protecting Personnel at Hazardous Waste Sites*, Butterworth-Heinemann, Boston, pp. 404-421.
2. Majthenyi, C. (1994) *Guide to Environmental Legislation in Ontario*, Southam Information and Technology Group, Don Mills.
3. The Canadian Institute (1994) *The Fundamentals of Environmental Law & Regulation*, Canadian Institute Publications, Toronto.
4. Andrews, L. P. (1990) *Worker Protection During Hazardous Waste Remediation*, Van Nostrand Reinhold, New York.
5. Levine, S.P., Turpin, R.D. and Gochfeld, M. (1994) Protecting personnel at hazardous waste sites: current issues, in W.F. Martin and S.P. Levine (eds.),

Protecting Personnel at Hazardous Waste Sites, Butterworth-Heinemann, Boston, pp. 1-22.

6. Haines, R. and Bardsley, D. (1992) *The Education and Training of Personnel Involved in the Handling and Monitoring of Hazardous Wastes*, Office of Official Publications of the European Communities, Luxembourg.
7. Haines, R. C. (1989) *A Study on the Safety Aspects Relating to the Handling and Monitoring of Hazardous Wastes*, Office of Official Publications of the European Communities, Luxembourg.
8. Dusenbury, M.R., Yarbrough, K.R., Benson, B.E. and W.F. Martin (1994) Federal government programs and information gathering, in W.F. Martin and S.P. Levine (eds.), *Protecting Personnel at Hazardous Waste Sites*, Butterworth-Heinemann, Boston, pp. 23-55.
9. NIOSH/OSHA/USCG/EPA (1985) *Occupational Safety and Health Guidance Manual for Hazardous Waste Site Activities*, US Government Printing Office, Washington.
10. Roughton, J. (1995) Protecting the hazardous waste worker, *Pollution Engineering* **27** (6), 88-91.
11. Martin, W.F. (1994) Site health and safety plans, in W.F. Martin and S.P. Levine (eds.), *Protecting Personnel at Hazardous Waste Sites*, Butterworth-Heinemann, Boston, pp. 422-536.
12. Task Force on Hazardous Waste Site Remediation (1990) *Hazardous Waste Site Remediation*, Water Pollution Control Federation, Alexandria.
13. Dahlstrom, D.L. (1994) Occupational health and safety programs for hazardous waste workers, in W.F. Martin and S.P. Levine (eds.), *Protecting Personnel at Hazardous Waste Sites*, Butterworth-Heinemann, Boston, pp. 56-80.
14. Hall, S.K. (1992) Health surveillance of hazardous materials workers, *Pollution Engineering* **24** (9), 58-62.

15. Melius, J. M. (1994) Medical surveillance for hazardous waste workers, in W.F. Martin and S.P. Levine (eds.), *Protecting Personnel at Hazardous Waste Sites*, Butterworth-Heinemann, Boston, pp. 183-193.

16. Schwope, A. D. and O'Leary, C. (1994) Personal protective equipment, in W. F. Martin and S.P. Levine (eds.), *Protecting Personnel at Hazardous Waste Sites*, Butterworth-Heinemann, Boston, pp. 219-257.

17. Schwope, A.D., Costas, P.P., Jackson, J.O. and Weitzman, D.J. (1987) *Guidelines for the Selection of Chemical Protective Clothing*, American Conference of Government and Industrial Hygienists, Cincinnati.

18. Forsberg, K. and Keith, L.H. (1989) *Chemical Protective Clothing Performance Index Book*, John Wiley & Sons, New York.

19. Cheremisinoff, P. N. (1989) Respiratory protection, *Pollution Engineering* **21**(7), 81-85.

III.5

EXPERIENCE WITH IN SITU TREATMENT SYSTEMS: AN OVERVIEW

ARTHUR S. KURZYDLO

Conestoga-Rovers & Associates

8615 West Bryn Mawr Avenue

Chicago, Illinois 60631

Introduction

In situ treatment systems form a very important class of remedial actions for soil and groundwater that are available today. The in situ methods rank high as popular and cost effective treatment technologies. Based on the information from the U.S. Environmental Protection Agency's (EPA) Vendor Information System for Innovative Treatment Technologies (VISITT) and Superfund Innovative Technology Evaluation (SITE) program (EPA, 1993) one-third of all innovative technologies are in situ treatments. Additionally, the current trend in site remediation is to focus on release control and prevention of further spread of contaminants and not on comprehensive clean-up programs. Thus, in situ technologies are becoming more attractive to consultants and regulators.

Before any treatment technology is proposed for the site, an array of basic information needs to be collected. Most important are site characteristics, waste characteristics, technology limitations, and identification of impacted media. Site characteristics include information on soil and groundwater.

E.A. McBean et al. (eds.) Remediation of Soil and Groundwater, 251–270.

The most important parameters for soil are:

- stratigraphy
- hydrology
- geotechnical parameters
- chemical composition
- surface topography
- engineered features

Whereas, for groundwater characterization the following are required:

- aquifer properties
- geochemical environment
- hydrologic properties
- chemical composition

To adequately characterize waste, information on volume, mobility, concentration and toxicity of every class of chemicals is required.

Once the site and waste are characterized, a proper treatment system can be selected. In general, the treatment options can be divided based on the waste's toxicity, mobility, and physical form. To reduce toxicity, possible technology classes include (EPA, 1989):

- Biological
- Thermal
- Chemical
- Physical

To reduce waste's mobility possible technology classes include (EPA, 1989):

- Containment
- Solidification

Currently, the trend in remediation is to include technology selection in the early stages of site and waste characterization. This will allow for better use of resources, focus the investigation, and result in time and cost savings. Therefore, a good understanding of available technologies, including their limitations, is essential to the success.

The list of in situ treatment systems is very long and is continuously expanding. Biological treatment technologies are represented by in situ bioremediation and its many variations. Physical treatment includes Soil Vapor Extraction (SVE), soil flushing, steam stripping, and air sparging. Chemical treatment includes oxidation-reduction, chemical precipitation, or electro-acoustic soil decontamination, and is usually applied in conjunction with other methods. Also, thermal treatment like soil warming and vitrification are being used as part of a treatment train. Stabilization and solidification methods such as lime based pozzolan or portland cement pozzolan processes are often combined with containment or encapsulation.

Each technology can be applied to the problem alone or in combination with other methods, often termed as a treatment train. The design of a treatment train is performed within a remedial program.

Design of a Remedial Program

Each remedial program is composed of several steps necessary for successful remediation. Following is the list of steps that are usually, but not always, performed for a remediation project:

1. Initial Site Investigation
2. Risk Assessment/Further Investigation
3. Feasibility Study (Biofeasibility Study)
4. Remedial Investigation
5. System Design
6. Source/Free Product Removal
7. Operation
8. Maintenance

Initial site investigation is usually performed in phases. Phase I covers extensive historical assessment of the site. Information like historical use, geological and hydrogeological maps, surface conditions, and activities around the site are gathered. Also, the type of contaminants, size of the contaminated area, location, and magnitude of contaminated sources are investigated. Phase II includes designing a sampling program based on the Phase I contamination hypothesis. Also, a first selection of the appropriate remedial technology is made. Sampling requirements of the selected technology are included in the program. In this phase, the hydrogeologic and geologic parameters like direction and rate of groundwater flow, the specific yield of the aquifer, the depths to the water table, soil classification, soil permeability, and pumping rate that can be sustained in the aquifer are measured.

After the initial investigation, the risk assessment and further investigation are performed to provide information necessary for feasibility studies and technology selection. The risk assessment process is usually comprised of four components: selection of indicator analytes, exposure assessment, toxicity assessment, and risk characterization. Further investigation provides information on the concentration, total amount, and special distribution of the contaminants. Also in this stage, sampling data is statistically evaluated and mapped on the previously defined geological and hydrogeological situation. A preliminary relationship between the contaminant sources and the present situation is developed.

A feasibility study (biofeasibility study for bioremediation) is performed to determine the applicability and potential for success of selected technology at a given site.

Remedial investigations are performed to confirm results of the previous investigations and feasibility study. At this stage it may be necessary to obtain data for evaluation of alternative engineering options.

The system design stage provides a detailed design for the project.

Next steps like source/free product removal, operation, and maintenance are self explanatory.

Environmental problems in Central and Eastern Europe require technologies that can offer simplicity, low cost, and flexibility. Therefore, only a limited number of technologies are presented here.

Bioremediation

Bioremediation is the natural degradation of foreign substances and contaminants by bacterial populations already present in the soil or those added to it. This technique offers partial or complete destruction of the contaminants. Partial destruction usually occurs because some chemicals may not be available to the microorganisms, due to transport limitations or because the chemicals may be biotransformed to other organic chemicals rather than completely mineralized. Bioremediation offers significant benefits including:

- low cost, due to the elimination of excavation, transportation, and disposal costs
- ability to apply the technology where physical limitations like structures inhibit removal
- low exposure of workers to hazardous compounds
- contamination that is sorbed to the aquifer matrix and dissolved in the groundwater can be treated simultaneously

Currently in the United States, over 160 field applications of bioremediation are registered with the EPA (Chaparian, 1995). A variety of contaminants including gasoline, diesel fuel, PNAs, BTEX, solvents, chlorinated hydrocarbons, pesticides, and heavy metals are treated at those sites. A limited number of well-documented, field demonstrations and the possibility of generating undesirable intermediate compounds are among the most important problems with the implementation of this technology.

Other limitations include a longer time for remediation than other methods, and the requirement of fair weather and mild temperatures. At the same time, these and other limitations are site-specific and bioremediation can perform better than other methods. For example, in very tight soils and clays bioremediation was shown to clean-up the site in two years as compared to 30 to 50 years required for air stripping or pump-and-treat technology (Bishop, 1992).

Temperature limitation is a more complex issue. Different microorganisms can function effectively within different temperature ranges, from 4 ^{0} C to 50 ^{0} C. Here, the issue is the rate of degradation. Microorganisms that can operate at lower temperatures degrade contaminants more slowly than those that can perform at higher temperatures. Practical solutions to that problem include heating up the air that is pushed into the ground or heating up the groundwater.

1. *Biological Transformation*

Bioremediation is possible because the biological transformation remediated by microorganisms occurs in the soil and groundwater. In general, an organic compound is oxidized (loses electrons) by an electron acceptor, which in itself is reduced (gains electrons). Under aerobic conditions oxygen commonly acts as the electron acceptor and under anaerobic conditions microorganisms use organic chemicals or inorganic anions as alternate electron acceptors. For aerobic heterotrophic respiration the reaction can be illustrated as:

$$\text{Microbes} + \text{Contaminants} = CO_2 + H_2O + \text{biomass}$$

here O_2 is the electron acceptor.

2. *Bioremediation Requirements*

Following are basic requirements for bioremediation to take place:

1. Organisms. Presence of appropriate organisms, capable of degrading the specific contaminates is needed.

2. Energy source. Organic carbon is required as the organisms energy source.

3. Carbon source. Since 50 % of bacteria's only weight is carbon, it is required for new cell growth.

4. Electron acceptor. They must be present for the reaction to occur. Typical acceptors include O_2 , NO_3 , SO_4 , and CO_2 .

5. Nutrients. The required nutrients for bacteria growth include nitrogen, phosphorus, calcium, magnesium, iron, and trace elements.

6. Favorable conditions. They include the correct temperature, pH, salinity, hydrostatic pressure, and no presence of toxic materials and radiation.

As explained earlier, a well designed remedial program follows a number of basic steps. In the case of bioremediation, many practitioners indicate that a well prepared and executed biofeasibility study is a critical part of the remediation program.

During this study, the direct measurement of the microbial capacity to degrade the contaminates is performed. Also, the concentration of naturally occurring microbes in the soil is measured. Additionally, information like biodegradation kinetics, biodegradation endpoint, and biodegradation range of concentration are obtained. Data from the laboratory kinetic studies need to be extrapolated to actual subsurface environments.

The most common extrapolation is through use of Monod's hyperbolic saturation function (Bedient, et al., 1994). Growth rate calculated from Monod kinetics is used to obtain the reduction of contaminant concentrations. In addition to Monod kinetics, first-order decay kinetics and instantaneous reaction models are used for biodegradation modeling.

The effect of site specific factors like soil moisture, pH, nutrients, temperature, oxygen supply, toxicity, and contaminant concentration is always included in design.

The most popular approaches for in situ bioremediation include stimulation, augmentation, and vegetative uptake. Figure 1 presents a process diagram for in situ bioremediation. Biostimulation consists of adding nutrients, such as nitrogen and phosphorous, as well as oxygen, to the microbial environment to stimulate activity leading to digestion of target contaminants. Bioaugmentation involves adding microbes to soil or groundwater that are able to degrade a specific contaminant. Most biotreatment practitioners employ some form of biostimulation, whereas some resist the use of bioaugmentation (Bishop, 1992). Well designed biofeasibility and a thorough understanding of the site conditions are required for successful design. Use of the commercially-available microorganisms should be supplemented with engineering design. Quite often, site conditions are different from the conditions under which the organisms were grown.

PROCESS DIAGRAM FOR IN SITU BIOREMEDIATION

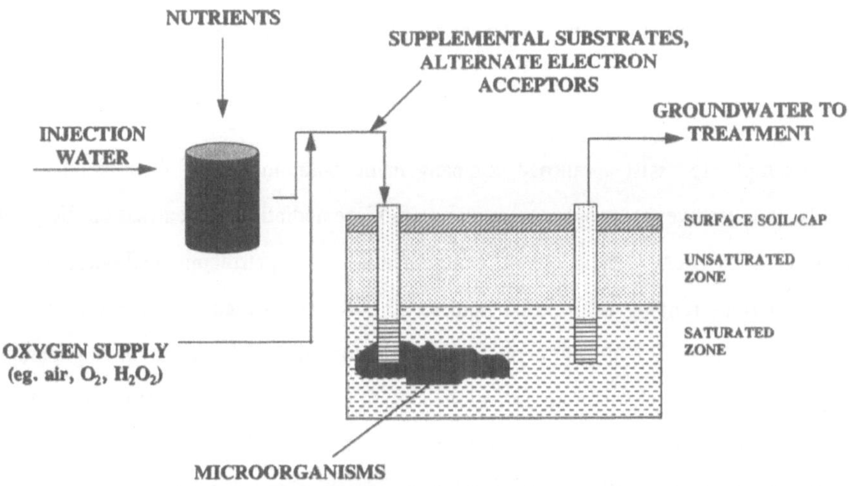

Figure 1. In situ bioremediation

Currently in soils with low level of contamination, plants are being used for stimulating microbial activity. Statistically significant reduction in PAHs was shown for soil with prairie grass vegetation (EPA, 1991).

3. *Example Applications*

Use of the bioremediation technology starting from 1985 shows exponential growth with 38 % annual growth rate. Current advances in soil vapor extraction and genetically engineered microbes will probably increase that rate.

To date, Conestoga-Rovers and Associates (CRA), like many other consultants, employed bioremediation either alone or in combination with other technologies (treatment trains) on many projects. Innovative treatment trains used or proposed by CRA include, in situ flushing followed by bioremediation, soil vapor extraction followed by bioremediation, bioremediation followed by solidification/stabilization and chemical treatment followed by bioremediation.

CRA has utilized bioremediation at a variety of sites ranging from chemical manufactures, railroad yards, lumber yards, former manufactured gas plants to gas stations.

For example, to assist a railroad company in decontamination of their maintenance yards, an extensive research program on in situ bioremediation was carried out by CRA and TreaTek-CRA. Soils were contaminated with petroleum hydrocarbons in concentrations ranging from 5,000 ppm to 60,000 ppm. Based on the results of the program, a field demonstration phase showed more than 85 percent biodegradation within 28 weeks. A diagram for an in situ bioremediation of unsaturated soils is presented on Figure 2. Currently, the project is in the implementation stage. Similarly, following the biofeasibility stage an in-situ bioremediation using bioaugmentation was employed at a major refinery. Approximately 1,600 m^3 of oil contaminated soil to a depth of 8 m was treated. Groundwater was extracted through one central and 10 surrounding treatment zone wells. Two vacuum pumps, operating at approximately 30 m^3/h, forced liquid from an 8 m deep central well to an aerated 50 m^3 reactor tank. Water was pumped from the tank to gravel infiltration trenches at the perimeter of the treatment area. Overall mean oil hydrocarbon concentration was reduced form 185 ± 49 ppm to 26 ± 6 ppm (86 % reduction) within 15 weeks.

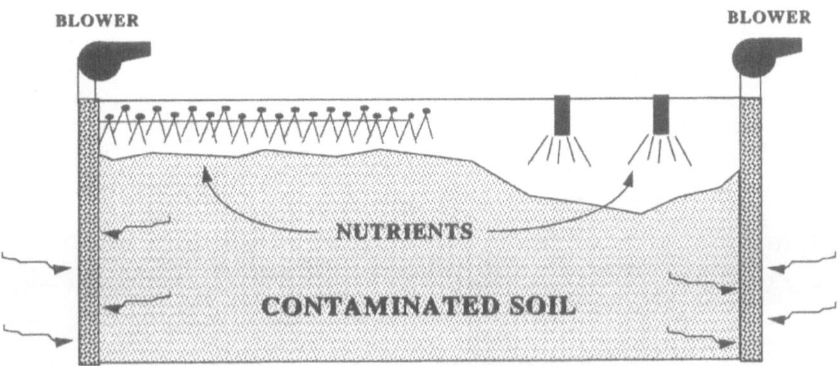

Figure 2. In situ bioremediation of unsaturated soils

Following are recent examples of in situ bioremediation employed in Central and Eastern Europe. Based on published information, in situ bioremediation was used in combination with other methods at several former Soviet military installations in Hungary and in the Czech Republic. At Vac-Mariaudvar base in situ land farming with nitrogen and phosphorus fertilizer addition followed by rye grass seeding was successfully employed. The site was contaminated with diesel fuel (10,000 ppm), lubricating oil (100,000 ppm) and gasoline (7,000 ppm) (Toth, 1995). In Vysoke Myto, Czech Republic, in situ bioremediation technology utilizing bioaugmentation with indigenous (naturally occurring) organisms for surface decontamination was engaged. The site was contaminated with gasoline, lubricating oil, and PCB's. Only hot zones were remediated, and contaminated concentrations ranged from 350 ppm to 59,620 ppm (Kastanek, 1995).

Soil Vapor Extraction

Vapor extraction or gas control method has been often referred to as soil vapor extraction (SVE), forced air venting, or in situ air stripping. The SVE system involves extraction of air from the vadose zone. Figure 3 presents a process diagram for a typical soil vapor extraction system.

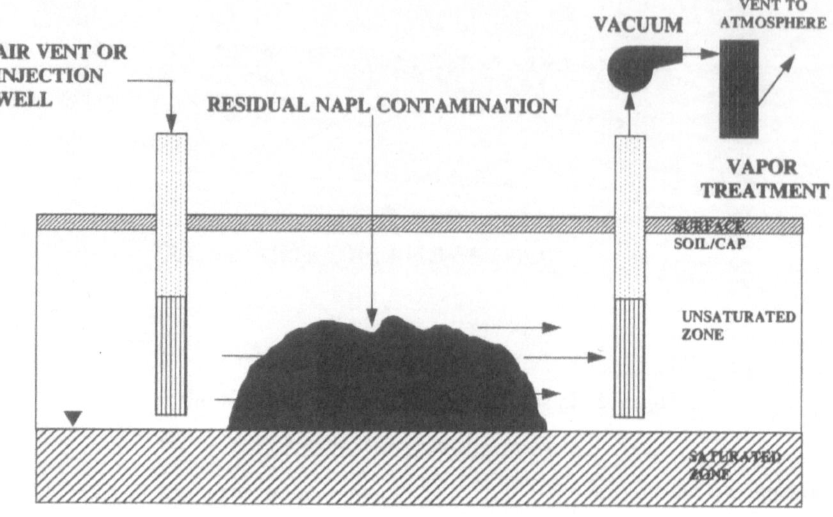

Figure 3. Typical soil vapor extraction system

The principle of vapor extraction is based on Henry's Law constant or air/water partition coefficient (K_h) and the concentration of a chemical in air compared to the concentration of the chemical in water (C_a/C_w).

If the K_h is high, the chemical has a greater tendency to partition into the air phase of the subsurface. Vacuum extraction keeps the system out of equilibrium by forcing the chemical to move from a high concentration zone (aqueous phase) into a low concentration zone (air phase).

The method involves pumping air into injection wells in soil and withdrawing it, along with volatile components, through extraction wells. Then, the air is cleaned by moving it through activated carbon. Vacuum extraction, bioventing, can also enhance biodegradation of volatile and semi-volatile chemicals in the soil, by providing oxygen to the soil for use by microorganisms.

Several chemical and site characteristics influence SVE system performance or its applicability.

1. *Chemical Characteristics*

The pure phase chemical vapor pressure measures the chemical's tendency to be in the air. For SVE to work, the vapor pressure should be greater then 14 mm of mercury at 20° C (EPA, 1992). For the compounds dissolved in the water phase, Henry's law constant is used for the chemical's tendency to be in the air. A minimum acceptable value for the Henry's Law constant is 0.01. Following are values of vapor pressure and Henry's Law constant for different compounds.

Compound	*Vapor Pressure (mm Hg) at 20 C*	*Henry's Law Constant (dimensionless)*
Benzene	95.2	0.24
Trichloroethylene (TCE)	57.9	0.42
Perchloroethylene (PCE)	14.3	0.34

2. *Site Characteristics*

Some of the site characteristics that influence the SVE system performance and applicability include soil moisture content, soil texture, distribution of contaminants, and immiscible fluids.

Soil moisture content may significantly tamper the SVE system effectiveness. Moisture competes with the contaminants for the surface area of soil and forces them into the water phase. A range of 2 to 5 percent of soil moisture content is considered a good low optimum for use with SVE (EPA, 1992).

Clay textured soils, because they are less permeable than sandy soils, are considered less applicable for SVE systems. If distribution of contaminates covers areas with clay soils it is important to place monitoring wells nearby to assess effectiveness of the system. Due to their low permeability, clays tend to diffuse contaminants over long period of time.

When non aqueous fluids are present in the subsurface, Raoult's Law should be used. Raoult's Law states that a chemical will volatilize in proportion to its concentration and its vapor pressure.

One form of a SVE system is bioventing. It is an innovative technology for the treatment of vadose zone contaminated by a variety of petroleum products. Bioventing stimulates the in situ biodegradation of the hydrocarbon contaminants by providing oxygen to the microorganisms. This technology utilizes a lower air flow rate than a typical SVE system. When combined with the soil warming methods it significantly increases biodegradation.

3. *Example Applications*

Following are examples of the use of SVE/Bioventing systems. Among historic applications was the use of SVE to remediate volatile organic compounds (VOCs) and semivolatile organic compounds (SVOCs) including PCBs, PAHs and some pesticides in Waukegan, Illinois (EPA, 1989 b). Another historical and successful application was the removal of volatile organic compounds in San Juan, Puerto Rico (EPA, 1989 b). Recently, a bioventing system was used to remove and biologically treat a gasoline spill at the Hill Air Force Base in Ogden, Utah. Bioventing using venting wells followed by a water separator and activated carbon treatment was employed at Vac-Mariaudvar, former Soviet military base (Toth, 1995) and at Tokol airbase in Hungary (Reiniger et al., 1995). Examples of CRA's experience with SVE and bioventing involve turnkey remedial services with the SVE system as the main component. At a site in Clay, New York the SVE system was successfully utilized to clean up the soil contaminated with chlorinated solvents. SVE system with soil washing and biological treatment was used by CRA at Dock Street Coal Gas Plant site in Stamford, Connecticut. Soil VOC levels were reduced to less than 0.5 ppm and SVOC levels to 500 ppm. CRA in cooperation with TreaTek-CRA provided a SVE system at an operating rail yard in Bakersfield, California. The petroleum hydrocarbon contaminants were located at a depth of 19 meters.

Contaminant Systems

Containment is often used in association with other treatment methods, like bioremediation, SVE, or pump-and-treat. The main objective is to contain either soil, groundwater, or both and prevent contamination from spreading. In most cases the containment system consists of a vertical barrier, cap and some type of groundwater management system.

Several different types of vertical barriers are currently being used in remediation. Among the most popular barriers are slurry walls, steel sheet pile walls, vibrating beam walls, deep soil mixing type walls, jet grouted walls, and composite walls.

There are four basic types of slurry cutoff walls: (i)soil-bentonite, (ii)cement-bentonite, (iii)plastic concrete, and (iv)concrete. The process of creating the slurry cutoff wall is usually divided into three steps: (i) excavation and slurry placement, (ii) slurry mixing, (iii) backfilling.

In the case of the soil-bentonite slurry wall, the bentonite slurry holds the walls during excavation. Then the excavated material is mixed with bentonite and used as a backfill. The soil-bentonite backfill forms a low permeability, highly plastic cutoff wall.

In the cement-bentonite type of wall, cement is added to a fully hydrated bentonite-water slurry. Then the cement-bentonite-water mixture is used as an excavation support; and after settling, it creates the wall.

The plastic concrete wall adds plastic to the cement-bentonite-water aggregate thus forms a much stronger wall. The plastic concrete is placed into the excavation through a tremi method. The concrete-type walls use structural concrete backfill placed with the tremi method.

The conventional steel sheet pile wall, when combined with a sealant, creates an impermeable barrier. This method has been successfully employed on several sites in the United States over the last couple of years. CRA is one of the first firms, and in the case of former Manufactured Gas Plant sites, the first consultant to design and implement a containment system with the sealed steel sheet pile wall. A wall measuring over 250 m in length and 8 m in depth was installed at the site in 1994. The wall was constructed on the embankment between the existing industrial building and the river. In cooperation with a groundwater interceptor drain, the wall prevents contaminants from entering the river and allows for collection and subsequent treatment of groundwater.

Deep soil mixing technology offers not only the method for creating a vertical barrier, but an in situ fixation technique. The method was demonstrated to treat soils contaminated with arsenic and heavy metals in concentration of up to 5,000 ppm. Soil was mixed with appropriate reagents using multi axis auger equipment and allowed to solidify.

A typical cap for the containment is made of asphalt, concrete or geomembrane. The selection depends on future use and economics.

Funnel and Gate In Situ Groundwater Treatment

Recently, a very interesting concept of creating a low permeability vertical barriers (funnel) across plumes with a gap (gate) is successfully being field tested. The method allows for a passage of the plume through the gate where reactive medium is positioned or released so that the plume water is treated. Figure 4 presents the concept of a funnel-and-gate in situ groundwater treatment system. The reactive medium can be solid phase or liquid and is controlled by the release system. The method provides source control, passive treatment and low maintenance cost.

CONTAMINATION PLUME BEING REMEDIATED AS IT MOVES PASSIVELY THROUGH A PERMEABLE REACTION

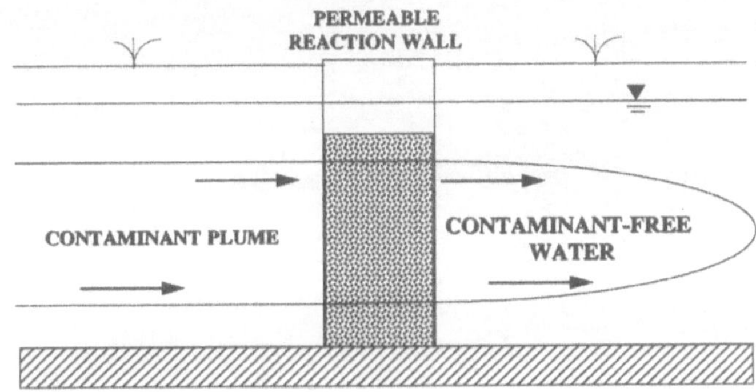

Figure 4. Funnel and gate in situ groundwater remediation

Air Sparging/Biosparging

This technology involves air injection below the water table to promote remediation/biodegradation and volatilization. In the physical process, volatile contaminants are transformed from the aqueous phase to the gaseous phase and removed via the injected air stream. The biological process involves delivery of supplemental oxygen to promote aerobic respiration. Figure 5 presents the process diagram for air sparging.

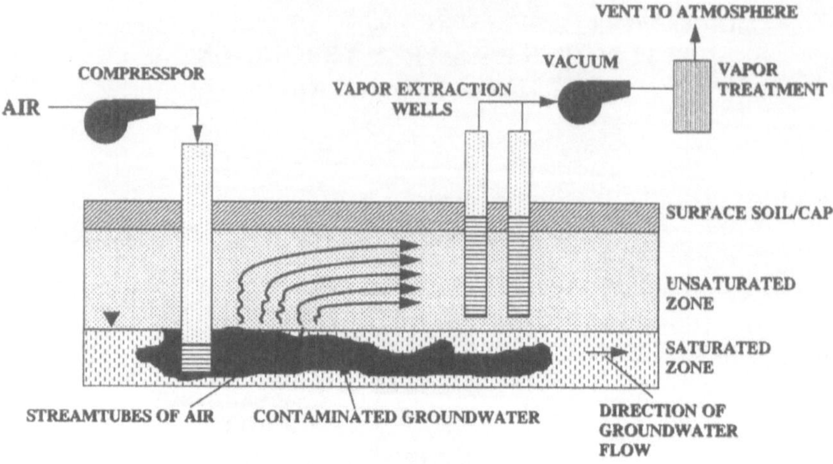

Figure 5. Process diagram for air sparging

Dynamic Underground Stripping

Dynamic underground stripping combines in situ steam sweeping, electrical resistance heating, and liquid and vapor extraction to mobilize and recover contaminants from the subsurface. Field testing was performed at Lawrence Livermore National Laboratory, Livermore, California (Yow et al., 1995). First, the site was heated for 12 weeks using electrical resistance heating method. Then, for five weeks steam sweeping was performed. Steam was delivered to the injection wells from a skid mounted 10 M Btu boiler. Liquid and vapor effluents were extracted through two wells near the center of the plume. During this phase 6,400 liters of gasoline was removed. The second phase of steam operations commenced after 3 months and continued for six weeks. In this phase a periodic steam injection was used. More than 18,500 liters of gasoline was removed. The process zone was kept desaturated and periodically depressurized. By January of 1994 the gasoline recovery rates had fallen to about 40 liters per month, suggesting almost complete removal.

9. Radio Frequency Heating

To treat contaminants at the former Manufactured Gas Plants (MGPs), researchers at the Illinois Institute of Technology Research Institute (IITRI) developed the use of radio frequency (RF) energy. The RF energy heats up tarry wastes which are then being pumped out. The liquid can be used in a variety of boilers and no excavation is required. Figure 6 presents the diagram for radio frequency heating.

References

1. Bedient, P. B., Rifai, H. S., and Newell, C. J., 1994, Groundwater Contamination, Transport, and Remediation, PTR Prentice Hall, Englewood Cliffs, New Jersey.
2. Bishop, J., 1992, "Biotechnology Coming of Age, Expanding Horizons in Environmental Area", Haz Mat World, May.
3. Chaparian, M., 1995, "Guaranteeing the Success of Bioremediation", Remediation Management, 1 (1), 17-19.
4. Kastanek, F., and Demnerova, K., 1995, "Biodegradation of Petroleum Hydrocarbons After the Departure of the Soviet Army", in Clean-up of Former Soviet Military Installations, Eds.: Herndon, R. C., Richter, P. I., Moerlins, J. E., Kuperberg, J. M., and Biczo, I. L., NATO ASI Series, Springer-Verlag.
5. Reiniger, R., and Horvath, Z., 1995, "Environmental Problems at Former Soviet Military Bases in Hungary", in Clean-up of Former Soviet Military Installations, Eds.: Herndon, R. C., Richter, P. I., Moerlins, J. E., Kuperberg, J. M., and Biczo, I. L., NATO ASI Series, Springer-Verlag.
6. Toth, J., 1995, "Results of Remediation Technologies Applied at Vac-Mariaudvar, A Former Soviet Military Installation", in Clean-up of Former Soviet Military Installations, Eds.: Herndon, R. C., Richter, P. I., Moerlins, J. E., Kuperberg, J. M., and Biczo, I. L., NATO ASI Series, Springer-Verlag.
7. U. S. EPA, 1989a, "Corrective Action: Technologies and Applications", EPA/625/4-89/020, Center for Environmental Research Information.

8. U. S. EPA, 1989b, "The Superfund Innovative Technology Evaluation Program: Technology Profiles", EPA/540/5-89/013.

9. U. S. EPA, 1991, "Site Characterization for Subsurface Remediation".

10. U. S. EPA, 1992, "RCRA Corrective Action Stabilization Technologies", EPA/625/R-92/014, Office of Research and Development.

11. U. S. EPA, 1993, "Profile of Innovative Technologies and Vendors for Waste Site Remediation", EPA/542/R-94/002, Office of Solid Waste Emergency Response.

12. Yow, J. L., Aines, R. D., and Newmark, R. L., 1995, "Demolishing NAPLs", Civil Engineering, 65 (8), ASCE, New York.

III.6

INCORPORATING RISK INTO DECISION-MAKING FOR CONTAMINATED SITE REMEDIATION

E.A. MCBEAN
Conestoga-Rovers & Associates Limited
651 Colby Drive
Waterloo, Ontario
Canada N2V 1C2

Introduction

The mismanagement of chemicals causing extensive environmental contamination is ubiquitous. These chemicals create concern with respect to both the health of humans and for the environment. Consequently, there is an immediate and growing need to determine if, when, and to what degree contaminants should be remediated. Depending on the chemicals being remediated and the media in which these chemicals reside, the choice of technology for remediation differs but, regardless, the need exists for considering the merits of site remediation.

The urgency for cleanup will be dictated by regulations at the local, regional and national levels but a very important concern is one of "how clean is clean enough?" One approach to the problem of determining the extent of necessary cleanup involves the utilization of risk assessment. Specifically, legislation in North America currently involves a focus toward remediation to the point of some specified level of risk

E.A. McBean et al. (eds.) Remediation of Soil and Groundwater, 271–283.

(typically 10^{-4}, 10^{-5} or 10^{-6} where, for example 10^{-6} interprets as a probabilitiy of death of one in one million). Site investigation translates to determining if the risk associated with a site exceeds the level of risk and, when this situation is estimated to exist, to identify site remediation options to reduce the risk to an acceptable level.

The initiatives for meeting the pre-determined levels of exposure risk are precipitated by the interest in reducing the environmental risk. A number of reasons explain the focus toward risk reduction associated with hazardous wastes including:
(i) people don't understand many of the characteristics of hazardous wastes and there is a fear of the unknown;
(ii) there are historical incidents that have precipitated enormous news coverage, thus heightening public concern; and,
(iii) the knowledge that somebody else will pay the economic costs associated with contaminated site remediation quickly persuades many people that no expense for risk reduction is too large.

As a result of these types of reasons, there is extensive pressure from governmental regulators in North America to have the risks associated with hazardous wastes decreased to very small magnitudes. On the other hand, as members of modern society, people are exposed to a variety of risks on a daily basis; when the risks are voluntary and, at least to some extent they understand the risks, the risks are accepted (e.g. smoking cigarettes and taking airplane flights). Thus, there are differences in attitude and acceptance as a function of the source of the exposure risk. The values in Table 1 indicate why 10^{-6} is considered acceptable.

Table 1. Risks Which Increase the Chance of Death by One in One Million

-Smoking of 1.4 cigarettes

-Drinking 1/2 liter of wine

-Spending 1 hour in a coal mine

-Travelling 6 minutes by canoe

-Travelling 10 miles by bicycle

-Flying 1000 miles by jet

-Drinking City of Miama drinking water for 1 year

-Eating 40 tablespoons of peanut butter

-Eating 100 charcoal broiled steaks

Ref. Wilson (1990)

Risk assessment in relation to environmental phenomena is a process that seeks to estimate the likelihood of occurrence of adverse effects due to exposures to chemical, physical and/or biological agents in humans and ecological impacts within an ecosystem. However, to determine the health-risk to a receptor, a risk assessor has to conduct an exposure assessment to determine chemical exposures of the receptor at a specific location, usually referred to as a "point of compliance". Such assessments of exposure must reflect all relevant exposure pathways. Examples of potential exposure pathways are described below; the complexity of incorporating all of these pathways is apparent..

EXPOSURE PATHWAYS

- INGESTION OF CHEMICALS IN DRINKING WATER
- INHALATION OF AIRBORNE (VAPOR PHASE) CHEMICALS IN THE SHOWER
- DERMAL CONTACT WITH CHEMICALS IN WATER
- INHALATION OF SOIL EMISSIONS

There are three methods through which the chemical exposures at a point of compliance can be obtained. The first, environmental sampling, is the most obvious method to obtain these data. Often through sampling it will be determined that insufficient data exist to use this source of risk assessment. In most cases then, reliance must be placed on a second method which utilizes a combination of sampling and contaminant transport modeling. The acceptability of this second method arises since the existence of data allows model calibration. With a model effectively calibrated, greater confidence exists in predictions of future contaminant migration. The third method of risk assessment estimates exposures for the situations of little or no data (other than source data), strictly through fate and transport modeling. In the absence of any chemical field data, computer modeling is used to estimate the receptor exposure concentrations.

The next level of difficulty associated with many environmental problems is the availability of modest data. For example, only limited knowledge frequently exists regarding the quantities and specifics of contents of solid wastes disposed at a site that needs remediation. Thus, it is difficult to estimate the exposure risks for nearby residents and/or onsite construction workers that may be encountered during site remediation. As a result, one response that is frequently adopted is to utilize a succession of worst-case assumptions. For example, we may know only that organic chemicals were disposed of at the site but to be conservative we will assume that all of the organics were benzene, a highly volatile chemical. As well, since the wind direction to be encountered during a particular time period during site remediation cannot be predicted definitively, the assumption may be made that the direction is selected to maximize exposure of the residential population-at-risk. Neither of the above assumptions are likely correct but such an approach can be utilized to estimate the maximum feasible exposure risk. If the maximum feasible exposure risk does not violate acceptable exposure levels, then an analyst knows that the concern is minimal. The downside to this conservative approach is that the maximum feasible exposure risk may exceed allowable exposure concentrations; if this is the result, care must be utilized to ensure that this conservative approach is recognized as conservative and not as a probable event.

In a typical scenario in which there is an exposure, contaminants may be transported via one or more media (including air, soils/sediments, surface water and ground water) to potential receptors (through, for example, inhalation, dermal contact and ingestion). The exposure assessment aspect of the risk qualification must then involve the characterization of the physical and exposure setting, including contaminant distributions leading from sources at a hazardous site to the points of exposure, the identification of chemical intakes for all potential receptors, and significant pathways of concern. If the exposure risk is unacceptable, then the exposure and risk assessment must be completed for each remediation alternative to allow a comparison between the different alternatives (to establish the cost-effectiveness for risk reduction).

Quantitative Evaluation of Carcinogenic Adverse Health Effects

The assessment of carcinogenic risk is accomplished by first calculating the lifetime average daily adsorbed dose (LADD). It is estimated by averaging the daily intake over a human lifetime, as per the following equation.

$$LADD = \frac{DIxEFxED}{365xAT} \qquad (1)$$

where LADD = chronic daily absorbed dose (mg/kg.day)

DI = daily intake for a specific pathway (mg/kg.day)

EF = exposure frequency (days/year)

ED = exposure duration (years)

AT = averaging time (years)

In Equation (1), the exposure frequency, EF, corresponds to the number of daily exposure events that occur within a year. The exposure duration, ED, is the number of years that the receptor is exposed to the conditions. The averaging time in Equation (1) is equal to the length of a human life.

Carcinogenic risks are quantitatively defined as the incremental probability of an individual developing cancer over a lifetime as a result of exposure to a potential carcinogen. This is often termed the *excess individual lifetime cancer risk.* The equation that describes this risk is:

$$Risk = LADDxSF \quad (2)$$

where Risk = a unitless probability (e.g. 3×10^{-5}) of an individual developing cancer
LADD = lifetime average daily absorbed dose (averaged over 70 years) (mg/kg/day)
SF = slope factor expressed in $(mg/kg.day)^{-1}$

Equation (2) is a linear low-dose cancer risk equation. It is termed low-dose cancer risk because relatively low intakes (compared to those experienced by test animals) are most likely from environmental exposures. Further, it is assumed that the dose-response relationship will be linear in the low-dose portion of the multi-stage dose-response curve. The slope factor is linear under this assumption and therefore is directly related to intake.

Quantitative Evaluation of Non-Carcinogenic Adverse Health Effects

The evaluation of a term called the chronic daily intake (CDI) is used in the assessment of non-carcinogenic effects from chronic exposure. It is computed by averaging the daily intake over the exposure period, as:

$$CDI = \frac{DIxEF}{365} \quad (3)$$

where CDI = chronic daily absorbed dose (mg/kg.day)
DI = daily intake for a specific pathway (mg/kg.day)
EF = exposure frequency (days/year)

Non-carcinogenic toxicity is quantified through a term called the hazard quotient. It is evaluated by comparing an exposure level over a specified time period (usually a lifetime of 70 years) with a reference dose derived for a similar period. This ratio of exposure to toxicity is mathematically described as:

$$Hazard.Quotient = \frac{CDI}{RfD} \quad (4)$$

where CDI = chronic daily intake

RfD = reference dose.

Note that both CDI and RfD are expressed in the same units and represent the same exposure period.

This quotient assumes that if the level of exposure is below the RfD value, then it is unlikely for sensitive populations to experience adverse health-effects. Therefore, if the hazard quotient remains below unity, no adverse health-effects are likely to be observed. It should be noted that the level of concern does not increase linearly as the quotient approaches and passes unity. This is because the slope of the dose-response curves is non-linear. Additionally, the nonlinearity is different for each chemical.

Aggregate Risk

There are likely to be several non-carcinogenic and carcinogenic risks that the receptor is exposed to, due to different chemicals at a hazardous site. As a result there will be aggregate risks to the receptor. These risks will be from several chemicals along an individual pathway, as well as multiple pathways. Non-carcinogenic risks along one pathway are additive for chronic exposures. The chronic non-carcinogenic hazard index for multiple toxicants along one pathway is:

$$Chronic.Hazard.Index = \frac{CDI_1}{RfD_1} + \frac{CDI_2}{RfD_2} + ... + \frac{CDI_i}{RfD_i} \quad (5)$$

where CDI_i = chronic daily intake for the ith toxicant (mg/kg.day)

RfD_i = chronic reference dose for the ith toxicant (mg/kg.day)

For carcinogenic risk along one pathway, the effects are assumed additive due to their probabilistic nature, and are expressed as:

$$Risk_T = \sum Risk_i \qquad (6)$$

where $Risk_T$ = the total cancer risk, expressed as a unitless probability

$Risk_i$ = the risk estimate for the ith substance.

Having described the aggregate risks along individual pathways, the total risk can be derived for all pathways. The total non-carcinogenic hazard index across multiple pathways is:

$$Total.Exposure.Hazard.Index = HI_{pathway.1} + HI_{pathway.2} + \ldots + HI_{pathway.i} \qquad (7)$$

where $HI_{Pathway\ i}$ = the total hazard index from all chemicals along pathway i

Similarly, the total cancer risk across multiple pathways is:

$$Total.Exposure.Cancer.Risk = Risk_{pathway.1} + Risk_{pathway.2} + \ldots + Risk_{pathway.i} \qquad (8)$$

where $Risk_{Pathway\ i}$ = the total risk from all chemicals along pathway i.

Uncertainties in Risk Assessment

Numerous sources of uncertainty exist throughout the risk assessment process. Each source potentially adds cumulatively to the next. It is important, therefore, to be aware of these uncertainties and their impact on risk calculations.

Uncertainty is inherent in any system that we attempt to model. The uncertainty occursas a result of our inability to completely define the components of the system. Uncertainty in risk assessment arises from limitations of data collection and technology performance. These limitations hinder the ability to accurately describe the source, the migration pathway parameters, the toxicity parameters, the effectiveness of the remedial technology, and so on. For example, our best estimate of risk might entail calculations which indicate acceptable soil slope-stability conditions during site excavation to remove contaminated soil. However, as an example of uncertainty, conditions other than those expected may occur and soil slope instability may initiate much higher exposure levels. Any derived value of expected exposure risk must be considered only that, a best estimate, and uncertainty exists in these estimates.

As evidenced from the comments above, risk assessment can play a very important role in relation to environmental site remediation. However, it is imperative to recognize that estimates of exposure risk have inherently built into them, considerable uncertainty.

Specifically:

(i) characterizations of the source - in many circumstances, the characterization of the source is relatively poor in terms of quantity, types of constituents and degree of dispersement that has already occurred, associated with the source;

(ii) determination of the exposure risk of nearby residents, for example, must consider the examination of the various migrational pathways which the source materials may assume. Some example indications of migrational pathways were illustrated in Figure 1. These include the ingestion of chemicals in drinking water, inhalation of airborne (vapor phase) chemicals in the shower, dermal contact with chemicals in water, and inhalation of soil particulates. The uncertainties implicit in quantifying the various pathways associated with the combinations of potential/relevant pathways are obvious.

(iii) the next stage of risk assessment, involves estimation of the health-related impact of the exposure. It follows that any chemical is toxic, if the exposure is sufficiently large. For example, the conventional rating scheme for lethal doses in humans was detailed previously in Table 1.

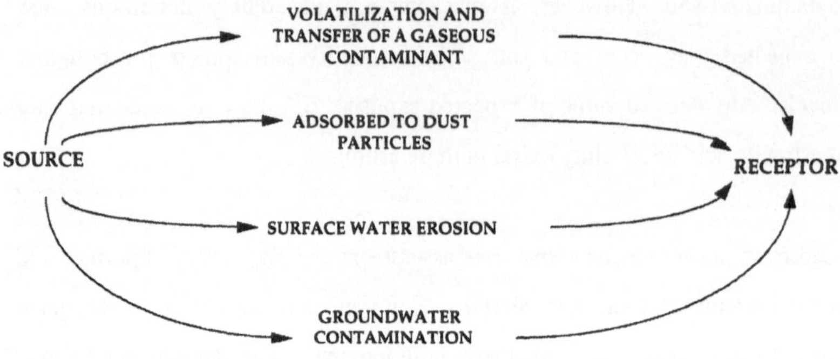

Figure 1. Migrational pathways.

Table 2. Conventional Rating Scheme for Lethal Doses in Humans

Toxicity Rating	Dose (mg/kg of body weight)	For Average Adult
Practically nontoxic	more than 15000	more than 1 quart
Slightly toxic	5000-15000	1 pint to 1 quart
Moderately toxic	500-5000	1 ounce to 1 pint
Very toxic	50-500	1 teaspoon to 1 ounce
Extremely toxic	5-50	7 drops to 1 teaspoon
Super toxic	less than 5	Less than 7 drops

However, what we need is to understand into which category a particular chemical fits. Acute toxicity is reasonably well understood since sufficient numbers of people have attempted and/or succeeded in committing suicide that information on acute toxicity is fairly well established. However, chronic toxicity, the low level but lengthy duration of exposure is much more difficult. In response, we turn to toxicologists and epidemiologists for answers. However, considerable elements of uncertainty exist within these venues. Toxicologists attempt to fix all environmental conditions and vary only one (or at most only a very few). However, to quantify carcinogenic risks set at levels of lifetime risk of 10^{-5}, for example, requires a study size of 100,000 or more animals. Since study subjects numbering at this level are infeasible, they adopt more severe exposures and then extrapolate the information. The concerns include the problems of extrapolation and in predicting the impact in humans when the test subject is some other species. Alternatively, the other primary source of information is from epidemiology studies. Epidemiologists observe the results from 'natural experiments'. These may involve the examination of reported sicknesses in workers associated with a particular type of manufacturing facility or residents who have been identified as having been consuming water with certain specific features. Therefore, this type of information-gathering does not have to extrapolate across species but the availability of information for a specific chemical constituent is a major limitation.

The result of concerns related to the dose-response is that uncertainties also exist. Dealing with uncertainties in risk characterization and risk management is not unique to the engineering aspects.

Monte Carlo Approaches to Risk Assessment

One of the most effective ways of incorporating some of the uncertainty aspects implicit in risk assessments is via Monte Carlo simulation. In other words, the features that contribute to the risk assessment are input to the model in the form of probability distributions that characterize the uncertainty of the feature. The stochastic nature of the variables that influence the leakage rates can be accounted for using a full distribution sampling technique. In order to represent the behavior of the system for a range of possible input parameter selections, the analysis requires that the model be solved for a number of realizations. The Monte Carlo sampling technique involves a random selection procedure from postulated parameter distributions, and requires a large number of realizations to represent the full range of possible input parameter selections. Models then accept individual possible values in accord with the probability distribution, and the model rerun.

A more rational use of Monte Carlo analysis is the Latin hypercube sampling technique introduced by Imam and Conover (1980). This method can provide better estimates than simple random sampling since it produces realizations from the entire probability distribution for the same number of simulations. The Latin hypercube technique involves the stratification of the range of each parameter into equally probable intervals. Then a uniform distribution is assumed for each section and a parameter value is randomly chosen from each section. The variance of a Latin Hypercube based sample mean is usually less than or equal to the variance of a Monte Carlo based sample mean for the same number of model simulations.

Conclusions

There is an immediate and growing need to determine if, when, and to what degree contaminants should be remediated in the environment. Risk assessment is a reasonable approach to consider for the rationalization/prioritization between the sites, to identify which sites should be given the highest priority for site remediation since the procedure is logically established to estimate the potential damage to human health and the environment for the alternative sites. Thus, in considering which of a number of hazardous sites should be remediated when there are limited funds, the existence of a hazardous chemical in the environment may not pose the most significant risk if there is no exposure pathway available; risk assessment thus possesses the ability for rationalization of priorities.

A significant concern with risk assessment is in relation to the implicit uncertainties in the estimates of the risk. The uncertainties in the estimates should be appropriately reflected in the interpretation of the risks.

References

1. Imam, R.L., and W. J. Conover, "Small Sample Sensitivity Analysis Techniques for Computer Models with an Application to Risk Assessment", Communication Statistitical Theor. Math., A9(7), pp. 1749-1842, 1980

2. Wilson, R., "Analyzing the Daily Risks of Life", in Readings in Risk, ed. T.S. Glickman and M. Gough, Resources for the Future, Washington, D.C., 1990.

III.7

LEACHATE TREATMENT SYSTEMS

M.S. CARVILLE & H.D.ROBINSON

Aspinwall & Company Ltd

Walford Manor,Baschurch

Shrewsbury,Shropshire

Great Britian, SY4 2HH

Introduction

Emplacement of solid wastes in landfill sites has been a common and accepted disposal practice for a number of years. Improvements in landfill management over time have controlled the impact of this method of disposal, and proposed EC legislation [1] indicates that increasing emphasis will be placed on sustainability for waste management. This will have the effect of making landfill a less attractive means of solid waste disposal.

The principle of sustainability being that the problems of one generation are not to be visited on succeeding generations, the conditions which determine why landfilling may be considered as less attractive are easily understood. Nevertheless, although the future for landfill may be uncertain, with a strong presumption against landfill in European policy, it is unlikely that there could ever be a total ban on that disposal route option.

E.A. McBean et al. (eds.) Remediation of Soil and Groundwater, 285–321.

It is now widely accepted among landfill scientists that the timescales involved in the complete degradation of wastes in a typical, large, high-density modern landfill site are likely to be centuries rather than decades. Despite this, many people in the waste disposal industry have not yet recognised the implications that this will have for landfill operations and management during and beyond the active tipping life of sites.

A simple calculation demonstrates the concept by considering only the "hydraulic retention time" of water through (or in) the landfill (Figure 1). Thirty metres of depth of wastes at a density of one tonne per cubic metre and with a moisture content of 33% by weight is equivalent to a depth of water of 10m spread through the mass of wastes. If the infiltration rate is limited to 100mm per year (readily achievable), then the mean time taken for infiltrating rainfall to make a single pass through the wastes before emerging from the base of the landfill is 100 years. This simplification makes no allowance either for the time taken to achieve the full field capacity moisture content or for the fact that complete degradation of wastes such as wood and paper into soluble products takes very many years.

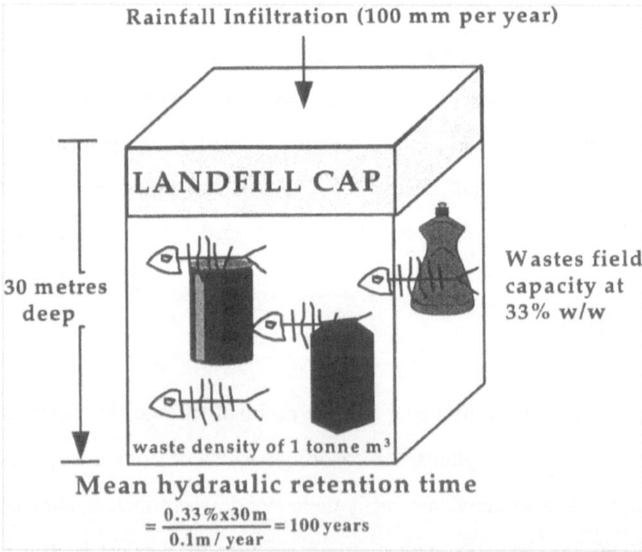

Figure 1. Simplified measurement of hydraulic retention time through a landfill

It is common knowledge, for instance, that newspapers have been unearthed from landfills over 30 years old but are still perfectly legible. Landfills during the 1950s and 1960s were operated following practices very different from those observed in the massive, high input, high density, deep engineered deposits of today. At a modern landfill, decomposition is expected to take even longer to complete. As an estimate, it may take 10 or 20 bed volumes of water to pass through before leachate is clean enough to discharge directly to controlled water.

Guidance used in landfilling practice in the UK today is presented in the UK Department of the Environment Waste Management Paper [2a] (WMP 26), "Landfilling Wastes", published in 1986 and updated recently by WMP 26B [2b] Its objectives included minimising the production of leachate, with the consequential effects of minimising treatment costs and reducing impacts on groundwater. High standards of cellular operation, use of intermediate cover, grading, capping and progressive restoration all contributed to the objectives set out originally in WMP 26. However, as stated above, such practices also contribute to prolonging the time for complete biodegradation of the emplaced wastes.

Landfill Classification

During the 1970s, landfills were operated under conditions of "controlled tipping of refuse", whereby refuse was deposited and compacted in shallow layers; the exposed layers were covered. At that time, the principles of "compaction" and "cover" were intended to control both odour emission and dispersal of light refuse, thereby preventing vermin and odour nuisance from occurring [3]. Typical sites used for controlled tipping included low lying land, marshland, saltings, and foreshores etc, where the sites could be improved by raising ground levels. At that time, it was accepted that polluting effluents might be produced (leachate), and there were examples where such effluents were allowed to " pass harmlessly into or through surrounding ground" [3]. These types of landfills were called "dilute and disperse" or "dilute and attenuate" sites, where permeation of the leachate into the water- bearing strata allowed it to be spread out over a large area, and thereby be rendered less concentrated. Landfills operated today generally preclude the entry of groundwater and the uncontrolled discharge of leachate and are classed as "containment sites".

Therein lies the problem, because the leachate, derived from water contacting the emplaced waste materials, is highly polluting. Where it is allowed to migrate uncontrolled from a site, it may contaminate (and, of course, at some sites, has contaminated) adjacent land and groundwater. The effective management of this wastewater is essential.

Leachate Generation

In order to manage leachate, it is of fundamental importance to know the quantities likely to be generated at any time during the operation of a landfill site, and post closure of the site. As stated above, good landfill management practices have historically been geared toward minimisation of water ingress to the emplaced wastes, and therefore, to the minimisation of leachate generated. Techniques employed include the control of groundwater to prevent it from entering the area to be landfilled, eg. by cut-off walls (Figure 2), the separation of surface water falling onto the tipping area from the wastes being deposited, the provision of temporary cover materials and the provision of a low permeability cap alongside progressive restoration of the landfilled area. Generally, the quantity of leachate being generated can be estimated from the amount of rain falling on the site after due allowance for runoff, evapotranspiration, groundwater ingress and cap (and temporary cover) permeabilities. By judicious choice of cell size and phasing of the landfilling operations, the quantities of leachate can be minimised.

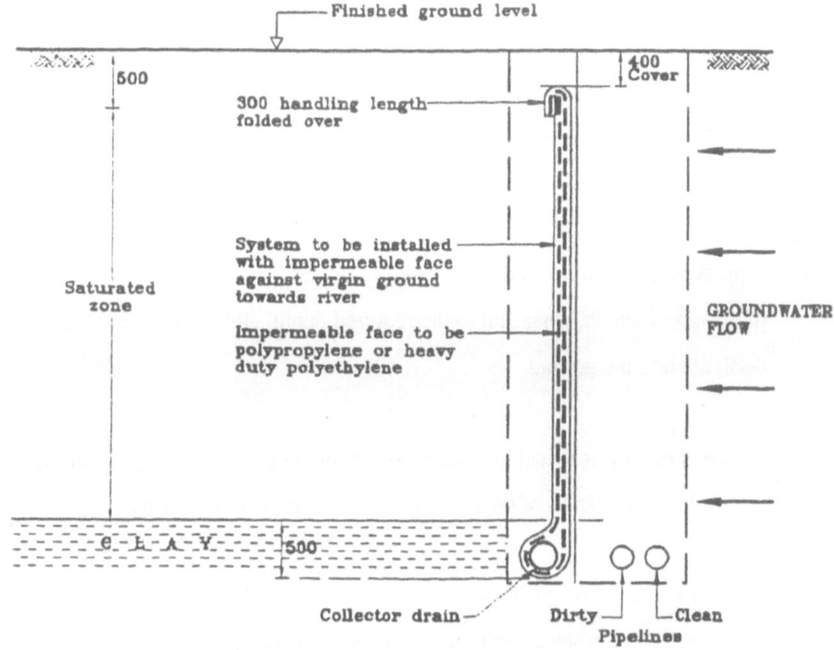

Figure 2. Typical groundwater cut-off installation detail

Leachate Disposal Routes

The generally recognized disposal routes for leachate are:

(i) to land

(ii) to air

(iii) to sewer

(iv) to controlled water

The chosen route will depend on a number of factors which will include the characterisation of the leachate in terms of quality, the land availability, the need for and extent of treatment before discharge. Typical land disposal routes include:

(i) recirculation, where collected leachate is pumped back across the landfill;

(ii) irrigation, where leachate is sprayed onto adjacent areas of land or trickle irrigated through a series of small diameter pipes across the designated area of land;

(iii) grass plot irrigation, where leachate is pumped across specially designed areas of grassland which, when full to the required depth, allow the leachate to dissipate into the ground.

Disposal routes to air might include release of ammonia gas after stripping, or emission of off gases after incineration of concentrate from reverse osmosis plants.

The latter two routes have greater potential when the leachate is relatively diluted, or when the leachate is pretreated and required polishing treatment.

For disposal either to sewer or to controlled water, e.g. surface watercourse, undertakers and/or regulatory authorities generally require some form of pretreatment prior to discharge.

Leachate Quality

One of the major factors affecting leachate quality is the age of the landfilled wastes. This is due to the microbiological and chemical processes that occur within the landfill. These processes were described in detail by Robinson [4] and are identified briefly here. The three major phases of decomposition are:

Phase 1. Aerobic decomposition rapidly (typically in less than a month) uses up oxygen which is present within the wastes.

Phase 2. Anaerobic and facultative organisms (acetogenic bacteria) hydrolyse and ferment cellulose and other putrescible materials, producing simpler, soluble compounds such as volatile fatty acids (with a high biochemical oxygen demand (BOD) value) and ammoniacal nitrogen (NH_3-N).

Phase 3 More sensitive and slower growing methanogenic bacteria gradually become established and start to consume simple organic compounds, producing the mixture of carbon dioxide and methane (plus various tract constituents) which is released as landfill gas.

The three phases can best be illustrated in graphical form. Figure 3 shows the variation of Chemical Oxygen Demand (COD) with time (in the leachate from Compton Bassett landfill site in the UK) [5]. Ammoniacal Nitrogen concentrations, also included on the Figure, show that reduction in COD is not due to dilution.

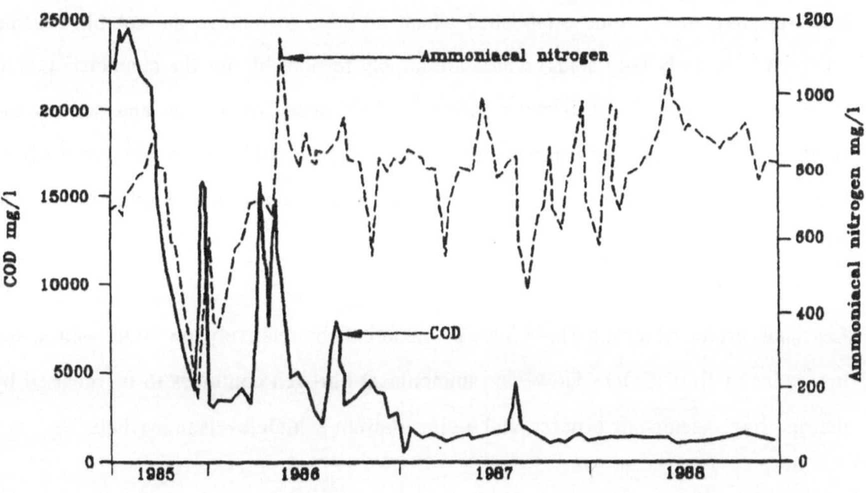

Figure 3. Landfill Leachate COD variation with time

Phase 1 is usually short, perhaps lasting only a few days or weeks, but it may persist for long periods, producing significant quantities of carbon dioxide, in shallow (<3m) deposits of waste where air can readily enter the waste, or if air is drawn into waste by pumping.

Phase 2 can last for years, or even decades. Leachates produced during this stage are characterised by high BOD values (commonly >10,000 mg/l); and high ratios of BOD:COD (commonly 0.7 or greater) indicating that a high proportion of soluble organic materials are readily degradable. Other typical characteristics are acidic pH values (typically 5 or 6), strong unpleasant smells and high concentrations of ammoniacal nitrogen (often 500 - 1,000mg/l). The aggressive chemical nature of such leachate assists in dissolution of other components of wastes, so leachates can contain high levels of iron, manganese, zinc, calcium and magnesium. Gas production is slow and consists mainly of carbon dioxide with lesser quantities of methane and hydrogen.

Although the transition from Phase 2 to Phase 3 can take many years, and may not be completed for decades, wastes have been known to reach Phase 3 in a few months. Bacteria gradually become established which are able to remove the soluble organic compounds (mainly fatty acids) which are largely responsible for the characteristics of Phase 2 leachates. These bacteria thrive in the absence of oxygen and convert the soluble organic compounds into methane and carbon dioxide which are given off as landfill gas, and therefore these soluble organic compounds are not found in leachate to the same extent.

Leachates produced during Phase 3 are characterised by relatively low BOD values, and low ratios of BOD:COD. However, ammoniacal nitrogen continues to be released by the first stage acetogenic process, and will be present at high levels in leachate.

It can be seen, therefore, that management of leachate has to account for the change in quality with time.

It is not surprising than that one of the most important considerations in leachate management is the assessment of its quality in terms of treatability. Leachate is basically a wastewater which, if unmanaged, will be a source of contamination both of the land and of the groundwater at, or adjacent to, the point(s) of its release. The composition of leachate depends ultimately on the nature of the wastes emplaced and, although the major pollutants, COD, BOD and NH_4-N are frequently prominent, other pollutants such as pesticides or other UK Red List substances [6] may be present in sufficient quantities so as to prove inhibiting for (biological) treatment.

Until recently, for proposed new landfill sites, the typical UK landfill leachate quality has been predicted using information provided in WMP 26. Table 1 shows the quality of leachate from domestic waste landfill sites, for both recent and aged wastes. In the case of existing sites it has been possible to take samples of leachate for analysis, and the latter have provided additional information in the classification of leachate. In particular, landfill sites where co-disposal has occurred provide some interesting data; leachate quality only rarely being significantly different from that found at sites which have received only domestic waste input.

Over the last five years, an extensive study of the composition of landfill leachate in the UK has been carried out by Aspinwall & Company, on behalf of the UK Department of the Environment. A paper [7] described the study methodology and provided examples of the type of data gathered during the project. The final report is expected in late 1995, and will contain analyses from some 72 sites, 30 of which have been analysed in detail. The changes in leachate composition with time, the value of a leachate quality database, the eleven categories of landfill studied, the selected analytical suite, heavy metal levels in leachate, and the composition of acetogenic and methanogenic leachates are discussed. Selected data is presented in six summary tables, two of which are reprinted here. Table 2 shows a summary of the composition of typical acetogenic leachates, while Table 3 presents a summary of the composition of the methanogenic leachates.

Table 1*. Typical composition of UK leachate from recent and aged domestic wastes at various stages of decomposition (all results in mg/l except pH value)

Determinant	Leachate from recent wastes	Leachate from aged wastes
pH-value	6.2	7.5
COD (Chemical Oxygen Demand)	23,800	1,160
BOD (Biochemical Oxygen Demand)	11,900	260
TOC (Total Organic Carbon)	8,000	465
Fatty acids (as C)	5,688	5
Ammoniacal-N	790	370
Oxidised-N	3	1
o-Phosphate	0.73	1.4
Chloride	1,315	2,080
Sodium (Na)	960	1,300
Magnesium (Mg)	252	185
Potassium (K)	780	590
Calcium (Ca)	1,820	250
Manganese (Mn)	27	2.1
Iron (Fe)	540	23
Nickel (Ni)	0.6	0.1
Copper (Cu)	0.12	0.3
Zinc (Zn)	21.5	0.4
Lead (Pb)	8.4	0.14

***Reference 2**

In a wider European context, a report is shortly to be presented to the 5th International Landfill Symposium, Sardinia 1995 by Hjelmar *et al* [8] which comments on the composition of typical leachates from EU landfill sites.

Quality Parameters

The important quality parameters which identify, first of all, a leachate and, second, the type of leachate or leachate contaminated groundwater are:

pH, COD, BOD_5, NH_4-N, electrical conductivity (EC) and chloride, iron and zinc

For leachate contaminated groundwaters, a comparison of their quality with that of the background groundwater will indicate the extent of any pollution. On site measurement of EC is an immediate identifier, and this can be confirmed also by analysis for ammoniacal nitrogen and chloride; noting that uncontaminated groundwater is generally free from ammoniacal nitrogen.

The change in COD and BOD_5, as stated earlier, is indicative of the process phase within a landfill, but one of the most important parameters for consideration in terms of treatment and disposal of leachate is ammoniacal nitrogen (NH_4-N). It can be seen from Tables 2 and 3 that the mean concentration of ammoniacal-N changes little with age: 922mg/l for recent wastes, 889mg/l for aged wastes. This is not surprising because the nitrogen is in its most reduced form as ammonium and therefore, under reducing conditions in an anaerobic reactor (the landfill), its form will not change. It is therefore a long term contaminant of concern. The toxicity of ammonia to the aquatic ecosystem emphasises the need for due consideration of significant removal of that parameter, whatever the strategy for treatment and disposal.

This does not, and should not, preclude the consideration of treatment of the other major parameters, ie:

- high concentrations of degradable and non-degradable organic materials
- concentrations of specific hazardous organics and inorganics
- nitrate ions
- sulphides
- odorous compounds
- suspended solids.

Table 2*. Summary of composition of acetogenic leachates sampled from large landfills with a high waste input rate, relatively dry (35 samples in all)

Determinand	Samples	Minimum	Maximum	Median	Mean	SD
pH-value	34	5.12	7.8	6.0	6.73	-
COD	35	2740	152000	23600	36817	32718
BOD_5	29	2000	68000	14600	18632	15643
ammoniacal-N	34	194	3610	582	922	802
chloride	34	659	4670	1490	1805	910
BOD_{20}	13	2000	125000	14900	25108	32870
TOC	24	1010	29000	7800	12217	10028
fatty acids (as C)	26	963	22414	5144	8197	6786
alkalinity (as $CaCO_3$)	24	2720	15870	5155	7251	4390
conductivity (μS/cm)	28	5800	52000	13195	16921	11602
nitrate-N	30	<0.2	18.0	0.7	1.80	3.41
nitrite-N	25	0.01	1.4	0.1	0.20	0.30
sulphate (as SO_4)	24	<5	1560	608	676	549
phosphate (as P)	21	0.6	22.6	3.3	5.0	5.47
sodium	26	474	2400	1270	1371	631
magnesium	28	25	820	400	384	196
potassium	28	350	3100	900	1143	760
calcium	31	270	6240	1600	2241	1656
chromium	26	0.03	0.3	0.12	0.13	0.08
manganese	26	1.40	164.0	22.95	32.94	37.29
iron	32	48.3	2300	475	653.8	566.2
nickel	24	<0.03	1.87	0.23	0.42	0.48
copper	24	0.020	1.100	0.075	0.130	0.216
zinc	34	0.09	140.0	6.85	17.37	29.56
cadmium	24	<0.01	0.10	0.01	0.02	0.03
lead	24	<0.04	0.65	0.30	0.28	0.16
arsenic	19	<0.001	0.148	0.010	0.024	0.039
mercury	15	<0.0001	0.0015	0.0003	0.0004	0.0004
heavy metals (b) excluding Zn	24	0.34	2.57	0.95	1.03	0.56

Notes:

(a)Results in mg/l except pH-value and conductivity.

(b)Represents the sum of concentrations of chromium, nickel, copper, cadmium, lead, arsenic and mercury

*** Reference 7**

Table 3*. Summary of composition of methanogenic leachates sampled from large landfills with a high waste input rate, relatively dry (29 samples in all)

Determinand	Samples	Minimum	Maximum	Median	Mean	SD
pH-value	29	6.8	8.2	7.35	7.52	-
COD	29	622	8000	1770	2307	1527
BOD_5	29	97	1770	253	374	378
ammoniacal-N	29	283	2040	902	889	396
chloride	29	570	4710	1950	2074	9870
BOD_{20}	24	110	1900	391	544	459
TOC	29	184	2270	555	733	470
fatty acids (as C)	29	<5	146	5	18	29
alkalinity (as $CaCO_3$)	29	3000	9130	5000	5376	1664
conductivity (μS/cm)	25	5990	19300	10000	11502	3890
nitrate-N	27	0.2	2.1	0.7	0.86	0.53
nitrite-N	27	<0.01	1.3	0.09	0.17	0.26
sulphate (as SO_4)	16	<5	322	35	67	83
phosphate (as P)	28	0.3	18.4	2.7	4.3	4.3
sodium	29	474	3650	1400	1480	691
magnesium	29	40	478	166	250	308
potassium	29	100	1580	791	854	760
calcium	29	23	501	117	151	106
chromium	28	<0.03	0.56	0.07	0.09	0.11
manganese	29	0.04	3.59	0.30	0.46	0.66
iron	29	1.6	160	15.3	27.4	32.8
nickel	29	<0.03	0.60	0.14	0.17	0.13
copper	27	<0.02	0.62	0.07	0.13	0.15
zinc	29	0.03	6.7	0.78	1.14	1.30
cadmium	27	<0.01	0.08	<0.01	0.015	0.02
lead	27	<0.04	1.9	0.13	0.20	0.35
arsenic	27	<0.001	0.148	0.010	0.024	0.039
mercury	23	<0.0001	0.0008	<0.0001	0.0002	0.0002
heavy metals (b) excluding Zn	27	0.15	2.78	0.51	0.61	0.49

Notes:

(a)Results in mg/l except pH-value and conductivity.

(b)Represents the sum of concentrations of chromium, nickel, copper, cadmium, lead, arsenic and mercury

*** Reference 7**

Leachate Treatment

The objective of leachate treatment at all sites containing landfilled wastes is to achieve the required standards for discharge, whether to sewer, land, watercourse or other controlled water. A variety of technologies is available to assist in the management of leachate, some of which perform the function of changing the form of the compounds in the leachate, while others change the compounds altogether. Some are used as pretreatments, while others are used as polishing treatments.

The technologies may be groups under the following headings:

(i) physical processes
(ii) physical - chemical processes
(iii) chemical processes
(iv) biological processes

Physical Processes

The technologies associated with physical treatments are reverse osmosis and evaporation.

Reverse Osmosis.
Reverse osmosis removes suspended and colloidal materials, ammoniacal nitrogen, heavy metals, and most dissolved solids. The process also reduced COD and BOD. It may be suitable for application to leachates with a high inorganic loading and low volumetric flow rates.

The process **does not** treat or degrade any contaminants, but, using ultra filtration membranes (operated at elevated pressures of 40 Bar or more (and elevated temperatures)), is able to concentrate soluble constituents of leachate into a brine or concentrate which can comprise 25 to 40% of the volume of the influent leachate, and a permeate which can achieve high standards of purity. Various means have been used for the disposal of this concentrate.

Recirculation of leachate back into the landfill has been the simplest, cheapest, and generally adopted option, but may not be acceptable in the longer term as a landfill where leachate contaminants are simply being returned into the wastes will not achieve accelerated completion. Should it be possible to tanker the concentrate off site, the overall trade effluent costs associated with what is basically concentrated leachate are likely to be unaffected, although transport costs will be less because of the reduced volume.

Evaporation.

The development of evaporation as a technique for leachate treatment has been comparatively recent. Evaporation can be carried out in two ways: either by natural evaporation under atmospheric conditions or in an accelerated evaporation process. Evaporation under atmospheric conditions will depend on the application eg whether directly to land or to lagoon. These options are undesirable as a means of disposal of the leachate because of land requirements and because of the uncontrolled release of volatile organics from the liquid (leachate). Small quantities of ammonia gas may be released depending on the pH of the liquid body.

Accelerated evaporation in flash evaporation/distillation has been extended to the "treatment" of leachate in recent years. Studies in Switzerland [9] and the USA [9] have shown the evaporation of leachate can be achieved successfully. The leachate described in [9] contained ammoniacal nitrogen in concentrations generally of greater magnitude than those contained in UK leachates (NH_4^+ -N, 2000 mg/l), while [9] used a leachate with NH_4^+ -N at 200mg/l and a Total Kjeldahl Nitrogen (TKN) of 400mg/l. The former indicated the desirability of removing the ammonia prior to evaporation, although it also stated that by reducing the pH, ammonium salts would be formed which, after evaporation, would be tied up with the condensate from the distillation process. The lack of a condensate analysis did not permit an assessment of the complete nitrogen cycle.

The principle of the evaporation system is to heat the leachate to a temperature of 60 - 80°C [9], or 100°C [10] to remove water vapour while inhibiting the release of metal components. The evaporate can then either be condensed or incinerated; both alternatives having been used. If the evaporate is condensed it can then be discharged or treated further prior to discharging in accordance with the discharge consent limit. In order to inhibit the release of ammonia during the evaporation stage, it is then essential to bind the ammonia into the other components of the leachate or remove it prior to evaporation.

If the evaporate is to be incinerated the ammonia can be volatilized with the water vapour and incinerated [10].

Low grade heat used in the evaporation process might be sourced from combustion of landfill gas, provided that the quality of the gas is sufficient highly. Predicted estimates of gas quality and quantity at specific landfills can indicate whether these would be sufficient for an evaporation process.

The concentrate can be passed for further evaporation leaving a "mush" [9] which can readily be landfilled. Typical volumes of such material can be estimated from these studies as less than 20% of the original leachate volume. It must be emphasized, however, that those studies were on a pilot scale operating at 20 m^3/d, compared with projected volumes of leachate generated from modern landfill sites of up to 500 m^3/d.

The studies by Hughes et al [10] indicate successful incineration of the evaporate, leaving air emissions as the form of "discharge of treated leachate". It is worth noting that the site was located in a remote arid region of the US. The concentrate is recycled either to the landfill or to the boiler to mix with new leachate. As the concentrate is recycled, there will probably be a buildup of the metals within it, to the extent that either there is precipitation within the leachate tank, the boiler or the evaporator. The report makes no mention of this aspect, although it comments on the requirement for intermittent use of the plant to allow for maintenance operations.

While the system may have benefits in the use of landfill gas as the source of heat energy, the technology appears still to be largely unproven, and may not be recommended for large sites without further study.

It must be remembered that, similar to reverse osmosis technology, evaporation as a process does not "treat" the leachate, but concentrates the ammonium in a different form.

Physical-Chemical Processes

The main physical-chemical processes involved in the treatment of leachate are air stripping of ammonia, activated carbon adsorption, and chemical oxidation.

Air Stripping of Ammonia.

The chemistry of air stripping of ammonia is well known and the technology associated with it is well proven. Basically, ammonia is readily soluble in water and dissociates as the ammonium ion over a wide range of pH in accordance with the following reaction:

$$NH_4^+ + OH^- \text{---->} NH_3 + H_2O$$

At pH values greater than 11.5, the balance of the reaction is towards the right, that is, ammonia is present as the dissolved gas. Severe agitation of the leachate will then release the ammonia gas into the air stream and then into the atmosphere. This is best carried out in a stripping tower where the leachate is introduced into the top of a packed tower and air is blown in from the bottom to give a countercurrent effect as the two streams meet.

The ability of a dissolved gas to be transferred from the liquid phase (leachate) into the gaseous phase (air) depends on the Henry's Law constant and the partial pressure of the gas in the air above the liquid [12]. Henry's Law constant is used in the following formula:

$$\frac{C_2}{C_1} = [1 + \frac{Hc\ Aw}{RT}] - 1$$

where

C_1	=	Initial ammonia concentration in the leachate;
C_2	=	Final ammonia concentration in the leachate;
Hc	=	Henry's Law constant;
Aw	=	the volumetric air/leachate ratio;
R	=	the universal gas constant;
T	=	the temperature.

The formula is applicable for ammonia in solution only on the assumption that pH correction has ensured that the ammonia is not in its dissociated form. Lime or other alkali therefore is required to raise the pH to a value of 11.5. Steiner et al [11] demonstrated that a dosing rate of 6g/l will achieve this.

The leachate discharged from the stripping tower will have a high pH and will be required to be neutralised before either being discharged or passed for further treatment. It has been shown [10] that 10 meq acid/litre of leachate are required to reduce the pH from a value of 10.0 to a value of 7.0.

A consequence of this process is that solids are produced which require to be settled and disposed of. These solids also include solids from the lime itself. In the laboratory trials settlement was found to be slow and would require the addition of polyelectrolyte to accelerate the process and therefore to reduce the size of a settlement tank.

The end products of air stripping of ammonia are ammonia gas and sludge solids. The ammonia gas can be re-absorbed for further treatment [15] or it can be reacted with phosphorus or sulphur to form usable fertilizers. In practice, the ammonia is often released to the atmosphere. It is necessary to determine the effect of stripped ammonia on its receiving medium (air) and to ensure that the final ammonia concentrations are below the prescribed limits.

Guidance from the UK Health and Safety Executive give long term limits of ammonia as 17 mg/m^3 [16]. In addition, the effects of ammonia on vegetation adjacent to the emission source are described in the guidance manual commissioned by Her Majesty's Industrial Air Pollution Inspectorate [17]. This states that ryegrass is affected by concentrations greater than 100ppm ammonia (68 mg/cu m). The USEPA Process Design Manual [15] indicates that concentrations in excess of 280 mg/m^{3}have caused eyes, nose and throat irritations.

Sludge arising from the stripping process will also require disposal. The chemistry of the reaction (with lime) to provide a pH of 11.5 will produce metal hydroxides, principally of iron, but also from heavy metals such as chromium and cadmium which can co-precipitate with the iron.

Activated Carbon Adsorption.

Granular activated carbon (GAC) is a highly porous material with a high surface area to volume ratio. GAC (and powdered activated carbon - PAC) have been used to adsorb residual quantities of organic materials from leachates that have previously had the majority of their organic contaminants removed using other treatment methods.

Suspended solids must be removed from the leachate prior to treatment, to prevent blockage of the carbon filter. This can be achieved by several means including plate separators and pressurized sand filters.

The activated carbon can be regenerated after it has become completely saturated with adsorbent. The regeneration cycle cannot be undertaken in situ. In situation where small volumes of GAC (often in modular units) have been used, the GAC may be disposed of by incineration rather than sent for regeneration.

This method of leachate treatment can be used in an effluent polishing situation to reduce COD loading, non-volatile organics and hazardous organics. The treatment can be highly effective with up to 99% removal attainable, but is generally very expensive if significant quantities of residual COD require treatment.

Chemical Oxidation.

Oxidation of leachate by the addition of oxidising agents and pH adjustment may be used for the removal of sulphides, sulphite, formaldehyde, cyanide, and phenolics. The principal use of this type of treatment is in situations where odours caused by sulphides are a particular problem.

The performance of the process depends on the reaction time and on the oxidising agent chosen. In addition to hydrogen peroxide, other reagents which may be used include calcium and sodium hypochlorite, ozone, and chlorine gas with caustic soda. Caution should be exercised in the use of oxidising agents to ensure their safe handling.

Treatment may be carried out as a batch or continuous process, using dilute solutions of the oxidising agents. A ratio of hydrogen peroxide to soluble sulphide of unity, at a neutral pH, with a contact time of about ten minutes is usually adequate to remove sulphides.

Organic material may also be removed by oxidising agents such as ozone, although high dosages are often required to bring about significant reductions in COD. Ozone has been used in wastewater treatment plants to control odour, improve suspended solids removal, oxidise pesticides and improve biodegradability of other organic compounds.

Coagulation, Flocculation and Settling.

The addition of reagents to leachate, followed by mixing and settlement, might occasionally be useful either before or after other methods of treatment. The reagents can result in a reduction in suspended solids, heavy metals, turbidity, colour and some organic loading concentrations.

Reagents which have been used in the process include lime, sodium and magnesium hydroxide, ferric chloride and sulphate, and polymeric coagulants.

Chemical Processes

Chemical processes which may be used in the treatment of leachates include breakpoint chlorination and ion exchange. These processes are particularly relevant for the removal of ammonia.

Breakpoint Chlorination.

Breakpoint chlorination is an alternative method of oxidizing ammonia to nitrogen gas, the technology of which is proven. The principle on which the method operates is that sufficient chlorine is added beyond the point where chloramines are formed and all other chlorine demand is satisfied to the point of complete oxidation of the ammonia.

The typical chemical reactions are:

$NH_3 + HOCl \rightarrow NH_2Cl + H_2O$

$NH_3Cl + HOCl \rightarrow NHCl_2 + H_2O$

$\underline{NHCl + HOCl \rightarrow NCl_3 + H_2O}$

overall, $NH_3 + 3HOCl \rightarrow NCl_3 + 3H_2O$

Two observations can be made:

(i) 2 mol of hypochlorous acid (or 1 mol of chlorine gas) are required for every 2 mol of ammonia oxidised; and

(ii) the reaction produces hydrochloric acid, therefore reducing the pH.

The molar ratio of $Cl_2 : NH_4^+$ of 3 : 2 gives a weight ratio of 7.5 : 1. In practice, this latter ratio has been found to be about 10 : 1 minimum, with 18 : 1 also being experienced [15], therefore, a significant amount of chlorine must be added daily.

The operation has been found to be very sensitive to pH. At pH values lower than 6, nitrogen trichloride tends to be the final oxidised product. At pH values greater than 8, nitrate is found to be the final oxidised product. At pH values around 7, no other nitrogenous materials are usually found, however, occasions have occurred when such materials (less than 10%) have been found in the treated liquid.

The resistance to a reduction in pH produced by the reaction is provided to a certain extent by the natural alkalinity of the leachate. However, the alkalinity required to neutralise the acid formed during the reaction is large. It is estimated that 23,500 mg $Ca(OH)_2$/litre of leachate with no natural alkalinity will require to be added to maintain effective pH control.

The reaction time associated with breakpoint chlorination has been found to be very quick. With close pH control, the recommended design contact time is one minute [15] and the contact tank should be designed to resemble as close as possible a plug flow reactor.

Ion Exchange.

Ion exchange technology has been applied in the waste industry. Fundamentally, the process exploits the affinity of natural and synthetic resins for divalent cations, such as calcium and magnesium, and a typical example of such application is in the water softening process. Similarly, Iron (II) and Manganese(II) can also be exchanged but control of the oxidation state is necessary to prevent Iron (III) and Manganese (III) being formed,

In terms of wastewaters, both industrial and leachate, the use of ion exchange may be limited to the removal of ammonia. Notwithstanding the affinity of resins for divalent cations, a natural zeolite, Clinoptile, is known to favor the ammonium ion [18].

Because of the fouling of the resin by the precipitation of calcium carbonate and residual organic matter from, eg, aerobic treatments, pretreatment using filtration and/or activated carbon technologies employed, ion exchange technology tends not to be economically viable [13].

The natural ion exchange capacity (CEC) of indigenous soils has sometimes been used to polish leachates . In such instances, the natural soils, eg peat, must be tested for effectiveness first because the variability of characteristics of those soils includes their CEC.

Biological Treatment Processes

Biological treatment systems may be aerobic or anaerobic. Ground formations which air cannot reach will become anaerobic as oxygen contained in the original formations becomes used up. Landfills may be classed under such ground formations. More importantly, because landfills contain degradable materials, they become active anaerobic reactors, and once they have become methanogenic, they also become very efficient at reducing the organic content within the "reactor". Unless there is an essential need to reduce COD and BOD concentrations while a landfill is in the acetogenic phase, there is generally no effective use to which a separate anaerobic treatment process can be put. If the latter does become an important consideration, quite often it is efficient to recirculate leachate through the waste layers to accelerate the development of methanogenesis.

Aerobic Processes

Since one of the most important contaminants to be removed from leachate is ammoniacal nitrogen, it is not surprising to find that aerobic processes have significant use in the treatment of leachates. These processes can take one of two forms:

(i) attached growth processes;

(ii) suspended growth processes;

Attached Growth Processes

Trickling or Percolating Filters. The use of this technique as a single-stage treatment for high strength leachates is limited, because increasing organic and inorganic loadings tend to cause clogging, due to the build-up of slimes (microbial growth) and the precipitation and build-up of inorganic (especially iron) salts. The accumulation of these materials restricts liquid flow, causing ponding.

Problems of clogging by inorganic salts might be avoided by physical-chemical pre-treatment, or by a long retention time in an open lagoon prior to treatment, which enables precipitation or settlement of iron and calcium salts.

Rotating Biological Contactors.

RBCs comprise a series of closely-spaced discs, 3-4 metres in diameter, mounted on a common horizontal shaft. This is set in a tank such that the surface of wastewater passing through the tank almost reaches the shaft. The shaft continually rotates at 1 or 2 rpm, and a layer of biological growth (some 2 to 4mm thick) is soon established on the large surface area of the discs. This growth assimilates the contaminants in the wastewater; aeration being provided by the relatively low powered rotation which exposes the discs to the atmosphere after contacting them with the wastewater. Typically, a 10m long by 4m diameter module contains about 10,000m^2 of surface area.

A full-scale rotating biological contactor with an installed surface area of about 30,000m^2 was constructed [19] at Pitsea landfill site in Essex, UK, in 1984/5. That plant used landfill gas or propane to heat the influent leachate to optimum temperature for treatment and was primarily designed to nitrify ammonia from a concentration of about 300mg/l in strongly methanogenic, but relatively weak, leachate from a co-disposal landfill. It was treated at a rate of up to 400m^3 per day. RBC plants would perform poorly if used to treat high strength acetogenic leachate, due to inorganic and biological fouling. Very high effluent recycle rates (10 or 20 times treatment rate) would be needed to avoid toxic inhibition by ammonia of the microbial films and the plant would be relatively susceptible to variations in influent quality and flow rate. These type of plants tend to lack flexibility in operation because they are designed to treat a specific wastewater and can not readily be adapted to treat a significantly different wastewater.

Suspended Growth Processes.

In the use of suspended growth processes in aerated lagoons or tanks for aerobic leachate treatment, aeration encourages the formation and growth of suspended biological flocs, which breakdown and metabolise the polluting components of the leachate. Mean periods of retention from 10 to 20 days are capable of greater than 90% removal of COD and ammoniacal nitrogen.

Extended aeration treatment plants have been shown to be robust, both biologically and mechanically [5]. Mechanically, extended aeration plants can be engineered to require little maintenance, and to provide automated discharge of treated leachate as appropriate to a specific discharge consent.

The microbial flocs are resistant to shock loads. They can acclimatise to the presence of toxins and metal ions as well as to high ammoniacal-N and chloride levels, partly because the large volume of an extended aeration system enables them to dilute rapidly the incoming leachate doses.

The extended aeration plants developed for leachate treatment differ in their operation from conventional activated sludge processes (which were initially developed for treatment of domestic sewage and have been installed at some landfills). The short residence time of the conventional activated sludge plants can reduce COD values but they provide only limited removal of ammoniacal nitrogen.

To ensure sufficient phosphate levels for microbial growth within an extended aeration process, occasional addition of phosphoric acid is generally required. Regular inputs of alkali, preferably sodium hydroxide, may also be needed to counteract reductions in pH which occur during the nitrification process.

Extended aeration treatment systems, if operated correctly to a daily cycle in accordance with recent leachate treatment research [20], have been found to be a most flexible form of leachate treatment. They can readily cope with a wide range of flows and strengths of leachate. Easily altered parameters such as volume of leachate in the lagoons, and retention times, makes the plants resistant to shock loads (due to the large volume of aerated lagoon into which these increased concentrations or flows of leachate are discharged).

Trials carried out by Aspinwall & Company to determine the effectiveness of the aerobic treatment of methanogenic leachates, where COD : Nitrogen ratios are high, show that full nitrification is possible for ammoniacal nitrogen concentrations of over 2,000 mg/l [21]. Further trials were carried out to determine the effectiveness of the nitrification process at low temperatures, and up to 350mg NH_4-N/l were treated in the process to provide effluent NH_4-N concentrations of less than 1.0mg/l at 7°C. The results of the latter trials were used to design a full scale plant in Cambridgeshire, UK which is consistently nitrifying a leachate quantity of up to 200m^3/d with an influent concentration of 300mg NH_4-N/l : effluent NH_4-N concentrations are still less than 1.0mg/l.

In nitrate sensitive areas, it is sometimes necessary to ensure that, having nitrified the ammoniacal nitrogen to nitrate nitrogen, the latter is denitrified prior to discharge. This is particularly relevant when effluents are being discharged to surface watercourses. An extension of Aspinwall nitrification trials included a denitrifying step, using the techniques of pre-denitrification and post-denitrification. The layout of the units used in the trials are shown in Figures 4 and 5 respectively. Both pre-denitrification and post-denitrification trials demonstrated that stable treatment could be obtained with hydraulic retention times (HRT) less than 25 days. Mean HRTs in the anoxic reactor of 7 to 8 days was adequate to allow full denitrification to occur within it. Process loading rates for nitrification and denitrification derived from the trials were lower than the maximum rates reported in the literature.

Typical performance data for full scale plants designed by Aspinwall are reported in Table 4 below.

Table 4*. Performance of a wide range of full-scale leachate treatment plants installed by Aspinwall & Company on landfill sites in England, Ireland, Scotland and Wales since 1982

Leachate Treatment Plant	Ref	Discharge route	COD in	COD out	BOD_5 in	BOD_5 out	Ammoniacal-N in	Ammoniacal-N out
Full-scale:								
Bryn Posteg	(4)	STW	9750	210	7000	37	175	0.9
Whiteriver	PC	STW	4200	170	2300	18	290	2.0
Compton Bassett	(5)	STW	1646	268	710	10.4	601	3.4
Chapel Farm	(6)	STW	10500	491	4150	3.0	525	0.3
Walpole Drove	PC	WC	20600	825	9800	33	330	9.3
Summerston	PC	WC	4360	623	1480	6.0	569	0.8
Greengairs	(7)	WC	12100	133	4500	3.5	592	0.2
Harewood Whin	(8)	STW	11300	282	4300	25	229	1.0
Fiskerton	PC	STW	1640	378	660	16	414	<0.1
Borth	PC	WC	439	177	35	10	286	0.6
Bennadrove	PC	WC	1085	227	492	5.5	152	1.1
Gairloch	(9)	WC	602	140	-	-	62	0.4
Deep Moor	PC	STW	1560	147	811	3.0	262	0.5
Pilot-scale:								
A (Hong Kong)	(10)	STW	2560	791	167	35	2563	2.1
B SE England	CIC	-	2310	846	177	17	1000	0.6
C SW England	CIC	-	426	185	38	9	288	0.5
Pilot-scale I	(11)	-	841	307	102	15	302	<0.3
Pilot-scale II	This	-	615	354	70	<2	350	<0.3
Full-scale 19/4	This	WC	705	278	42	5	386	2.0

Notes: (a) All results in mg/l
(b) STW = sewage works; WC = watercourse
(c) PC = personal communication; CIC = commercial in confidence

***Reference 22**

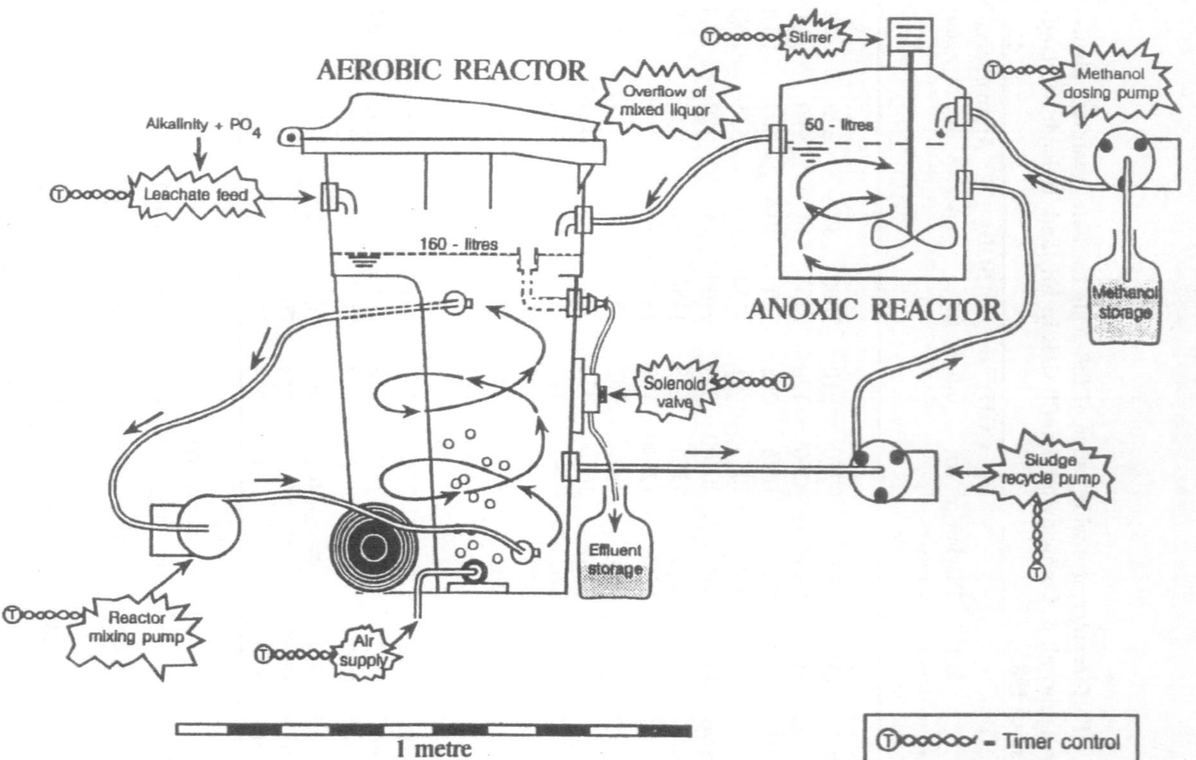

Figure 4. Design of the Experimental Unit used to achieve Pre-dentrification

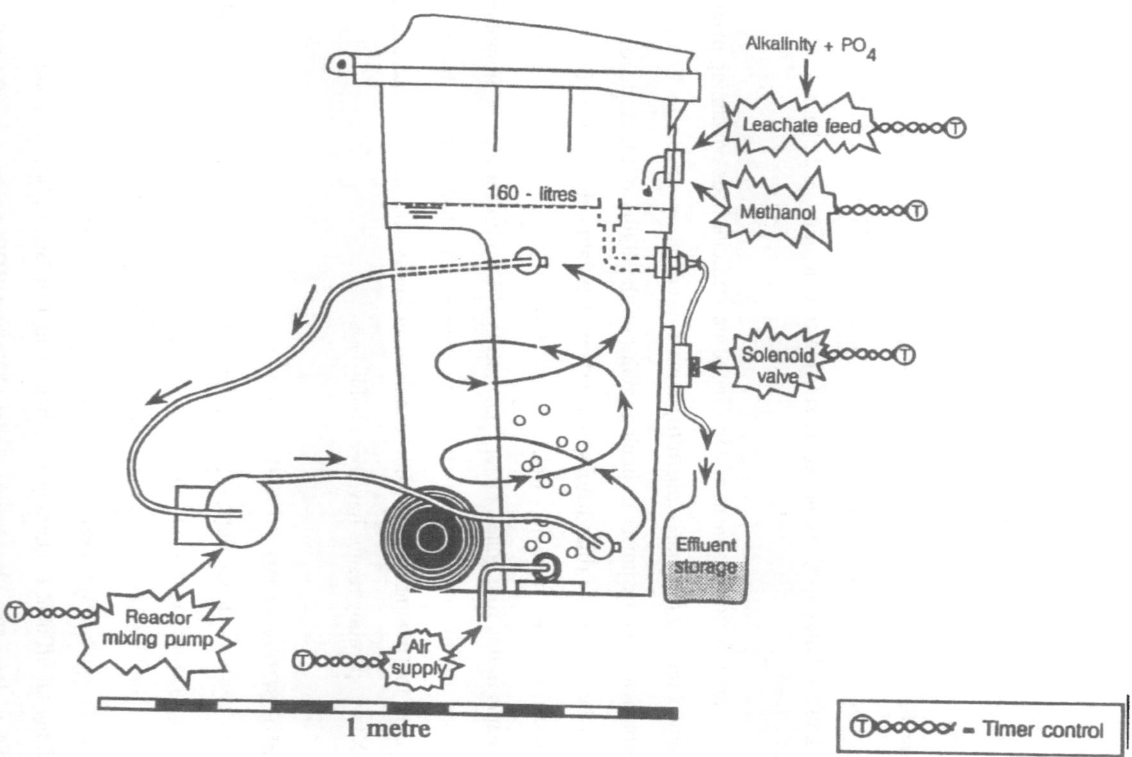

Figure 5. **Design of the Experimental Unit used to achieve Post-denitrification**

Reed Bed Treatment Systems

The treatment of industrial and domestic wastewaters by passage through beds containing the common reed (*Phragmites australis*) has been widely practised in recent years with varying degrees of success. Although many workers have demonstrated good removal of organic components of effluents, poor removal of ammoniacal nitrogen is a common finding, and this is likely to limit the value of reed beds for treatment of raw leachates. Nevertheless, such systems have considerable potential for secondary polishing of leachates that have been pretreated in aerobic biological plants.

In recent years effluent treatment systems based on artificial beds containing plants of the common reed (*Phragmites australis*) have caused much interest in Europe and North America. The treatment technology relies on the ability of the reeds to transfer oxygen to their extensive rhizomatous root system, stimulating the growth of bacteria in the surrounding soil medium, which break down organic substances and other contaminants in this root zone. Other constituents of the effluent may be immobilised or absorbed by the plants themselves, and recent reed bed systems have found applications for treatment of organic chemical effluents and landfill leachates [23] although few operating results have been published.

Removal of BOD and COD

Reed bed systems have achieved widespread success in obtaining removal of degradable organic compounds in screened/degritted domestic sewage and in other wastewaters, and loading rates of up to 5m^2 of bed area per population equivalent have been widely quoted for design purposes by many workers as being capable of maintaining an effluent containing less than 20mg/l of BOD [24]. Research into the treatment of leachates also indicates that COD and BOD removal is effective as a polishing treatment, ie when values were low [25].

Removal of Ammoniacal Nitrogen

The European Water Pollution Control Association (EWPCA, 1990) [32] has concluded that nitrification of ammoniacal nitrogen to nitrate nitrate has not generally occurred in temperate reed bed treatment systems, because of oxygen limitations, but noted that it has been reported in some "polishing" treatment schemes. In a summary of UK reed bed performance, Findlater et al (1990) [26] concluded that neither soil beds or beds containing coarse granular media, removed significant amounts of nitrogen, and individual authors [27] have discounted their use as an option in cases where an ammonia standard must be met to protect a receiving watercourse. Other authors [24][28] have also found conditions not to be conducive to any significant ammoniacal nitrogen removal, with little nitrification and subsequent denitrification occurring. This is no doubt due to the nitrification step being limiting, and therefore there is little nitrate nitrogen to denitrify. It is accepted that nitrification is temperature-sensitive, and slow below 5°C. In spite of this, workers at Portsmouth Polytechnic [29] have had some success in achieving nitrification in reed beds treating settled domestic sewage, at reduced carbonaceous oxygen demand loadings.

Nitrate Removal

Nitrate however can readily be removed if present in the influent applied to a reed bed, eg, the effluent from an aerobic reactor, and it is well-known that wetlands have high denitrifying capacity [30].

Pilot Reed Bed

A research programme was undertaken on behalf of the UK Department of the Environment to assess the effectiveness of reed bed treatment on effluent from a landfill leachate which had been pre-treated using an aerobic suspended growth process. The design of the bed followed the general principles established, for instance, in EWPCA 1990 and used a gravel medium with a horizontal surface; the latter is considered important because it allows full control of weed growth, and the gravel medium kept hydraulic gradients low, thus ensuring that the bed was loaded across its full width and length. A typical layout of the bed is shown in Figure 6 and recent results of the polishing process are reported in Table 5, and shown on Figures 7 to 9 [31].

Table 5. Pilot Trial Reed Bed Performance

	SS		BOD		COD	
DATE	**INFLUENT**	**EFFLUENT**	**INFLUENT**	**EFFLUENT**	**INFLUENT**	**EFFLUENT**
6/2/91	76		32.0		236	
19/3/91	86			5.5	527	156
10/4/91	67				246	
22/4/91			4.5		202	
5/6/91	52		12.0		190	
13/6/91	56		>39		162	
8/7/91	42		30.0		158	
9/7/91	24		18.0		121	
5/8/91	19	13	14.5		101	87
15/8/91	50	13	41.0	1.0	220	338
28/8/91	27	26	41.0	2.0	154	128
23/9/91	81	9	8.0	0.5	150	114
3/10/91	68	11	7.0	<0.5	147	116
22/10/91	58	3	3.5	3.0	182	120
4/11/91			5.0	2.5	214	148
26/11/91	114	35	36.0	17.0	613	370
5/12/91	170	28	34.0	5.0	448	240
16/12/91	66	24		9.5	255	260
10/1/92	90		10.0		407	
16/1/92	68	6	10.5	11.0	360	319
26/1/92					547	
10/2/92	60	13	9.5	5.5	311	318
20/2/92	76	9	5.0	4.5	435	366
14/3/92	23	14	3.0	8.0	264	296
31/3/92	96	17	14.0	5.0	326	294
7/4/92	82	11	40.0	18.0	334	270
1/5/92	244	21	37.0	2.0	406	266
13/5/92	38	22	26.0	1.5	279	232
18/5/92	108	37	14.0	0.5	361	288
9/6/92	56	30	21.0	16.0	360	279
30/6/92	54	7	13.5	1.0	320	270
17/7/92	33	9	6.0	3.0	256	253
30/7/92	20	7	3.5	0.5	260	243
11/8/92	103	8	41.0	4.0	296	238
2/9/92			60.0		373	
16/9/92	62	27	28.0	1.0	267	219
15/10/92	38	10	6.0	2.0	218	187
4/11/92			9.5	4.5	236	205
16/11/92	13	5	5.0	1.0	184	183
3/12/92	20	7	8.5	0.5	147	135
13/12/92	21	10	4.0	1.5	134	124
22/12/92	33	7	8.5	2.5	146	126

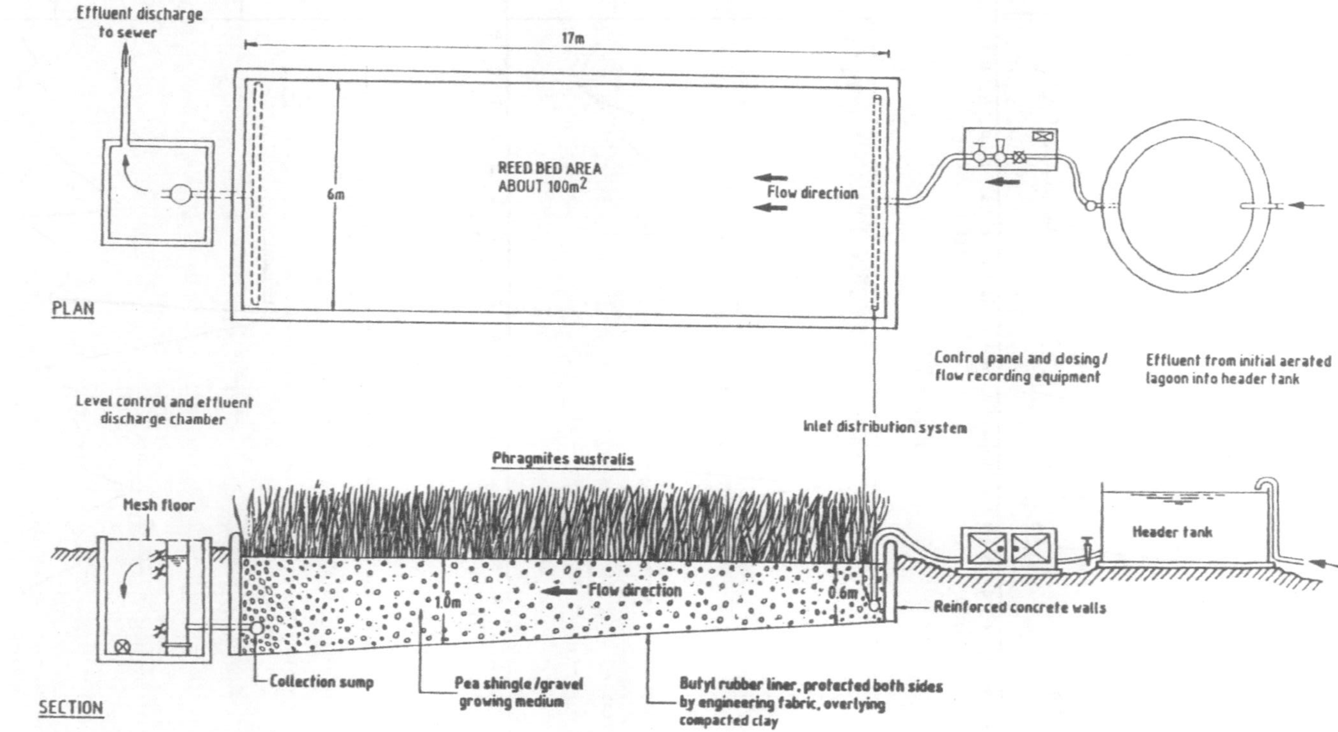

Figure 6. Sketch of Experimental Reed Bed Polishing System (Not to scale)

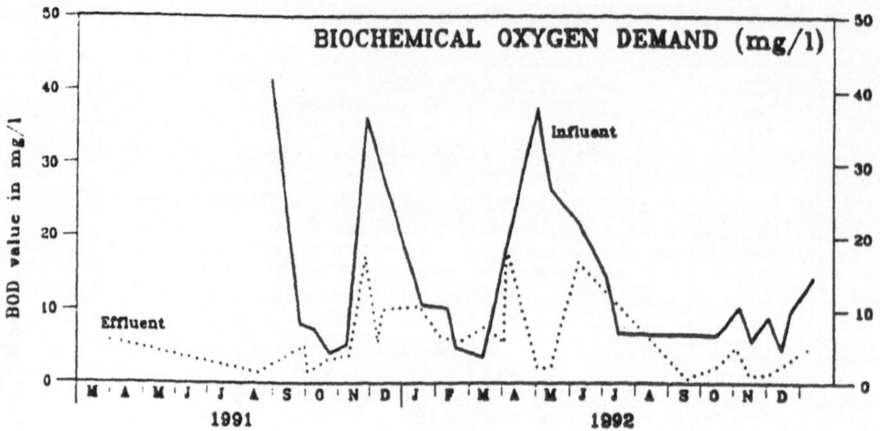

Figure 7.

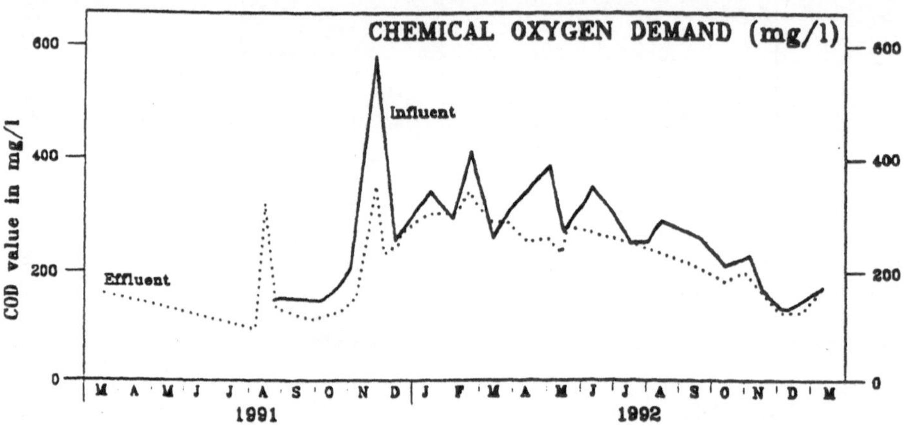

Figure 8.

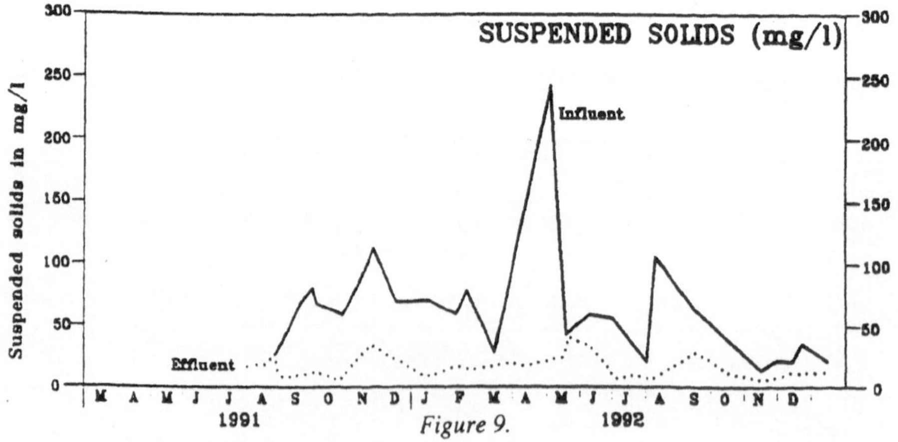

Figure 9.

References

1. Proposal for Council Directive on Landfill of Waste COMM (93) 275 Final. Brussels, June 1994.
2. UK Department of the Environment Waste Management Paper 26 Landfilling Wastes, A technical memorandum for the disposal of wastes on landfill sites, HMSO, 1986.
3. UK Department of the Environment, Report on the Working Party on Refuse Disposal, HMS, 1971.
4. Robinson HD, Development of Methanogenic Conditions within Landfills, 2nd International Landfill Symposium, Sardinia 1989.
5. Robinson HD, On-Site Treatment of Leachates from Landfilled Wastes Journal IWEM, 1990, Vol 4. No 1.
6. Environmental Data Services Ltd, Dangerous Substances in Water - A Practical Guide, 1992.
7. Robinson HD, Gronow J, A Review of Landfill Leachate Composition in the UK, 4th International Landfill Symposium, Sardinia 1993.
8. Hjelmar et al, Review of Composition of EU Landfill Leachates, 5th International Landfill Symposium, Sardinia 1995.
9. Hofstetter UP, Treatment of Leachate with the new Autoflash Evaporator, 3rd International Landfill Symposium, Sardinia 1991.
10. Hughes JL et al, A Technology Development for Leachate Management by Evaporation, 14th Annual Madison Waste Conference, University of Wisconsin, Madison, 1991.
11. Steiner et al, A Demonstration of a Leachate Treatment Plant, US International Technical Information Service, Springfield, Va, PB/269/502, 1977.
12. Eckenfelder Jr WW, Industrial Water Pollution Control, 2nd Edition. McGraw-Hill, 1989.
13. Metcalf & Eddy Inc, Wastewater Engineering Treatment, Disposal, Reuse, McGraw-Hill, 1979.

14. Robinson HD and Maris PJ, Treatment of Leachate from Domestic Wastes in Landfills, Report for Leicestershire County Council No 162 - m/A1623C/CO.2100, July 1981.
15. USEPA, Process Design Manual for Nitrogen Control, EPA - 625/1-75/00.
16. UK Health & Safety Executive, Occupational Exposure Limits, 1992.
17. Taylor HJ, Ashmore MR and Bell JN, Air Pollution Injury to Vegetation, Guidance Manual, Commissioned by HMIP.
18. Mercer BW et al: Ammonia Removal from Secondary Effluents by Selective Ion Exchange, J WPCF, Vol 42, 10 1969.
19. Knox K, Two Stage Control and Treatment of a Methanogenic Leachate, 2nd International Landfill Symposium, Sardinia 1989.
20. HK Government Environmental Protection Department: NENT Landfill Leachate Disposal Study: July 1990.
21. HK Government Environmental Protection Department: SENT Landfill Denitrification Trials: March 1992
22. Aspinwall Report to Local Authority 1995.
23. ENDS, Report No 190, 1990.
24. Bayes CD, Bache DH & Dickson RA, Land Treatment Systems, Design and Performance with Special Reference to Reed Beds, Journal IWEM 1989 3(6).
25. Findlater BC, Hobson JA & Cooper PF, Reed Bed Treatment Systems, Performance Evaluation, Proc International Conference on The Use of Constructed Wetlands in Water Pollution Control, Cambridge, UK, 1990.
26. Christian JNW, Reed Bed Treatment Systems, Experimental Gravel Beds at Gravesend, Proc International Conference on The Use of Constructed Wetlands in Water Pollution Control, Cambridge, UK, 1990.
27. Wolstenholme R & Bayes CD, An Evaluation of Nutrient Removal by the Reed Bed Treatment System at Valleyfield, Fife Scotland, Proc International Conference on The Use of Constructed Wetlands in Water Pollution, Cambridge, UK, 1990.
28. Butler JE et al, Gravel Bed Hydroponic Systems used for Secondary and Tertiary Treatment of Sewage Effluent, Journal IWEM 1990 4(3).
29. Gersberg RM, Elkins BV & Golman CR, Nitrogen Removal in Artificial Wetlands, Water Research 17, 1009-1014.

30. Bowmer K, Nutrient Removal from Effluents by an Artificial Wetland, Influence of Rhizosphere Aeration and Preferential Flow Studies using Bromide and Dye Tracers, Water Research 21 591-599.

31. Robinson HD: The Treatment of Landfill Leachates using Reed Bed Systems:4th International Landfill Symposium S. Margherita di Pula, Sardinia 1993.

32. European Water Pollution Control Association (EWPCA), 1990. "European Design and Operations Guidelines for Reed Bed Treatment Systems". Report prepared by the EC/EWPCA Emergent Hydrophyte Treatment Systems Expert Contact Group (editor P F Cooper), August 1990, ix + 25 pp.

III.8

COMPARISON OF ALTERNATIVE REMEDIATION APPROACHES UTILIZING TIME-RISK CURVES

EDWARD A. MCBEAN & FRANK A. ROVERS
Conestoga-Rovers & Associates Limited
651 Colby Drive
Waterloo, Ontario,
Canada N2V 1C2

Introduction

Extensive societal pressures are being directed toward decreasing the risks to which people are exposed. Although these risk-reduction initiatives are being experienced throughout many aspects of society, the initiatives are particularly relevant in relation to the remediation of hazardous waste sites.

In addressing risk considerations associated with hazardous waste sites, risk assessment seeks to estimate the likelihood of occurrence of adverse effects due to exposures to chemical, physical and/or biological agents in humans and ecological impacts within an ecosystem. The goal of risk management is then to identify that option which best balances the benefits of an activity against a real or perceived risk, with the costs of eliminating that risk. Time-risk curves are a useful addition to the risk management process, in terms of identifying the initiative with the most effective risk-reduction consequences while reflecting the temporal variations of the risk.

E.A. McBean et al. (eds.) Remediation of Soil and Groundwater, 323–328.

Basis of Risk-Time Curves

The use of cost-times in the selection between remediation alternatives has been standard practice for many years. Typically, the curves themselves have only been considered peripherally since, using a discount rate, the temporal variabilities of the costs are modified to a present value and/or translated into an equivalent annual cost. The equivalencing concept is utilized to allow comparisons between alternatives by removing the individual temporal variations, placing the comparisons between the alternatives onto an equal basis.

Risk-time curves are of a similar nature, except that the curves indicate the changing levels of risk, as a function of time, for each of the remedial alternatives. The flexibility and easy comprehensibility of the results to a non-technical audiences, have made the approach a particularly useful one.

As an example of risk-time curves, consider the remediation of ground water contaminated with solvents. In this example, the contamination is the result of improper disposal where solvents were simply dumped into lagoons which in turn leaked into the soil profile for a number of years. The contamination exists over considerable depths of soil profile (40 m to bedrock) as depicted schematically in Figure 1 In considering the potential remediation options, at least the following two possibilities exist, namely:

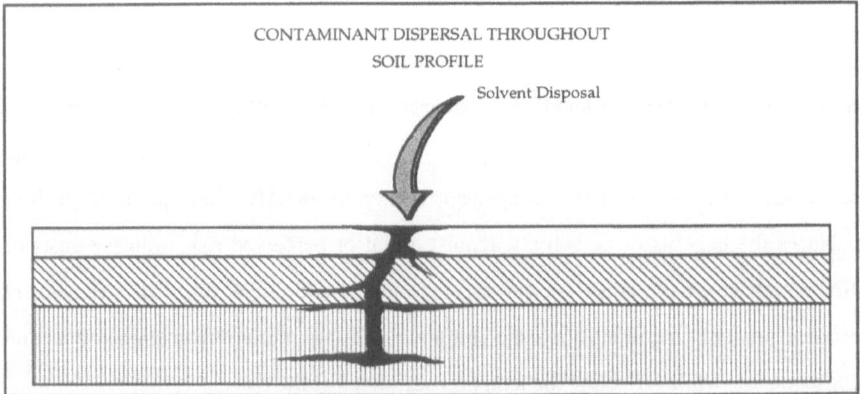

Figure 1. Contaminant dispersal throught soil profile

(i) Option 1 - Excavation and Disposal - Excavation of the contaminated soil, temporary stockpiling on site, with eventual removal and disposal in a secure landfill site. The impacts in terms of risk include volatile emissions during excavation and from the stockpiles of soil, and the potential for backfill recontamination from upwelling of contaminants from the bedrock. The excavation, soil stockpiling, and removal of the soils at the site can pose significant risks to on-site workers resulting from accidental chemical exposures, physical hazards, heat stress, or other conditions commonly encountered during excavation. An indication of the resulting risk-time curve is indicated in Figure 2, in which the curve shows the best estimate of the cancer risk with time. The 'status-quo' or do-nothing option indicates the risk versus time for the do-nothing approach. The status quo curve illustrates a continuing gradual increase in risk as the contamination continues to migrate over time. Significant temporal variation of risk versus time in the excavation and disposal option is apparent. The period of elevated risk is lengthy (5 to 10 years) since the excavation and removal of large quantities of soil will take a lengthy period of time.

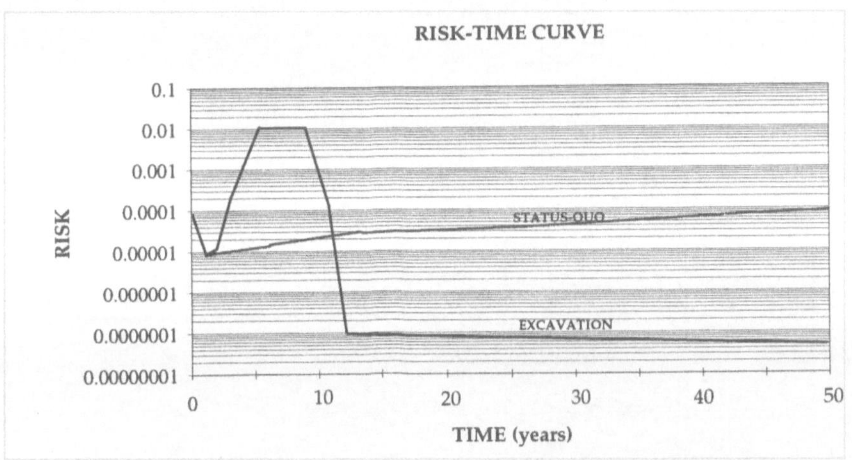

Figure 2. Risk-time curve - excavation status-quo

(ii) Option 2 - Inplace Containment - Placement of an engineered cap over the contaminated soils, in combination this with a passive gas collection system. These two actions (capping and gas collection) or collectively referred to as the inplace containment approach, would effectively contain the volatile organic chemical emissions. The implementation of a comprehensive monitoring program would be required to ensure that the inplace containment system continued to function as designed, but the inplace containment option would involve a small fraction of the costs implied in Option 1. The resulting best estimate, or expected value, of the risk-time curve is depicted in Figure 3. Again, the status-quo curve has been included in Figure 3 for purposes of showing relative comparisons.

A comparison of the risk-time curves and related information for Options 1 and 2, and the implications of the two approaches, demonstrate several relevant features:

(i) the economic costs associated with Option 1 are substantial, in comparison with those of Option 2. However, the inplace containment option involves "writing off" the

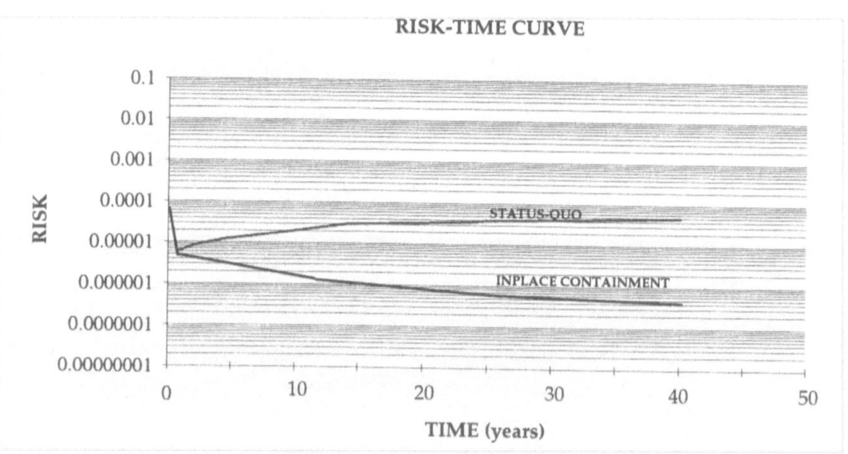

Figure 3. Risk-time curve - inplace containment vs. status-quo

value of the contaminated groundwater resource and soil environs. The economic opportunities foregone by allowing the contaminated soil to remain contaminated essentially in perpetuity, must be evaluated in economic terms to complete the economic analyses.

(ii) The inplace containment approach of Option 2 involves significantly less exposure risk, in comparison with that of Option 1. There is no disturbance of the soil and thus by placement of the cap, there is no increase in exposure pathways for the chemicals in the ground. The risk-time curves are useful in indicating to both technical and nontechnical individuals alike, the temporal variability of the expected risks.

(iii) The construction activities involved in Option 2, with placement of a gas collection system and an engineered cap are fully understood, and thus the uncertainties of risk-time predictions are likely very small. Alternatively, the excavation of a 40 m deep pit to remove the soils may result in problems such as slope instability and/or the excavation of previously-unexpected chemicals. The result is that an upper-bound risk-time curve reflecting uncertainty could well entail a curve that is several orders of magnitude higher than that depicted in Figure 2.

In all likelihood, the best option for the hypothetical site as briefly described in the example above, involves inplace containment. The advantages of inplace containment over excavation and removal of the contaminated soil include lower costs, lower risk and less uncertainty of the risks. These advantages must be compared with the opportunities foregone as a result of leaving the aquifer contaminated virtually in perpetuity. The uncertainty of the risks associated with the excavation option, with a nonzero probability of a very large exposure risk, also argue strongly in support of inplace containment as the most appropriate contaminant remediation option for the site described herein.

The Generality of Concerns Associated with Risk Assessment and Management

The principles and concerns associated with risk assessment and management, as detailed in the context of contaminated site remediation, are general and have merit in application to a wide range of environmental and water-related issues. There remains the need for extensive methodological development and technical analyses to resolve the questions regarding the appropriateness of worst-case assumptions and the most effective way of dealing with uncertainty in risk assessments and risk management. The use of risk-time curve methodology is but one of a number of methodologies which has merit in providing insights into problems of selection between a number of potential remediation alternatives.

III.9

ENVIRONMENTAL TECHNOLOGIES - HUNGARY

AMERICAN EMBASSY COMMERCIAL SERVICE

Bajza u. 31

1062 Budapest, HUNGARY

Overview

Hungary experiences all the environmental problems common to Central and Eastern European countries. Air, water and soil contamination are severe. Although Hungarian environmental regulations are among the strictest in the world, they are not uniformly enforced. While much attention is directed toward the need for environmental protection, insufficient resources are actually devoted to it.

Nonetheless, Hungary's environmental situation has demonstrated significant positive strides in recent years. Some success is attributable to the country's shift from high sulfur coal power generation to other less polluting technologies, particularly nuclear, which provides half of Hungary's electricity needs. Industrial contamination has dropped considerably, more due to the closure of inefficient plants than genuine environmental concerns. Subsidies on hazardous or energy-inefficient agricultural, industrial and residential inputs have been reduced or eliminated, causing more rational usage of these inputs. Crossborder pollution issues are slowly being recognized and addressed. In the whole, the outlook for improved air quality is somewhat upbeat. The

E.A. McBean et al. (eds.) Remediation of Soil and Groundwater, 329–337.

country's water and soil contamination problems appear to be more serious long-term issues.

Market Size

Hungarian statistics on the environmental sector are virtually unobtainable. Since the industry includes may diverse products, services, and technologies, it is extremely difficult to estimate market size or trade in environmental goods and services.

This young industry has evolved in Hungary in response to the transition to the market-oriented economy in the early 90's. The country's spending on environmental projects is low, less than 1 percent of GDP. IN 1994, the total amount was about $150 million, about the same as in 1993. Of this, 60% was for water/wastewater, 15% for waste management and 10% for air pollution control. In the future, the private sector is expected to exceed the governments expenditures in meeting environmental pollution control and abatement standards.

Market Analysis

Air Pollution Control

Ambient (background) air quality in Hungary has demonstrated notable improvement in recent years, although much of the improvement is the result of reduced economic activity, rather than actual investment in environmental technologies. Still, positive strides have been made to correct inefficient and highly polluting energy practices, particularly the burning of low-grade coal. Changes in home heating systems to natural gas and agricultural practices have also had a positive impact on air quality.

The Ministry for Environmental & Regional Policy (MERP) estimates that manufacturing emits 40% of air pollutants, transportation emits another 40% and 20% is accounted for by residences (home heating units).

Sulfur dioxide (SO2) and airborne particulate emissions from thermal power stations declined by a third as Hungary underwent a major transition from coal to nuclear power generation during the last decade. Average annual concentrations of airborne lead (PB) and cadmium (CD) also decreased significantly during the 80's and 90's due to reduced lead levels in gasoline, and the introduction of less environmentally-damaging industrial technologies.

Levels of other gases such as nitrogen oxides (NO2, NO3) and ozone have demonstrated less improvement. While carbon monoxide (CO) emissions by industry continue to fall, CO produced by the transportation sector registered a jump between 1990 and 1992. This is due largely to the exponential growth in the number of automobiles ion the country now exceeds 2 million. For environmental reasons, the government now prohibits the importation of vehicles more than six years oil into the country. The average age of automobiles in Hungary is nearly 10 years.

The Country's relatively energy-efficient public transportation sector is quickly losing ground as private automobiles proliferate. In Budapest, transportation has become the predominant source for nitrogen oxides, carbon monoxide, hydrocarbons and airborne lead. Most air pollution originates in and is concentrated around urban areas.

Hungary uses approximately two-thirds more energy for a required task than the European OECD average, producing more pollutants. According to the Ministry for Environment & Regional Policy, sever air pollution and other environmental damages affect almost half of Hungary's population, of which about 30% live in seriously polluted places.

Water/Wastewater Treatment

Rivers provide Hungary with two-thirds of its available water resources and groundwater makes up to other third. Thus water quality is determined by the neighboring countries. Most of Hungary's groundwater resources are comprised of bank-filtered wells in alluvial formations along the rivers - aquifers which are highly susceptible to inorganic contamination. Artesian and karstic waters also serve as valuable water resources.

The growth of water demand by industry, agriculture and municipalities has grown dramatically in recent decades. Today, 70-75% of this demand is from industry, 15% from agriculture and less than 10% from municipalities. While industry is the largest polluter, there is substantial contamination from agricultural run-off (agrichemicals).

The water supply of Hungarian households has improved significantly between 1990-1994. While 97% of Hungarians have access to potable water only 43% are connected to a sewage system. There is now a significant "utility gap" between the percentage of the population served with community water and those connected to sewage treatments plants. One major result is that water supplies are becoming more polluted, mostly with untreated sewage. Furthermore a survey of Hungary's wastewater treatment facilities revealed that an estimated two-thirds of the country's wastewaters is discharged following only mechanical treatment and only one-third is treated biologically.

Investment of %5-6 billion would be required for Hungary's sewage system to reach EU standards with 67% of households being connected to a sewage system. Two-thirds of the money would go for pipeline expansion while the rest would finance the construction of new and upgrading of existing treatment plants.

In Budapest, with 20% of Hungary's population 90% of the households are connected to the sewage system. However, in other cities 50-60% are connected, while in villages with 45% of the country's population - only 4-5% of the households are connected to the sewage system. Government plans call for 57-58% of households nation-wide to be connected to the sewage system in the next few years.

The government recently approved a 10-year wastewater treatment program for Budapest and 22 cities. The program calls for investment estimated at $2 billion. According to the program the central government will provide 25% of the total cost of expansion/upgrading activities and 35% for construction of new facilities. The remainder has to be financed by the local governments. The World Bank is negotiating a $170 million loan with the Ministry for Transportation and Water Management to assist the municipalities.

Waste Management and Contaminated Soil Remediation

Hungary generates yearly some 85 million tons of waste, of which 4-5 million tons are municipal solid waste and the rest in industrial waste. The management of municipal waste is the responsibility of the local government's. The country's only major incinerator is in Budapest. Most waste is disposed of in landfills which do not meet environmental standards. Only 10% of the waste is recycled.

Approximately 2-3% of wastes generated annually are hazardous. One-third is "red mud", i.e. waste form the production of aluminum from bauxite. Actual installed capacity to treat hazardous wastes wither by incineration or by final disposal only exists for 10% of the total volume. Recycling programs have been increasing slowly and storage space is reaching capacity. Hungary has one hazardous waste incinerator in Dorog with 25,000 ton/year capacity owned by the French Sarp Industries, one final disposal site in Aszod and four regional temporary storage sites. Some companies have constructed on-site incineration facilities.

The Ministry for Public Welfare is responsible for the management of medical/hospital waste. Some medical hospital waste is being incinerated in Dorog and for the disposal of the rest regional incinerators are being planned. Debrecen is the first being constructed with a Danis government grant. The smaller incinerators in operation are being upgraded or phased out for not meeting environmental standards.

After the departure of Soviet soldiers in June 1991 the cleanup tasks of their 171 army bases was huge. Environmental damage was estimated at 1 billion. After an initial surge of activity and the cleanup of the 20 most contaminated bases, further activities have now closed considerably due to a lack of funds. For the most part, the Hungarian government is focusing on containment, rather than true cleanup of these sites. The State Assets Privatization Agency (APV) has been recently assigned the tasks to privatize the former Soviet military bases along with other assets in their present status.

Market Opportunities for U.S. firms

The new Act LIII of 1995 on the General Rules of Environmental Protection was passed by Parliament on May 30, 1995. Decrees and regulations for details are expected to follow soon. With stricter enforcement of the new anti-pollution law, more environment-related investments are anticipated in all areas: air, water/wastewater, hazardous/municipal waste.

Business opportunities are abundant, however financing of environmental projects is a problem. Because of budget restrains and economic restructuring, the near-term opportunities are limited. But the prospects for long-term market growth are strong.

Central government allocations are funneled either through the Central Environmental/Water Funds or, for certain wastewater projects, directly to the municipalities (up to 25-40% of the cost of the projects). Industry-related projects are financed by the companies themselves.

The best way for foreign companies is to get involved in environmental projects is teaming up and subcontracting arrangements with local Hungarian companies. A list of local environmental consulting and engineering companies is available with the Embassy's Commercial Services.

Major Contracts In-Country In Environmental Areas:

1. American Embassy, Commercial Service
   American Embassy, Science & Technology
   Szabadag ter 12
   1054 Budapest, Hungary
   Tel: (36 1) 267 4400 Fax: (36 1) 269 9326
2. Ministry for Environment and Regional Policy
   Ms Eszter Szovenyi, U. S. Desk Officer (direct tel: 201 3764)
   Fo utca 44-50
   1001 Budapest, Hungary
   Tel: (36 1) 201 2846 & 201 5580 Fax: (36 1) 201 4133
3. Ministry of Industry and Trade
   Ms. Martha Hibbey, Department Head fir Environmental Protection
   Vigado utca 6
   1051 Budapest, Hungary
   Tel/Fax: (36 1) 118 4867
4. Ministry of Transport, Telecommunications and Water Management
   Ms. Agnes Pinter, Advisor, Water Management Development Department
   Ms. Ibolya Gazdag, Counselor, Water Management Department
   Dob utca 75
   1077 Budapest, Hungary
   Tel: (36 1) 141 4300 Fax: (36 1) 122 8695
5. National Authority for Environmental Control
   Mr. Robert Reiniger, Director
   Fo utca 44
   1011 Budapest, Hungary
   Tel: (36 1) 201 4133 Fax: (36 1) 201 4282

6. Central Environmental Fund
c/o Ministry of Environment
Ms. Bela Donath, Executive Secretary
Fo u 44
1011 Budapest, Hungary
Tel: (36 1) 201 4173 Fax: (36 1) 201 3653

7. National Water Authority
Mr. Bela Fenyvesi, General Manager
Marvany u 1
1012 Budapest, Hungary
Tel: (36 1) 201 1729 Fax: (36 1) 201 9332

8. KGI Institute for Environmental Management
Dr. Istvan Endredy, Director General
Alkotmany utca 29
1054 Budapest, Hungary
Tel: (36 1) 111 5826 Fax: (36 1) 132 8270

9. Regional Environmental Center for Central and Eastern Europe
Stanislaw Sitnicki, Executive Director
Miklos ter 1
1054 Budapest, Hungary
Tel: (36 1) 168 8685 Fax: (36 1) 168 7851

10. Vituki Center for Water Resources Development
Dr. Odon Starosolszky, Director General
Kvassay Jeno ut 1
1095 Budapest, Hungary
Tel: (36 1) 114 1620 Fax: (36 1) 134 1514

11. Hungarian Association of Water and Sewage Companies
Dr. Maria Papp, Secretary General
Sas u 24
1051 Budapest, Hungary
Tel: (36 1) 153 3241 Fax: (36 1) 112 3067 and 153 3241

12. Association of Environmental Consultants
    Ms. Anna Szekely, Secretary General
    Angol utca 12
    1149 Budapest, Hungary
    Tel: (36 1) 164 0026 Fax: (36 1) 163 2874

Competitors in the Market

1. BGT Hungaria (Subsidiary of Groundwater Technologies, U.S.)
   Tas vezer utca 28
   1113 Budapest, Hungary
   Tel/Fax: (36 1) 185 4523
2. COMCO-Martech Hungary (Subsidiary of Martech, U.S.)
   Patakhegyi ut 83-85
   1028 Budapest, Hungary
   Tel: (36 1) 176 5532 Fax: (361) 176 5388
3. Culligan Europe (Subsidiary of Culligan, U.S.)
   Kossuth u 34
   1221 Budapest, Hungary
   Tel: (36 1) 227 9000 Fax: (361) 228 1658
4. GIBB Consulting (Subsidiary of Law International Inc., U.S.)
   P O Box 433
   1371 Budapest, Hungary
   Tel/Fax: (36 1) 202 7244
5. KROFTA Hungary (Subsidiary of Krofta, U.S.)
   Abel Jeno utca 22
   1113 Budapest, Hungary
   Tel/Fax: (36 1) 166 0889

Part IV:

Case Study Applications

IV.1

IMPLEMENTATION TIME AND UNCERTAINTY INFLUENCES ON PREDICTION

M. CISLEROVÁ
Faculty of Civil Engineering
Czech Technical University
Thakurova 7, 166 29 Prague 6
Czech Republic

Introduction

The major cause of widespread groundwater pollution is the introduction of solid and liquid wastes into the subsurface or near-surface soil. Liquid waste and leachate from the solid wastes directly affect groundwater quality, although it may take many years for contaminated water to be flushed from a system because of the slow rate of movement. In other cases, however, the speed of the travel of contaminants from source through the subsurface may be surprisingly fast. The resulting deterioration in water quality may be serious in both cases such that the source becomes a hazard to human health and the environment. As a consequence, predicting the movement of pollutants in the subsurface is an urgent topic all over the world and represents a difficult and demanding engineering discipline.

E.A. McBean et al. (eds.) Remediation of Soil and Groundwater, 341–353.

Of interest is the character of groundwater pollution in this part of the world? Groundwater contamination is often a local rather than a regional problem. Originating at a point source, it would be unlikely to extend more then a few thousand meters. However, if there are many point sources, region wide pollution of groundwater may occur. Serious pollution of regional subsurface environments is caused by mining or by improper management in agriculture. As well, "famous" are large areas polluted by military activities. A common and continuing problem is groundwater contamination from improperly located and/or designed landfills. Contaminant plumes may grow for years, continued long after abandonment of such sites. The growing volumes of solid wastes, the lack of geologically suitable sites, and /or inadequate regulations on landfill location may result in the contamination from recently-opened sites. Often the regulation rules are followed just formally from designers and from owners as well. Only a small percentage of contamination spills is accidental.

Landfill leachate may contain extremely high concentrations of inorganic and organic materials of unknown composition and amount. The date of deposition cannot be always traced. The volume of leachate produced is a function of the quantity of water percolating through the refuse and the hydraulic properties of surrounding soils. Contaminants moving through soil and rock may be attenuated to some degree through dilution and dispersion, mechanical filtration, chemical reactions, volatilization, adsorption, biological assimilation, membrane filtration, and radioactive decay.

Processes to Predict

The maintaining source of movement is precipitation. The residence time of the pollutants in particular zones or reservoirs forming the subsurface porous media profile is dependent on that part of precipitation which infiltrates, on its amount and its spatial and temporal distribution. With respect to pollutants we focus on water near the earth's surface and how it moves within and between the various reservoirs, namely variably-saturated (vadose) zone representing soil profile and saturated zone composed of subsurface water aquifers. As can be expected there are many factors influencing flow processes, the principal characteristics being the hydraulic properties of each zone, generally defining the retention ability and the hydraulic conductivity. Fundamental knowledge of basic theoretical principles of flow and transport processes is contained, for example, Bear [1], [2], Freeze and Cherry [3], Kutìlek and Nielsen [4], and many others.

As water flows through the soil profile, solute is carried away at a rate equal to the flow velocity. This process is known as advection. In a porous medium, the flow paths are tortuous, so that different flow paths have varying lengths. Solute transported along a shorter path would arrive at a closing profile sooner than solute particle following a longer path. The result is mechanical dispersion. The third process commonly occurring is the diffusion within gaseous and liquid phases. To express the processes mathematically, advective, dispersive and diffusive fluxes have to be included in the macroscopic mass balance equation of solute species transported in a porous medium domain. If the solute is conservative, it means it does not react chemically with the other constituents present and is not adsorbed on the solid phase surface. This basic form of advective-dispersive equation is adequate to describe the process, assuming we are able to supply boundary and initial conditions over the solute transport domain.

A contaminant transported by natural flow exhibits dispersion both longitudinally and laterally. The shape of a contaminant plume is influenced by the character of the source. If there is continuous source, the contamination front moves by advection in the direction of flow. A point source, such as a contaminant spill will move by advection as a slug with the flow of groundwater. Due to hydrodynamic dispersion, the slug will expand in three dimensions to fill a larger volume, but with a lower solute concentration.

If there is a reaction between the solute and other groundwater constituents, or if adsorption of the solute on the solid phase occurs, the rate of advance of a contaminant front will be retarded. The plume of solute will spread more slowly and the concentrations will be lower than those of non-reactive solutes. Adsorption and chemical reactions affect various solutes at different rates. In the transport equation the additional terms have to be added in the form of sources or sinks to describe the chemical reactions. Each constituent has to be balanced separately.

Simulation Models as the Tools for Prediction

Modern and very efficient tools to simulate the contaminant movement are simulation models. Several simple methods are available as well, allowing estimation of solute travel times for simple systems, but for more complex situations or chemical reactions, we must have recourse to numerical models. Theoretical aspects of flow and solute transport modeling can be found in books of Bear and Verruijt [5], Kinzelbach [6], Dagan [7] and many others. The applicability of models for predictive purposes is presented. For example,. by Anderson and Woessner [8] on three frequently used codes of saturated subsurface flow and particle tracking.

Because advective transport and hydrodynamic dispersion depend on the velocity of groundwater flow, the mathematical simulation model has to solve at least two partial differential equations. One is the flow equation from which velocities are calculated, and the other is the solute transport equation, which describes the concentration of pollutant in groundwater. For the advective-dispersive equation, a number of analytical solutions exist for simple boundary conditions and steady-state flow. In majority, both equations are solved numerically, resulting in a powerful tool able to produce complex studies.

Accurate predictions of groundwater flow conditions is an essential first step in simulating contaminant transport in groundwater. Model analyses and predictions can lead to an improved understanding of an aquifer system and serve as an aid to making decisions or formulating policies. All predictions should be presented together with a realistic assessment of the confidence. Historically, models describing the flow can be divided into two groups. Models of unsaturated-saturated flow, applicable essentially for vadose zone, are based on Richard's equation and represent flow in subsurface generally. The use of this type of model for prediction is not yet routine. The larger group of models is dealing with regional horizontal flow in confined and unconfined aquifers where only saturated flow is considered. Here the applicability for the engineering practice is promising and this type of models may be used with relative confidence. Particle tracking analysis can be used routinely with flow models to track contaminant path, as a result of advective transport. To include two other processes that affect contaminant movement, namely dispersion and chemical reaction, again the advective-dispersive equation has to be solved.

Existing field-scale models can be classified from many aspects, but no classification seems to be universally accepted. As basic model categories, judged on different levels and therefore representing at best the fragments of a larger classification scheme Roth et al. [9] indicates:

- purely deterministic vs. stochastic models
- stochastic vs. deterministic models with spatially-variable parameters
- black-box models producing a system response to an input signal vs. models with variable degree of process resolution
- functional vs. mechanistic models
- physically based vs. empirical models
- research vs. management models
- rate vs. capacity models

A comprehensive review of flow and solute transport modeling is presented by van der Heijde et al. [10], or more recently Mangold and Tsang [11]. Many models are available commercially, although, the most powerful models are often proprietary.

In decision-making connected with required system interpretation and the aims of prediction, the purpose of the modeling effort has to be clearly identified to determine properly the magnitude of the modeling effort [8]. It includes decision-making as to whether the model is steady-state or transient, one-, two-, or three-dimensional, analytical or numerical, and the types of processes to be included and which simplifying assumptions are acceptable. The correct modeling strategy helps to prevent the model misuse in both directions - in the form of ignoring important aspects, or in the form of overkill.

Modeling can be defined as the art and science of collecting a set of discrete observations which represent our incomplete knowledge of the real world and produces predictions of the behavior of a system. Such predictions will be necessarily uncertain, as will our knowledge of the true behavior of the system [8].

Data for Models and Strategy of Sampling

The state and characteristics of the real world are incorporated into the simulation models described by a set of information termed the input data, such as the spatial distribution of soil and aquifer properties, or the time-histories of system stresses. These inputs are often highly variable in time and space. Some may be inherently uncertain such as the future time series of rainfall infiltration and subsequent recharge. The processes at work in the real system, including those induced by proposed management actions, are represented by the inputs to yield the outputs that characterize the behavior of the system. Outputs might be contaminant concentration distributions in space and time, travel times, mass losses, or exposure levels and duration at selected locations. Groundwater flow and transport systems input data are loaded by significant variability and uncertainty which remain in the outputs. It is evident that the quality of the results depends crucially on the quality of inputs. The major sources of uncertainty in the modeling process have to be reflected at several different levels [12], included during the sampling process, the process of parameter identification, input estimation and model validation and accuracy assessment.

What is the most common negligence? The hydraulic and dispersive properties are often improperly determined, neglect heterogeneity, ignore spatial and temporal variability, and assume different scales of data samples in contrast with the scale of the region. Theoretical assumptions are frequently ignored during the sampling and input evaluation process. Most contaminant concentration measurements are collected at or near a contaminated site after an indication that some problem exists. These measurements are usually limited and scattered. The real geometry of contaminant plumes can stay quite uncovered. Contaminant data are usually more difficult to interpret than hydrologic data because the compounds observed and their physical state depend on chemical and biological conditions in the subsurface environment.

Correctly done field sampling studies can help to reduce the major types of uncertainty, that means the lack of knowledge about the processes that control contaminant transport and incomplete knowledge of the spatially and temporally variable environmental factors that influence these processes.

The design of model-oriented field sampling programs requires a good understanding of subsurface physical, chemical and biological processes, of model and input estimating algorithms, and sampling technology. It makes sampling design a truly multidisciplinary effort that typically requires the active participation of several specialists.

Input and validation, the further phase of a model-oriented field sampling program, presume that the processes included in the model are those that control contaminant behavior at the site of interest. The sampling program should provide the data needed to obtain the most accurate model inputs and predictions possible, subject to resource constraints. Here the traditional sampling problem is addressed by classical statistics, as known from hydrology.

Heterogeneity, Variability and Scales

Geological systems, by their very nature, are complex, three-dimensional, heterogeneous, and often anisotropic. The greater the degree to which a model approximates the true heterogeneity as being uniform or homogeneous, the more the true variability in velocity must be incorporated into larger dispersion coefficients. The more accurately and precisely we can define spatial and temporal variations in velocity, the lower will be the apparent magnitude of dispersivity. The role of heterogeneity is not easy to quantify and much research is in progress on this problem.

The sampling program should recognize that contaminant dispersion is largely a manifestation of unknown hydrogeological heterogeneities. Dispersion and advection processes are interrelated and are dependent on the scale of the model. Because dispersion is related to the variance of velocity, neglecting or ignoring the true velocity distribution must be compensated in a model by a correspondingly higher value of dispersivity. The scale dependence of dispersivity coefficients (macrodispersion) is recognized as one of the limitations in the application of conventional solute transport models to field problems. Gelhar et al. [13], [14] show that most reported values of longitudinal dispersivity fall in a range between 0.01 and 1.0 on the scale of the measurement.

Preferential Flow

A special attention should be paid to preferential flow, the phenomenon not yet well described which plays an important role in the speed of contaminant propagation. In our engineering practice preferential flow is mostly ignored and neglected. Preferential flow greatly affects the migration of pollutants in the vadose zone in terms of both arrival time to the groundwater table and the amount of the discharged solute. During an infiltration event, water and solutes can preferentially flow into and through macropores and bypass a major portion of the soil matrix. Significant preferential flow - a feature of coarse soils, was detected in the majority of soils encountered in the Bohemia basin as products of weathering of crystalline rock (about 60% of all soils ([15], [16], and others). A number of experiments involving strongly adsorbing tracers have shown that preferential flow may occur with little or no solute retardation.

The conceptual problem of predicting macropore flow is similar to that of describing solute transport in saturated fissured and fractured media . In both cases, rapid flow in the macropores (or fissures) is accompanied by much slower infiltration or diffusion -controlled mass transfer into the soil or matrix. The major distinction is, that unlike the case for fissures and fractures in aquifers, flow and transport in macropores occur only when specific conditions are satisfied. Identification of these conditions and modeling of macropore flow and transport has been recently investigated. The occurrence of macropore flow depends significantly on the antecedent soil water conditions, hydraulic properties of the soil matrix, the rate of water input at the soil surface, and the spatial distribution and interconnectedness of the macropore sequences. The impact of such preferential flow on solute transport is further determined by the rate of diffusive mass transfer into the soil matrix and the sorptive properties of the macropore and matrix regions.

One of the major limitations in predicting macropore flow and bypassing is a limitation in characterizing the geometric and hydraulic features of macropores and in providing the values for the required model parameters. Our understanding of the relation of the preferential flow to observable soil features is quite poor [17]. With respect to predictability we face the uncertainty as to when macropore flow is dominant and how it should be modeled [9].

Multiphase Flow

Another special phenomenon yet is multiphase flow. For many common and dangerous contaminants (e.g. organic fluids), a multiphase flow and transport should be considered to predict their migration. Codes that simulate this problem exist but the application is not common. Their use is primarily concerned with the source (e.g. water or oil) quantity, the programs are relatively new and not tested (also they are largely proprietary) and insufficient data exist to employ multiphase principles.

Migration patterns associated with immiscible and miscible organic fluids are schematically described by Abriola [18]. If not remedied, the migration of such a immiscible organic liquid phase represents an acute or chronic source of pollution. Movement of the organic liquid through the vadose zone is governed by the potential of the organic liquid, which in turn depends upon the fluid retention and relative permeability properties of the air/organic/water/solid system. As an organic liquid flows through a porous medium, some is adsorbed to the medium or trapped within the pore space. This organic contamination held within the soil column by capillary forces (at its residual saturation) represents a chronic source of pollution because it can be leached by percolating soil moisture and carried to the water table.

If the organic liquid is lighter than water, it may migrate as a distinct immiscible contaminant (the acute source) within the capillary fringe overlying the water table. The soluble fraction of the organic liquid will also contaminate the water table aquifer and migrate as a miscible phase within groundwater. If the organic liquid is heavier than water, it will migrate vertically through the vadose zone and water table to directly contaminate the groundwater aquifer. It may also penetrate water-confining strata that are permeable to the organic liquid and, consequently, contaminate underlying confined aquifers. The organic contaminant may form a pool on the bedrock of the aquifer and move in a direction defined by the bedrock relief rather than by the hydraulic gradient.

Contamination of groundwater occurs by dissolution of the soluble fraction into groundwater contacting either the main body of the contaminant or the organic liquid held by specific retention with the porous medium.

Conclusions

With respect to routine engineering practice, there is an opinion that a general understanding of subsurface processes and reactions has grown significantly in recent decades, but the predictive capability regarding solute transport processes is not yet satisfactory. Our understanding of some processes and reactions is not sufficient for predictive purposes of complex, heterogeneous and anisotropic environments. Significant efforts are still needed to translate the research results into an accepted field-scale technology. Assessments of model accuracy and validity at the field scale are an important aspect of this translation from science to application. Interdisciplinary efforts of geologists, hydrologists, geochemists, geostatisticians, and others are essential to study and solve groundwater contamination problems with greater confidence.

Simulation models routinely are used only in the case of well defined flow problems, mostly concerning the flow in aquifers, with appropriate, carefully collected field data.

Studies should be carried out by experienced modelers. In practice however, the use of flow and solute transport models is booming, being often inadequate. Many of available computer codes and graphics packages are easy to use so that impressive-looking results can be readily produced using few data or data of poor quality. In that case, professional ethics require that modeling results should not be used for prediction [8].

A substantial problem is a lack of qualified engineers who have a basic understanding of all processes involved, and would be able to deal with the situation professionally. We need to introduce a high quality training in practical and theoretical disciplines connected with the subject to our universities. Until now there has been a resistance against introducing modern approaches. Resources need to be devoted both to continue fundamental research and to decrease the gap between the state of art and the state of practice. It is not a problem of subsurface hydrology only [19],[20].

References

1. Bear, J. (1972) Dynamics of Fluids in Porous Media, American Elsevier, New York.
2. Bear, J. (1979) Hydraulics of Groundwater, McGraw-Hill, New York.
3. Freeze, R.A. and Cherry, J.A. (1979) Groundwater, Prentice Hall, Englewood Cliffs, New Jersey.
4. Kutìlek, M. and Nielsen, D.R. (1994) Soil Hydrology, Catena - Verlag, Cremlingen-Destedt.
5. Bear, J. and Verruijt, A. (1987) Modeling Groundwater Flow and Pollution: Theory and Applications of Transport in Porous Media, D.Riedel Publishing Co, Dodrecht.
6. Kinzelbach, W. (1986) Groundwater Modeling: An Introduction with Simple Programs in BASIC, Elsevier Sc.Publishers, Amsterdam.
7. Dagan, G. (1989) Flow and Transport in Porous Formations, Springer-Verlag.

8. Anderson M.P, Woessner, W.W. (1992) Applied Groundwater Modeling: Simulation of Flow and Advective Transport, Academic Press, Inc., San Diego.

9. Roth, K., Fluhler, H.W.A. Jury and J.C.Parker (eds.) (1990) Field-Scale Water and Solute Flux in Soils, Birkhauser, Verlag, Basel.

10. van der Heijde, P.K.M., El-Kadi, A.I., and Williams, S.A. (1988) Groundwater Modeling: An Overview and Status Report, EPA/600/2-89/028, Ada, Oklahoma.

11. Mangold, D.C. and Tsang, Ch.F. (1991) A summary of subsurface hydrological and hydrochemical models, Rev.of Geophysics 29, 51-79

12. Mc Laughlin, D. and Wood, E.F. (1988) A distributed parameter approach for evaluating the accuracy of groundwater predictions, 2. Application to groundwater flow, Water Resour. Res. 24, 1048-1060.

13. Gelhar, L.W. (1986) Stochastic Subsurface Hydrology from Theory to Applications, Water Resour. Res. 22, 135-145.

14. Gelhar, L.W., Welty, C. and Rehfeldt, K.R. (1992) A Critical Review of Data on Field-Scale Dispersion of Aquifers, Water Resour. Res. 28, 1955-1974.

15. Cislerova·, M., äim_nek, J. and Vogel, T. (1988) Changes of steady-state infiltration rates in recurrent ponding infiltration experiments, Journal of Hydrology 104, 1-16.

16. Cislerova· M., Vogel, T. and Simunek, J. (1990) The Infiltration-Outflow Experiment Used to Detect Flow Deviations, in K.Roth, H.Fluhler, W.A.Jury and J.C.Parker (eds.), Field-Scale Solute and Water Transport through Soils, Birkhauser Verlag, Basel, pp. 109-117.

17. Hopmans J.W., Cislerova·, M. and Vogel, T. (1994) X-Ray Tomography of Soil Properties, in S.H. Anderson and J.W. Hopmans (eds.), Tomography of Soil-Water-Root Processes, SSSA Special Publ. No. 36, pp. 17-28.

18. Abriola, L.M. (1988) Multiphase Flow and Transport Models for Organic Chemicals: A Review and Assessment, EA-5976, Electric Power Research Institute, Palo Alto, Calif.

19. Nash, J.E., Eagelson, P.S., Philip, J.R. and van der Molen, W.H. (1990) The education of hydrologists, Hydrol. Sci. J. 35, 597-607.

20. Philip, J.R. (1991) Soils, Natural Science and Models, Soil Science 151, 91-98.

IV.2

CASE STUDY APPLICATIONS: A RESEARCH PERSPECTIVE

NEIL R. THOMSON

Department of Civil Engineering, and

Waterloo Centre for Groundwater Research

University of Waterloo

Waterloo, Ontario, Canada

N2L 3G1

Introduction

Over the last 15 years ground water and soil remediation technologies have been developed by ground water scientists and engineers mostly out the need to satisfy regulatory agencies. Given this short time to develop an experience base, many of the technologies that have been classified as standard (as opposed to emerging) remediation technologies are just now showing some performance problems. From a theoretical perspective, some of these technologies are very appealing due to their low cost (both capital and maintenance) and expected performance. Even results from laboratory columns and larger bench scale boxes tend to show that these technologies are good candidates for some sites. However, once implemented at a site it is the complex geological nature of the subsurface system coupled with the array of sources and associated contaminants that hinder the performance of the selected technology. The choice of a remediation system or systems at a particular site may vary depending on the desired remedial objective. These remedial objectives range from the control of an

E.A. McBean et al. (eds.) Remediation of Soil and Groundwater, 355–378.

aqueous phase plume at a property boundary, to the complete restoration of both the soil and ground water to a desired safe standard.

Obviously, the hydrogeology and contaminant characteristics of a site under consideration must be adequately understood before remedial options can be determined to satisfy the desired objective. However, the situation in which a complete understanding of the site hydrogeology and the contaminant characteristics exists is quite rare. With this imperfect knowledge of the site hydrogeology (in particular the spatial variability in permeability), and uncertainty regarding the contaminant source location, strength, composition, reactivity and historical presence, a remedial technology is selected. It is therefore important to develop a comprehensive program to carefully monitor the progress of the remedial schemes. This monitoring program should address concerns dealing with engineering parameters associated with the selected technology, the effective zone of clean-up, the in-situ mass removed or destroyed, and an estimate of the amount of mass remaining in the subsurface. This final concern has a direct impact on risk reduction.

This paper initially provides some details on various ground water contaminants and sources with a focus on non-aqueous phase liquids (NAPLs). Since most remedial technologies rely on the ability of the subsurface regime to transport fluids (e.g., water, gas, NAPLs, nutrient enriched mixtures, and surfactants), and to provide for adequate mass transfer between various phases, a brief discussion on chemical partitioning and mass transfer is provided. Following this discussion the role of variations in geological media on the performance of pump-and-treat systems and soil vacuum extraction systems are described.

Ground Water Contaminants and Sources

It is clear that worldwide ground water pollution problems date back to the 1800s. With the industrial developments in the 1900s, ground water problems increased due to the large increase in the production of a variety of chemical wastes associated with the production of pesticides, plastics, wood preserving compounds, chlorinated solvents, and petroleum-based products. Depending on the nature of these chemical wastes a number of options (e.g., landfilling, open dumping, surface impoundments, and deep well disposal) were available to each industry to handle their waste. However, poor waste handling and transportation practices, and faulty storage facilities have created a variety of ground water problems. Since some of these contamination problems occurred quite a long time ago, and even though the mass of contaminants released may be small, the ground water problem may be extensive due to the long period of time available for contaminant migration to occur.

A variety of compounds have been identified as ground water contaminants. These include inorganic compounds (e.g., arsenic, chromium, mercury, zinc, and nitrates), synthetic organic chemicals (e.g., benzene, 1,1-dichloroethane, vinyl chloride, and tetrachloroethylene), radioactive contaminants (e.g., strontium 90, cesium 141, and plutonium 239), and pathogens. The organic compounds have created the greatest potential for ground water contamination due to their widespread use within the industrial sector and their properties (i.e., significant solubility and high toxicity). Many of these organic chemicals are part of a class of chemicals called non-aqueous phase liquids (NAPLs). These compounds have solubilities that are in some cases three-orders of magnitude higher than regulated drinking water limits. Schwille [11] was one of the first to demonstrate that NAPLs can be easily divided into two classes depending on their density relative to water. Immiscible liquids both less dense than water (LNAPLs) and more dense than water (DNAPLs) infiltrate into the ground at a rate depending on the type of soil and the NAPL characteristics. Whether or not the NAPL reaches the water table depends mostly on the spilled volume and the retention capacity of the soil. If the retention capacity of the unsaturated zone is exceeded, the fate of the NAPL becomes largely dependent upon its density. A light immiscible liquid will tend to form a thin pool on the surface of the water table, whereas, a dense immiscible liquid will continue to sink past the water table leaving a residual in its wake. Upon reaching a layer which it cannot penetrate, lateral spreading will occur, not necessarily following the direction of ground water flow but the local topography of the layer. In all situations, the residual and pools of free product will contribute dissolved components to the ground water, probably for many decades. It has been estimated that a spill volume as small as 40 litres can contaminant significant volumes of water [7].

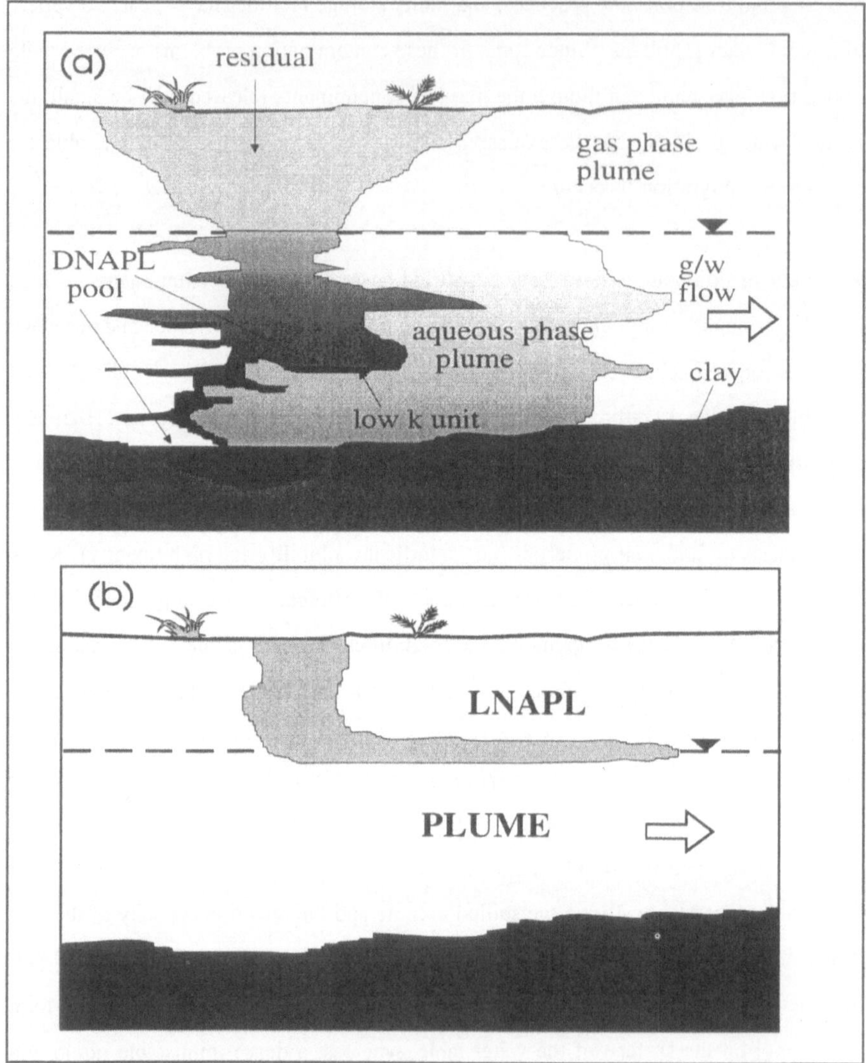

Figure 1. Schematic of (a) DNAPL and (b) LNAPL release in porous media.

A schematic of a DNAPL release which has had sufficient volume to penetrate the water table and form a pool both on a lower permeability lense and on a lower confining clay layer is shown in Figure 1(a). Also shown in this schematic is the aqueous and gaseous phase plumes which evolve from both the NAPL residual and pools. In contrast, a LNAPL release is shown in Figure 1(b) in which the NAPL has spread out within the capillary fringe and the top portion of the saturated zone. Recent findings on DNAPLs has shown that the movement of a DNAPL in porous media is controlled by the wetting properties of the DNAPL/soil system and the difference in interfacial tension. Slight spatial differences in the permeability in a relatively homogeneous sandy porous medium have been shown to cause the DNAPL to follow a tortuous path [9]. This somewhat heterogeneous nature of the location of the DNAPL is one of the factors that makes remedial technologies which focus on DNAPL removal ineffective.

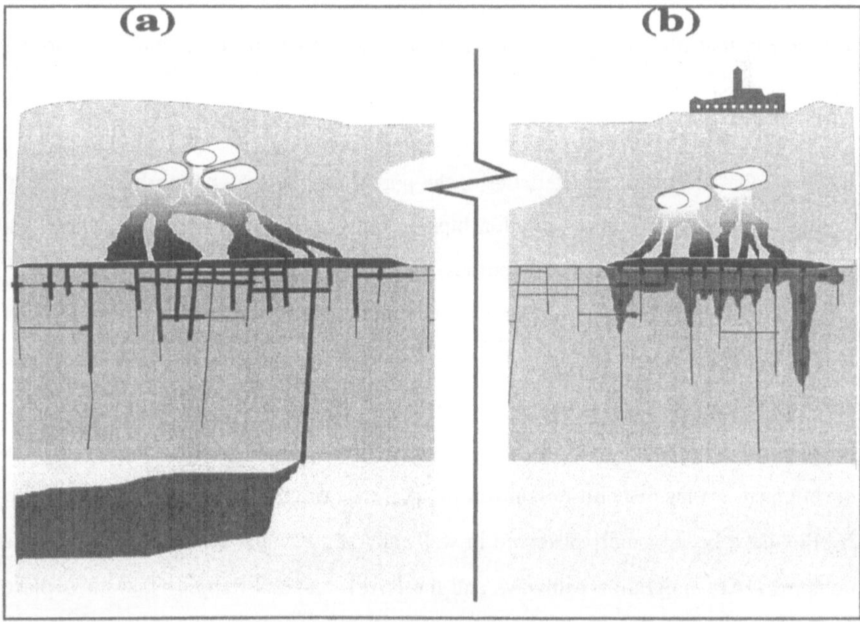

Figure 2. Schematic of DNAPL release in a fractured porous media system.

The prognosis for the remediation of a fractured porous media system contaminated by a DNAPL release is much worse than for a porous medium system. A typical fractured porous media scenario involving a DNAPL pooled above a fractured aquitard overlying a drinking water supply aquifer is shown in Figure 2. Figure 2(a) represents a situation where several fractures have been invaded by a NAPL which had sufficient pressure to overcome the entry pressure of the fracture. The actual migration in most of the fractures has been arrested by aperture constrictions; however, one long fracture spanning the entire depth of the aquitard has allowed NAPL to migrate into the drinking water aquifer. Clearly, the presence of a NAPL in this drinking water aquifer will result in the formation of a long-term aqueous phase plume which may cause the water in the aquifer to be unfit for consumption. The situation in Figure 2(b) is somewhat different than the one presented in Figure 2(a) since the DNAPL has not penetrated as deeply into the fracture network, but the fractured porous media system has the characteristics such that contaminant mass has diffused significantly into the matrix portion of the aquitard. The factors that play a role in NAPL migration and retention in fractures are not yet clearly understood.

It is clear that physical characteristics of the porous media (e.g., permeability, porosity, capillary pressure-saturation relationships, relative permeability relationships, and sorption capacity) or fractured porous media systems (e.g., fracture frequency, orientation, and aperture; matrix porosity and sorption capacity) and the type of DNAPL play a significant role in the remediation technology selected. Most of these properties are difficult to determine at actual field sites, but are important in trying to assess the location of the NAPL which in most cases represents a long-term source. In most situations it is very difficult to confirm the presence of NAPL at a site. At some sites NAPLs have been visually observed in wells and cores, while at other sites due to the complex NAPL migration pathways and the NAPL residual trapping that takes place, the presence of NAPLs cannot be confirmed. In these cases indirect methods which rely on some speculation must be used to determine the presence of NAPLs. For example, a general rule of thumb is that if the concentration of a constituent related to a NAPL is greater than 1% of the effective solubility, then NAPLs are most likely present. An additional complicating problem is that the actual physical properties of the subsurface NAPLs compared to the industrial grade counterparts differ due to the presence of complex mixtures, and the in-situ weathering process. At those sites where NAPL has been located and if attempts to remove the mobile NAPL through pumping are undertaken, a considerable amount of the NAPL will remain in the immobile or residual state and continue to contribute to the long-term ground water contamination problem.

Chemical Partitioning and Mass Transfer

At the pore scale in a porous medium containing NAPLs the contaminant mass may partition into the solid (sorbed) phase, the gas phase, and the water phase in addition to being present as a pure phase (NAPL). A schematic of this four phase system is shown in Figure 3. The total mass per unit volume of porous media associated with this system can be determined by

$$C_T = \rho_n \theta_n + C_w \theta_w + C_a \theta_a + C_s \rho_b \qquad (1)$$

where C_T is the mass of contaminant per unit volume of soil; ρ_n is the mass density of the pure phase NAPL; θ_n, θ_w, and θ_a is the NAPL content, moisture content and air content respectively; C_w is the aqueous phase concentration; C_a is the air phase concentration; C_s is the mass of contaminant sorbed per mass of soil; and ρ_b is the bulk soil density. The equilibrium concentration or vapour pressure of a compound in the air phase can be estimated from Raoult's law expressed as

$$P_{g(i)} = X_i P^o_{g(i)} \qquad (2)$$

where $P_{g(i)}$ is the vapour pressure of component i in the NAPL mixture, X_i is the mole fraction of component i in the mixture, and $P_{g(i)}{}^o$ is the vapour pressure of a NAPL comprised only of component i.

The equilibrium concentration in the water phase is estimated from the effective solubility which can be determined from

$$C_{w(i)} = X_i C^o_{w(i)} \qquad (3)$$

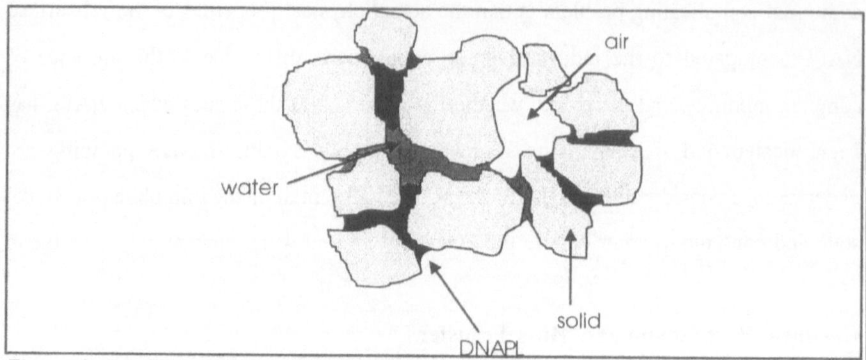

Figure 3. Schematic of pore scale system.

where $C_{w(i)}$ is the effective solubility of component i in the NAPL mixture, X_i is the mole fraction of component i in the mixture, and $C_{w(i)}{}^o$ is the solubility of the pure compound i in water. Assuming that the soil grains are water wet, then one way to estimate the mass of contaminant sorbed at equilibrium is through a linear isotherm of the form

$$C_s = k_d C_w \tag{4}$$

where k_d is the linear distribution coefficient. Since the gas and aqueous phase are in contact, the equilibrium relationship between the respective concentrations in each phase is controlled by Henry's law given by

$$C_g = HC_w \tag{5}$$

where H is the dimensionless Henry's law coefficient. Equation (1) through (5) can be used to determine how the various constituents will partition between the four phases if equilibrium conditions are assumed.

Various remediation technologies rely on the ability to reduce the equilibrium concentration in one of the phases (usually the water or gas phase) which causes the mass contained in the other phases to transfer in an attempt to re-establish equilibrium. The mass transfer from a single component NAPL to either the aqueous or the gaseous phase is usually assumed to be a function of a mass transfer coefficient, a driving force represented by a concentration difference, and the interfacial area of the NAPL. In general this can be expressed as

$$J_{n,\beta} = ka(C_{\beta}^{o} - C_{\beta}) \tag{6}$$

where $J_{n,\beta}$ is the mass flux from the NAPL to phase β (water or gas), k is the mass transfer coefficient, a is the NAPL-phase β interface area per unit volume of porous media. Since the mass transfer coefficient and the interfacial area are difficult to determine they are usually lumped as

$$K = k\,a \tag{7}$$

where K is the lumped mass transfer coefficient and reflects such factors as the pore scale NAPL geometry, fluid saturations, and physical fluid properties. Attempts to experimentally correlate the lumped mass transfer coefficient have produced relationships that relate the dimensionless Sherwood number to the Reynolds number, the NAPL saturation, soil porosity, and the mean grain diameter (for example see, [5],[10])). Unfortunately, most of this experimental work is based on a homogeneous distribution of NAPL within the porous media which may only apply to a small portion of an actual site.

Since contaminant mass must transfer between phases in order for mass to be removed from a ground water system, the assumption that equilibrium conditions are reached instantaneously has been a point of contention. Recent research has demonstrated that mass transfer rates from the NAPL to the aqueous phase, from the NAPL to the gaseous phase, and between the aqueous phase and the solid phase is kinetically-controlled (see for example, [5], [6], [10], [13]). The implication of these kinetically controlled processes is that any estimate of clean-up time assuming equilibrium conditions may severely underestimate the actual clean-up time due to mass transfer limitations.

Example 1: Pump-and-Treat Systems

The most frequently applied remediation technology involves pump-and-treat systems.

Ground water that contains either dissolved inorganic or organic compounds can be removed from the saturated ground water zone and treated on the surface before being released into a surface water body or injected back into the ground. For sites where NAPLs are present it appears that the most common use of pump-and-treat systems is to provide some form of hydraulic containment or capture of the dissolved phase plume. This technology relies on the removal of the aqueous phase; thus mass associated with the pure phase and the solid phase must be transferred into the mobile aqueous phase. Figure 4 presents a schematic profile of a saturated aquifer system containing residual and pooled DNAPL where a pump-and-treat system has been employed. For this pump-and-treat system to be effective, mass from the residual and the pooled DNAPL must partition into the mobile aqueous phase. Due to the low solubility of the NAPL, the existence of regions where there is very little direct contact of NAPL with the mobile aqueous phase, and the kinetically limited mass transfer between the NAPL and the aqueous phase, it would take a great many years before all of the NAPL were removed. Even after the NAPL has been removed, mass still remains in the aqueous and solid phases. For the sorbed mass, the sorption process is a nonlinear relationship [6], and for multi-component systems the sorption process has been shown to be competitive in nature [1]. In addition, since most sites have been contaminated for a long period of time, the less permeable zones act as locations in which mass can be sequestered. The release of mass stored in these less permeable zones is controlled by aqueous phase diffusion, a very slow process.

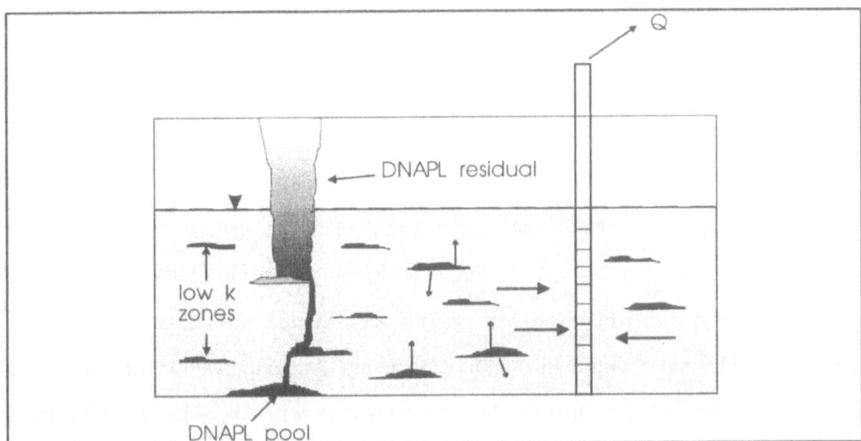

Figure 4. Pump-and-treat system with residual and pooled DNAPL in the presence of less permeable zones.

To illustrate the impact that heterogeneous media has on a pump-and-treat system a numerical model was employed to develop an aqueous phase plume within a homogeneous and heterogeneous aquifer system and then used to simulate the removal of this plume by a pump-and-treat system. The confined hypothetical 1.2 km x 2.6 km aquifer system is shown in Figure 5. For the homogeneous system a uniform hydraulic conductivity of 1×10^{-2} cm/s was used, while for the heterogeneous system a random hydraulic conductivity field with a mean of 1×10^{-2} cm/s, a ln-variance of 0.3 and an isotropic correlation length of 50 m was used. For each of these descriptions used for the geologic system, the same boundary conditions for flow, the same source location and rate, and the same pump-and-treat system design were employed. To create a steady-state flow field a hydraulic gradient of 0.005 m/m was applied across the system from the bottom to the top. The left and right-hand sides of the aquifer system were assigned no flow boundaries. The contaminant source with a strength of 60 mass units/day was applied at the source location shown in Figure 5 for a period of 10 years and then removed. A longitudinal dispersivity of 100 m, a transverse dispersivity of 10 m, an aqueous phase diffusion coefficient of 1×10^{-4} m^2/day, and a porosity of 0.35 were used to develop an aqueous phase plume for each system. Figure 6 presents the aqueous phase plume at 10 years for each system.

A pump-and-treat system comprising two extraction wells was employed. One extraction well pumping at 175 m^3/day was located near the source and the other well pumping at 125 m^3/day was located near the leading edge of the plume as shown in Figure 6. Figure 7 presents the fraction of mass remaining for each geologic description for an extraction period of 25 years. Notice that the mass recovered from both systems is identical for a period of ~5 years, and then a deviation in the rate of mass recovery is apparent between the systems. This deviation is a result of the nature of the heterogeneous system which gives rise to diffusion-limited mass recovery from those less permeable zones within the aquifer system. An extrapolation of these mass recovery profiles indicates that the heterogenous case will take ~25 years longer to remove the mass in the aquifer than the homogeneous case.

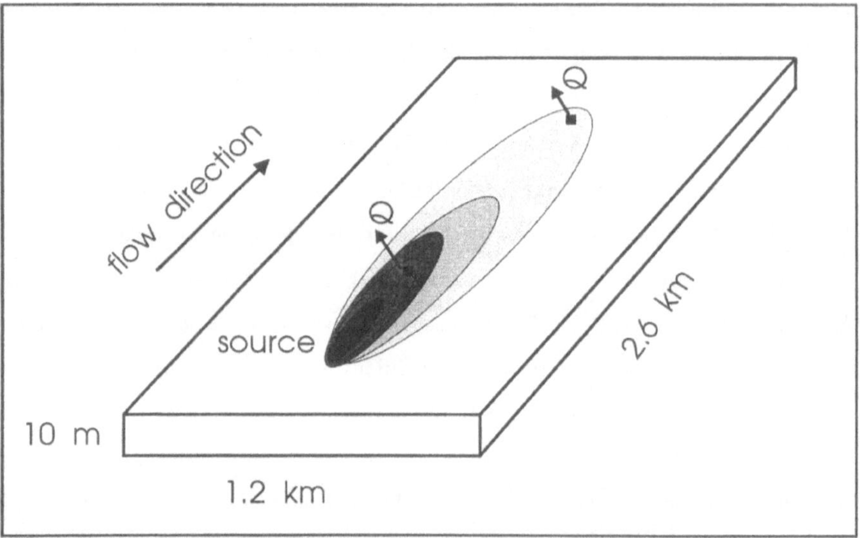

Figure 5. Confined aquifer system used for pump-and-treat example.

In recent years, a substantial effort has been directed towards a comprehensive review of existing pump-and-treat systems. A review of 16 pump-and-treat systems in the United States by Doty and Travis [4] indicated that for sites where *clean-up* time predictions were made, remedial operations involving pump-and-treat systems at 25% of these sites have lasted at least twice as long as predicted. Perhaps the most comprehensive review to date has been conducted by the National Research Council [8] in which operating data from 77 pump-and-treat sites in the United States were reviewed. At 69 of these sites the remedial objectives had not yet been reached although these objectives may possibly be reached in the future. For the remaining 8 sites the current success of the pump-and-treat operation indicates that site clean-up would be reached in a relatively short time (less than a decade).

This review by the National Research Council covered a spectrum of sites from those sites with a relatively homogeneous subsurface system containing a contaminant that was only in the dissolved form, to sites that are highly fractured and have DNAPLs present.

An example of a site that has heterogeneous geology contaminated with possible DNAPLs is one located in San Jose, California. At this facility, ground water

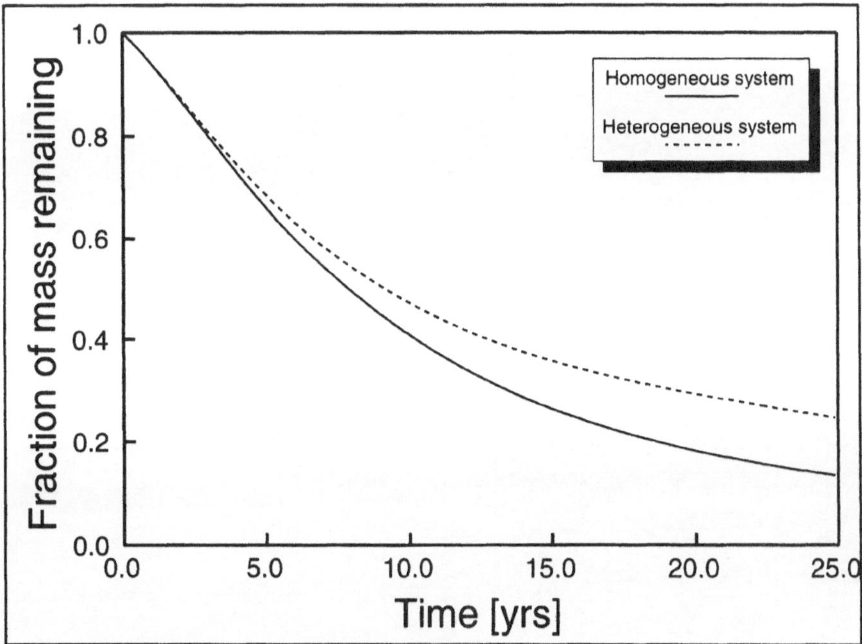

Figure 6. Fraction of mass remaining in system.

contamination was detected during a routine audit. Site investigations began immediately and it was determined that 1,1,1-trichloroethane, 1,1-dichloroethene and freon 113 were the primary contaminants. The geology at this site is composed of a complex layering of sand and gravel occurring between discontinuous less permeable units of silts and clays. The aqueous phase plume is ~4000 m long and ~450 m wide. Prior to the commencement of pumping, the estimated plume volume was ~$2x10^6$ m^3. The pump-and-treat system began operating in 1983 with 12 extraction wells (4 near the source area and 8 in the plume area). In 1990, an additional well was placed in each of the source and plume areas. By 1994, ~18 times the estimated plume volume had been removed and treated. The peak concentration in the plume has only been slightly reduced. Four times the estimated mass in the plume has been removed, indicating that mass continues to be released to the ground water either from the sorbed or pure phase. At this point it appears that, due in part to the complex geological environment, this pump-and-treat system will not restore the aquifer to drinking water standards.

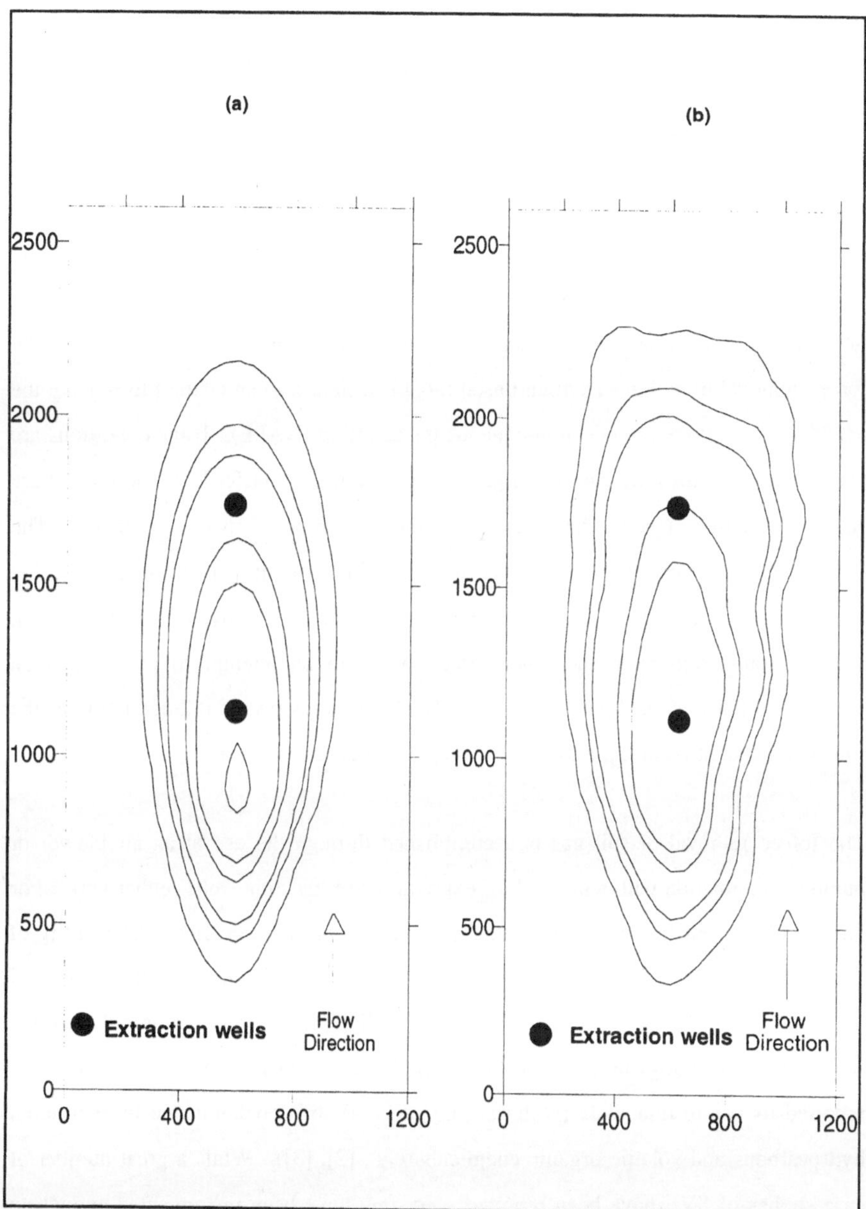

Figure 7. Aqueous phase plume at 10 years for (a) homogeneous system and (b) heterogeneous system (contour lines for 0.001, 0.005, 0.01, 0.05,0.1, 0.5).

If the source area at this site had been initially isolated from the aqueous phase plume, then pump-and-treat could have been used for plume control. However at this site it is apparent that nonlinear sorption and low permeability zones play an important role and therefore must not be ignored as important processes.

Example 2: Soil Vacuum Extraction

Soil vacuum extraction (SVE) is a physical treatment technology which involves the forced removal of soil gas from an unsaturated soil matrix in an attempt to remove the associated volatile phase of a non-aqueous phase liquid (NAPL). If the conditions are conducive, then this removal of soil gas results in a mass transfer from the pure phase NAPL, from the aqueous phase, and from the sorbed phase to the gas phase. The forced removal of the air phase results in a reduction in C_a in (1) and produces a dis-equilibrium between the four phases. Since some mass is removed from the system, mass is transferred from the other three phases in an attempt to re-establish an equilibrium between all four phases. Ideally, this process results in a reduction of the soil bulk concentration C_T, and hence a step towards site clean-up.

The forced removal of soil gas is accomplished through the use of an air blower or pump in conjunction with a network of extraction and injection wells (either vertical or horizontal) or trenches to create a suitable air phase pressure gradient. The utility of such a system allows for the enhancement of air flow through the subsurface zones of highest contamination. Soil vacuum extraction (SVE) has been a commonly applied remediation technology at many sites throughout North America. Several authors have reported its use to remediate (with varying success) soils contaminated by petroleum hydrocarbons and volatile organic chemicals (e.g., [2], [3]). While a great number of case studies of SVE have been reported, very few have been instrumented enough to gain insights into the processes involved and the efficiency of various venting configurations. In addition, few case studies have reported the degree of success of the remediation in terms of soil concentration. Success has mainly been judged in terms of mass removal.

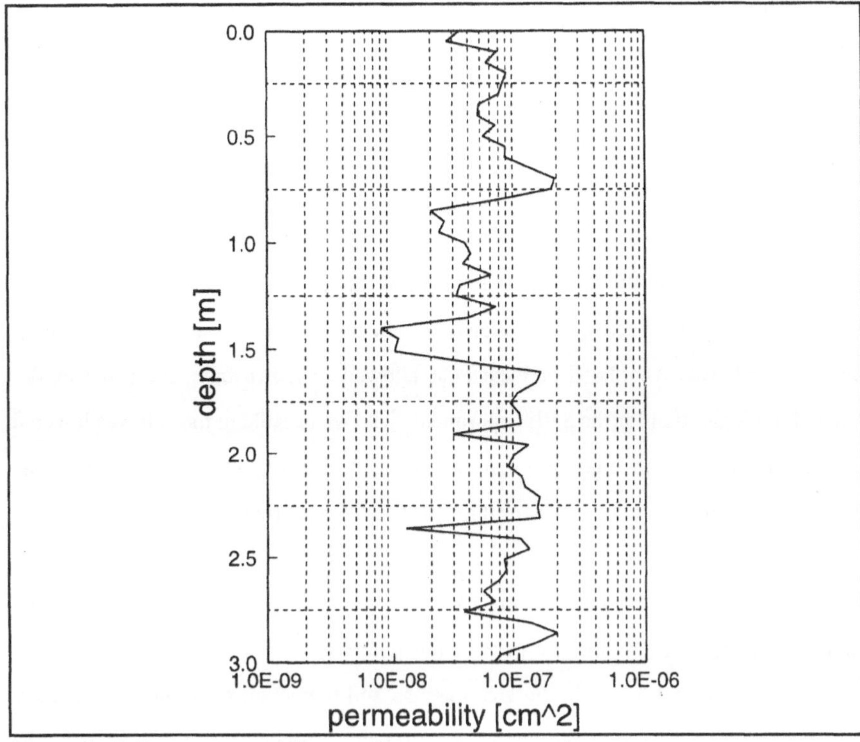

Figure 8. Soil core permeability profile.

A field trial investigating SVE at the field scale in naturally heterogeneous soil was performed at CFB Borden, Alliston, Ontario. The purpose of this field experiment was to investigate SVE at the field scale in a naturally heterogeneous soil in terms of both mass removal performance and soil clean-up performance. The experiment was performed in a block of a surficial sand aquifer measuring 9 m x 9 m in plan and approximately 3.3 m deep. The block was isolated by double sheet pile walls which were pneumatically driven through the aquifer and keyed into the clay aquitard, forming an impermeable base [12]. The sand in this aquifer is considered to be a fairly homogeneous, clean, well sorted fine-medium grained sand, and observations of core samples showed distinct horizontal layering varying from silt to coarse sand with occasional pebbles. Permeameter tests performed on core samples revealed variations on the millimetre scale with the layers typically differed in permeability by a factor of 1.7. Figure 8 presents the permeability results for one of the cores taken from the test cell.

In July 1991, seven hundred and seventy litres of perchloroethylene (PCE) was injected into the cell at a constant head. The injection occurred 60 cm below ground surface (b.g.s.). At the time of the injection the water level in the cell was maintained at 15 cm b.g.s., ensuring that the cell was fully saturated (the upper 15 cm was tension saturated). Once the spill was completed and the PCE allowed to redistribute, the cell was de-watered in preparation for the SVE experiment. The water table in the cell was lowered to approximately 3.0 m b.g.s. to create as much of an unsaturated zone as possible. While de-watering the cell, 174 L of pure phase PCE was removed.

To investigate the performance of the SVE system, it was necessary to carefully and comprehensively monitor the progress of the field trial. The parameters monitored were soil gas concentration, *in situ* gas pressure, soil temperature, volumetric moisture content, effluent concentration, flow rate, line pressure, barometric pressure, and ambient temperature. The spatial variability of the soil and NAPL distribution necessitated a large network of instrumentation to properly assess the remediation progress. Figure 9 shows the locations of the monitoring installations. In-situ soil gas concentrations and in-situ pressures were measured at seventeen multi-level sampling nests distributed throughout the cell. Nests S1 to S16 had two or three samplers each, installed to nominal depths of one, two and three metres, and nest S17 had six samplers installed at half-metre intervals. These multi-level sampling nests were also used as electrodes for time domain reflectometry (TDR), which is a method of non-destructively measuring the volumetric moisture content of the soil. Measurements made at each multi-level sampling nest allowed the average moisture content from 0 m to 1 m b.g.s., and from 1 m to 2 m b.g.s. to be determined. This provided an indication of the spatial variability of the volumetric moisture content. These measurements were complemented by moisture content profiles measured over 16 cm lengths at *in situ* TDR probes at locations TDR-1 and TDR-2 (see Figure 9). The subsurface temperature was measured at four nests. Each nest contained three thermocouples installed to depths of 1, 2, 3 m b.g.s. To monitor the mass removal performance of the SVE system, the volumetric flow rate and the effluent concentration were measured. Along with the gas pressure

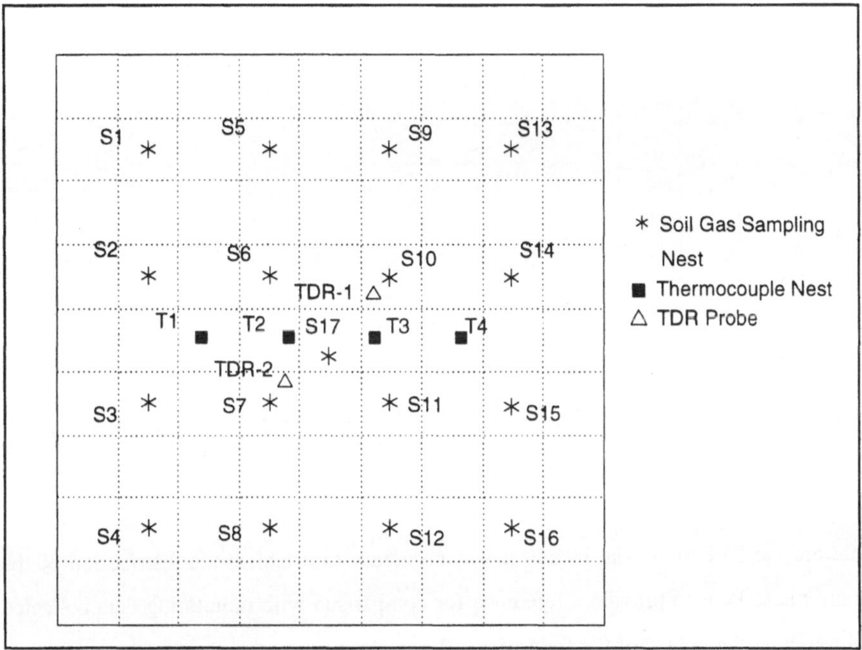

Figure 9. Location of monitoring installations.

throughout the piping manifold, these parameters provided an indication of the rate at which mass was being removed from the subsurface.

The vacuum system consisted of six wells arranged in two rows of three wells along opposite sides of the experimental cell. The wells were fully screened across most of the depth of the cell. To allow for some flexibility in the operation of the SVE system, each well was equipped with valves that allowed it to be used as either an extraction well or as a passive air inlet well.

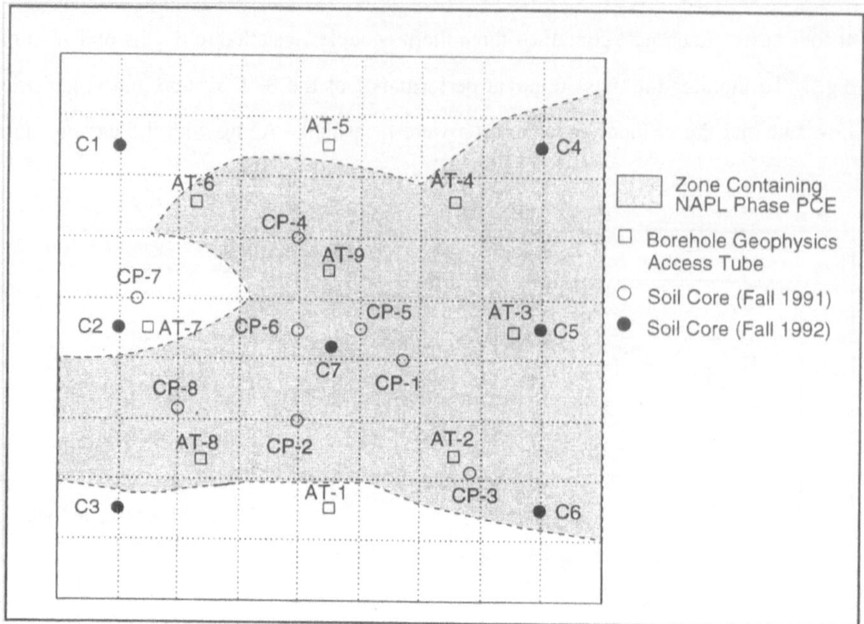

Figure 10. Estimate of the spatial extent of PCE pools at 0.8 m bgs.

Before the SVE field trial began, it was necessary to establish the distribution of the pure phase PCE to provide a reference for comparison with monitoring data collected throughout the course of the field trial. These data were obtained from three principle sources: ground penetrating radar sections, chemical analysis of soil cores, and soil gas concentration. The first two types of data were combined to produce figures indicating the approximate locations of pools of pure phase PCE. Several pools of PCE were located through the cell. As an example, Figure 10 shows the location of a large pool of PCE at 0.8 m b.g.s. The off-gas concentration profile was typical of other SVE remediation efforts reported in the literature in that it had three distinct stages delineated by changes in the slope. Figure 11 shows the off-gas concentration profile and the cumulative mass removed after ~540 days of operation. Notice that the effluent concentration decreased to very low levels even when there was a significant portion of the original mass remaining in the subsurface. The slope of the off-gas concentration profile was approaching zero at approximately 100 days, indicating that a steady state off-gas concentration had been reached at this point in the field trial. This decrease in off-gas concentration reflects a decrease in the mass flux out of the cell; however, the decrease in mass flux was not accompanied by low soil gas concentrations throughout the unsaturated zone in the subsurface. For example, the mass flux after 58 days of operation was calculated as 0.42 kg/day, even though a soil gas survey performed ten days earlier showed that there were areas within the subsurface that still contained very high soil gas concentrations (>50% of the saturated gas phase concentration of PCE). This disparity between the mass flux and the degree of contamination is evidence that heterogeneous soil conditions may have been limiting the performance of this vacuum system in terms of both mass removal, and soil clean-up. One reason for the slow mass removal was that some of the mass of PCE was located within the capillary fringe and the saturated zone and therefore is not directly assessable by the mobile air. However, this cannot completely explain the slow mass removal rate, since soil gas surveys revealed that there were zones well above the capillary fringe that exhibiting slow mass removal.

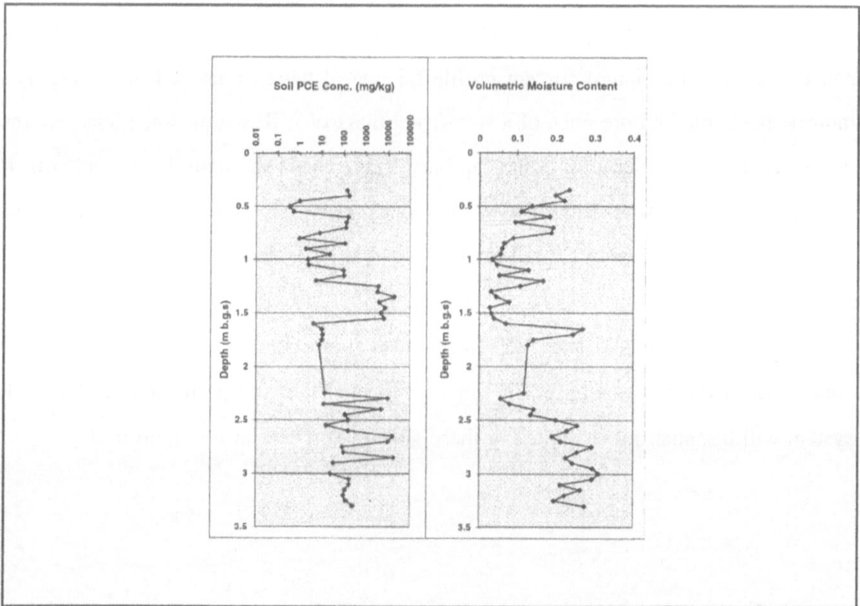

Figure 11. Bulk soil PCE concentration and moisture content.

The fact that this porous medium is only weakly heterogeneous would suggest that a SVE system should have easily removed all of the PCE from the unsaturated zone. Since this was not the case, it is interesting to investigate possible causes of this lack of performance. An analysis of the in-situ gas phase pressure data indicated that a sufficient vacuum was propagated throughout the unsaturated zone in the cell and therefore any mass transferred to the mobile gas phase was removed. The moisture content throughout the unsaturated zone ranged from a high of 30% down to 0.5%, decreasing from west to east. The moisture content on the west side of the cell was the highest measured at the ~1 m depth. The relatively high moisture content would have reduced the air permeability in these regions. If pure phase PCE was separated from the moving air by a zone with a high water content, then aqueous phase diffusion limitations would result in slow removal of this mass. After ~200 days after SVE system start-up, eight soil cores were collected throughout the cell at locations that coincided with consistently high soil gas measurements or were nearby the locations where previous soil cores were collected. Two sub-samples were collected every 0.05 m along each core. One sample was analyzed for bulk soil PCE concentration, while the other sample was analyzed for moisture content. Figure 12 presents a soil PCE concentration and moisture content profile for one of these cores. Clearly these data indicate the continued presence of a significant mass of PCE within some lenses in the vadose zone, and particularly in the capillary fringe above the underlying aquitard. If the remaining mass was all trapped within zones with high water saturation, then this decrease in the mass removal performance could be expected.

Since many contaminated sites will have much more complicated geology than the simple beach deposit at Borden, it can be expected that the performance of a SVE system will be much more limited at these sites than it was in this field trial.

Summary

Due to the complex nature of subsurface systems and the wide range of contaminants, the remediation of ground water systems is a difficult task. Our inability to characterize the heterogeneous nature of the subsurface system is perhaps the biggest problem. The existence of low permeability regions allows mass to be removed from the more mobile regions and then to diffuse slowly back as the concentration in the more mobile regions decrease. The presence of non-aqueous phase liquids which pose a long-term threat to drinking water supplies are difficult to detect and hence isolate and treat. Our understanding of the kinetics associated with the mass transfer processes are still evolving. These kinetics are essential to the design and implementation of various remedial technologies. Although the complete restoration of the ground water system to drinking water standards may not be feasible at many sites, the control of aqueous phase plumes and the removal of some contaminant mass from a subsurface system may still provide some important benefits.

References

1. Allen-King, R.M. (1995) personal communication, Dept. of Geology, Washington State University, Pullman WA.
2. Ardito, C.P., Billing, J.F (1990) Alternative remediation strategies: the subsurface volatilization and ventilation system. In *Proceedings of NWWA/API Conference on Petroleum Hydrocarbons and Organic Chemicals in Ground Water*, Houston, Texas.
3. Conner, R.J. (1988) Case study of soil venting. *Pollution Engineering*, July, pp. 74-78
4. Doty, C.B., and Travis, C.C. (1991) The effectiveness of groundwater pumping as a restoration technology, ORNL/TM-11866, Oak Ridge National Laboratory, Oak Ridge, TN.

5. Geller, J.T. and Hunt, J.R. (1993) Mass transfer from nonaqueous phase organic liquids in water-saturated porous media, *Water Resour. Res.*, 29(4), 833-845.

6. Grathwohl, P. and Reinhard, M. (1993) Desorption of trichloroethylene in aquifer materials: rate limitation at the grain scale., *Environ. Sci. Technol.*, 27(12), 2360-2366.

7. Mackay, D.M., Roberts, P.V., and Cherry, J.A., (1985) *Transport of organic contaminants in groundwater*, Environ. Sci. Technol., 19(5), 384-392.

8. National Research Council (1994) Alternatives for ground water cleanup, National Academy Press, Washington DC.

9. Poulsen, M.M. and Kueper, B.H. (1992) A field experiment to study the behaviour of tetrachloroethylene in unsaturated porous media. *Environmental Science and Technology*, 26(5), pp. 889-895.

10. Powers, S., Abriola, L.M., Dunkin, J.S., and Weber Jr., W.J. (1994) Phenomenological models for transient NAPL-water mass transfer processes, *J. Contam. Hydrol.*, 16, 1-33.

11. Schwille, F. (1988) Dense chlorinated solvents in porous and fractured media. *Lewis Publishers*, 146 pp.

12. Starr, R.C., Cherry, J.A. and Vales, E.S. (1992) A new type of steel sheet piling with sealed joints for groundwater pollution control. Presented at *45th Canadian Geotechnical Conference*, October 26-28, Toronto, Ontario.

13. Wilkins, M.D., Abriola, L.M., and Pennell, K.D., (1995) An experimental investigation of rate limited nonaqueous phase liquid volatilization in unsaturated porous media: steady state mass transfer., *Water Resour. Res*, 31(9), 2159-2172.

IV.3

IN-SITU AND EX-SITU REMEDIATION OF CHLORINATED SOLVENT SPECIES AT AN ACTIVE HEAVY MANUFACTURING FACILITY LOCATED IN THE MIDWESTERN UNITED STATES:A CASE STUDY

BRUCE C. CLEGG

Conestoga-Rovers & Associates

8615 West Bryn Mawr Avenue

Chicago, Illinois 60631

Overview

In November 1990, during the initial phase of an expansion and modernization program, a large manufacturing Facility located in the midwestern United States identified the presence of volatile organic compounds (VOCs) in soils beneath a production building (hereinafter referred to as Building 1). For over 75 years, the Facility has conducted heavy manufacturing operations related to its main business line in addition to many unrelated ancillary activities. One such ancillary operation was a drycleaning operation. As the most likely result of historical releases from this operation, tetrachloroethene (PCE) and various other related degradation products (namely, trichloroethene, 1,2-dichloroethene and vinyl chloride) were found in subsurface soils at concentrations ranging from approximately 10 µg/kg to over 1,000 mg/kg. Groundwater underlying the Facility was observed to contain a similar compliment of compounds.

Various features of the Facility and its setting may be similar to conditions encountered at Central and Eastern European production plants. Moreover, the soil remedies

E.A. McBean et al. (eds.) Remediation of Soil and Groundwater, 379–393.

selected as most appropriate for this site and the unique logistical problems associated with their implementation may have direct applicability at Central and Eastern European manufacturing sites and could be embraced in similar settings and circumstances.

As will be seen in the following, both ex-situ and in-situ soil vapor extraction (SVE) systems were effectively utilized to remedy site soils. Unique features of this remedy included the installation of subsurface piping and appurtenances in a manner that did not impede an aggressive plant modernization schedule. In addition, the construction and placement of a SVE blower and controls was performed in a manner and location that had no adverse impact to ongoing manufacturing operations.

Background

The plant occupies an area of approximately 260 hectares and is located adjacent to a large river system. The facility manufactures various heavy machine components and assembles equipment that is intended for use in the construction and agricultural sectors. Manufacturing operations have been conducted at this location for over 75 years. Ambient mean temperatures in the region range from -4°C to 24°C. Average annual precipitation is approximately 90 cm.

Building 1 occupies in excess of approximately 7.4 hectares and is located east of a drainage ditch, approximately 110 meters east of a large river. Although there was no evidence of a recent release or active source for VOCs, historical information indicated that a dry cleaning operation and other storage and recycling operations were previously located in this area. Upon discovery of the presence of organic contaminants, the Facility rapidly proceeded with the excavation of source materials (VOC solvent-impacted soil) under the oversight of local regulatory agency personnel. Soil excavated from the western portion of Building 1 prior to the discovery of the presence of VOCs had been taken to an exterior lot. Upon discovery of VOCs at Building 1, the soils staged in the lot were placed in specially lined roll-off boxes and transported to a second production building, Building 2, for treatment.

Out of concern for any remaining VOC residue, a soil sampling program within Building 1 was performed. The successful implementation of the investigative program indicated that some residual VOCs were present in the subsurface soils beneath Building 1.

The Facility implemented a plan to rapidly ameliorate contaminants identified in the soil and groundwater beneath the study area. In November 1994, the Facility commenced operation of a soil vacuum extraction system designed to remove VOCs present in subsurface soils beneath Building 1. In addition, treatment cells were specifically designed and constructed to treat soils excavated from Building 1 by ex-situ soil vapor extraction. Further details on both the ex-situ and in-situ VOC removal are provided in the following.

Building 1 Soil Ex-Situ Remedy

1. Soil Excavation and Staging

During expansion/modernization activities at Building 1, the Facility accumulated and staged approximately 12,000 tonnes of soil and debris in Building 2. All soils excavated from within Building 1 were secured in Building 2 due to the potential presence of residual VOCs. Soils were staged in a total of four soil staging areas within the limits of Building 2. The staging areas were lined with 1.27 mm high density polyethylene underliners and secured with plastic overliners. The staging areas were both situated on a concrete base and were enclosed beneath a fixed roof. The locations of these storage areas are as shown on Figure 1.

In general, closure of the soil staging area included construction of individual treatment units to remove VOCs from the soil using SVE. Off gasses collected from the treatment units are being treated through granular activated carbon (GAC) adsorbers prior to venting to the atmosphere. Once the soil meets the cleanup objectives (CUOs) identified in Table 1, the soil will be used as fill at the Facility. Further details

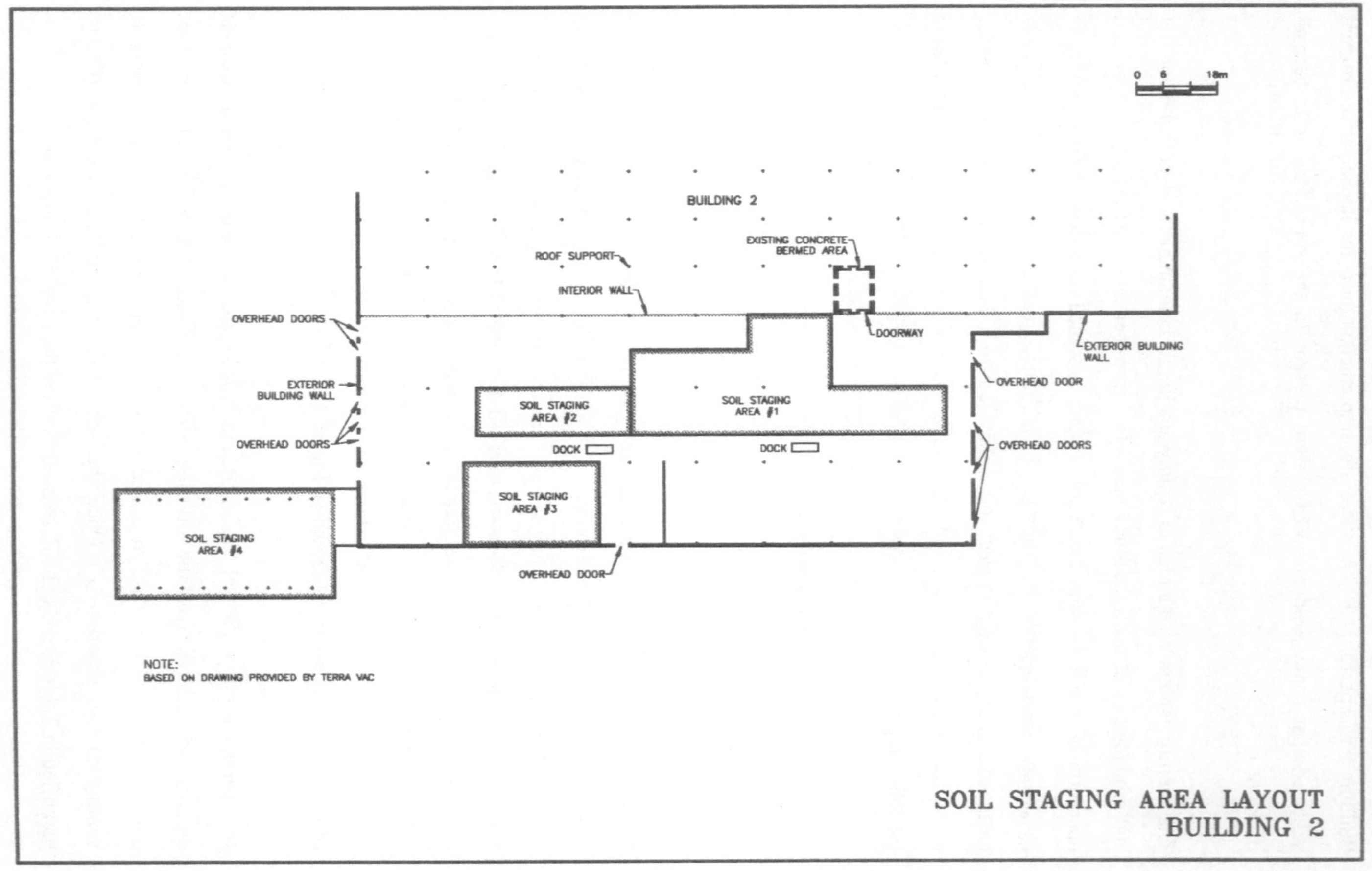
0 6 18m
BUILDING 2
ROOF SUPPORT
INTERIOR WALL
EXISTING CONCRETE
BERMED AREA
DOORWAY
EXTERIOR BUILDING
WALL
OVERHEAD DOORS
EXTERIOR
BUILDING WALL
OVERHEAD DOORS
SOIL STAGING
AREA #2
SOIL STAGING
AREA #1
OVERHEAD DOOR
DOCK
DOCK
OVERHEAD DOORS
SOIL STAGING
AREA #3
SOIL STAGING
AREA #4
OVERHEAD DOOR
NOTE:
BASED ON DRAWING PROVIDED BY TERRA VAC
SOIL STAGING AREA LAYOUT
BUILDING 2

Figure 1

pertaining to closure of the Building 2 soil staging areas are provided in the sections which follow.

TABLE 1. Building 2 VOC Cleanup Objectives for Soils

Compound	Concentration Objective (μg/kg)
Tetrachloroethene	25
Trichloroethene	25
cis-1,2-Dichloroethene	200

2. *Construction of Ex-Situ Treatment System*

2.1 Construction of Treatment Units

Treatment units were constructed of braced plywood walls lined with two layers of .5 mm high density polyethylene (HDPE) plastic welded into one continuous sheet in accordance with an approved closure plan. In some locations where soil staging areas were converted to treatment units, an existing 1.5 mm liner was incorporated into the treatment unit construction as the underliner. Construction of treatment Units identified as #. 1, 2, 3, and 4 was completed in January 1995.

2.2 Soil Screening

Screening of soil was performed to remove debris and to enhance the permeability of soils placed in the treatment units. In order to provide the greatest efficiency in soil handling operations, the soil screen was moved to various locations as the work progressed. Water was applied, on an "as required" basis, to minimize the generation of dust using an optimal application rate such that runoff from the soil staging areas was not generated. Final placement of soil within the treatment units was performed using a long stick excavator. Appropriate care was exercised during soil placement activities so as not to compromise the integrity of the HDPE liner or vacuum extraction wells. All soil screening activities were completed in January 1995. A total of approximately 6,900 cubic meters of soil was screened and placed in the treatment units.

Concrete debris removed from the soil was placed in a stockpile within a designated Exclusion Zone prior to triple rinsing of the concrete. Plastic sheeting removed from the screen was placed directly into roll-off boxes for disposal off site. The soil was removed from the plastic sheeting prior to placement in the roll-off boxes.

Triple-rinsed concrete and plastic sheeting were disposed of as a special waste in accordance with an Agency-approved design plan. A total of approximately 840 m^3 of concrete was triple rinsed and disposed off Site. In addition, a total of approximately 100 m^3 of plastic sheeting was disposed off Site.

2.3 Installation of Soil Vacuum Extraction Equipment

The installation of extraction wells was performed in conjunction with the placement of soil in the treatment units. Well screens were attached to 0.6 m by 0.6 m plywood bases and were set in place prior to placement of soil in the treatment units. Appropriate care was exercised when placing soil around the wells so as not to damage or displace the wells. Six wells were installed following placement of soil in treatment Unit No. 1. These wells were installed by excavating soils with an excavator, placing the well, and backfilling the excavation.

Once soil placement was completed within a particular treatment unit, a .15 mm polyethylene overliner was placed over the entire treatment unit. Seams between panels of polyethylene sheeting were sealed using tape. Outside edges of the overliner were fixed to the treatment unit walls. Penetrations cut in the overliner for well risers were also sealed with tape. Following installation of the overliner, manifold piping (including annular (pressure/vacuum measurement) ports, gate valves, and sampling ports) was installed.

2.4 Cleaning of Concrete Floor

At the completion of soil handling activities all concrete floors within the key work zones were cleared of large particulate matter then decontaminated using a high pressure hot water wash. Washwater generated during the cleaning operation was collected using a wet-vacuum unit and placed in drums or transferred to on-site temporary holding tanks. Collected washwater was disposed of at the Facility's Treatment Plant. Prior to disposal at the Treatment Plant, the rinse water was sampled and analyzed to verify conformance with the Treatment Plant's operating permit.

Soil Treatment

1. Overview

The vacuum extraction system commenced operation in January 1995. Air samples were collected throughout the startup and analyzed for VOCs to monitor the effectiveness of the SVE system operation.

When system operating trends were established, approximately two weeks after completion of startup, vacuum levels were recorded at piezometers three times a week and VOC concentrations and air flow were monitored monthly. After approximately two months of operation the frequency of the measurement of vacuum levels was reduced to once per month.

2. System Performance to Date

Vapor-phase VOC concentration measurements were collected to track the progress of the soil treatment. In order to monitor the quantity of VOCs being extracted from individual extraction wells, treatment units and the total system (Treatment units 1, 2, 3 and 4 combined), sampling ports were installed at various locations in the manifold piping. The sampling ports were used to collect vapor phase samples extracted from the soil.

VOC analysis was performed on Site using a portable gas chromotograph (GC) equipped with dual channel flame ionization detectors (FID). The GC was calibrated prior to use and operated in accordance with a Quality Assurance Plan. Throughout all of the VOC analyses performed, PCE comprised between 95-100 percent of the observed vapor phase VOCs being extracted.

Figures 2 and 3 indicate very clearly that a significant bulk mass removal occurred quite readily soon after system startup. However, after a period of approximately 200 days of continuous system operation, the system appeared to approach a steady-state condition with concomitant VOC removal rates and total mass removal approaching rate and mass asymptotes, respectively.

As an indicator of biological activity, carbon dioxide (CO_2) concentrations were monitored on a regular periodic basis. Figure 4 illustrates the recorded CO_2 concentrations at various points in time. As is evident from Figure 4, biological activity was initially very high but declined, more or less proportionately, with declining VOC mass and soil moisture content.

After approximately nine months of continuous operation, approximately 160 kgs of the monitored volatiles were removed. However, three additional months of continuous operation may be necessary to reach the established CUO's for each monitored parameter.

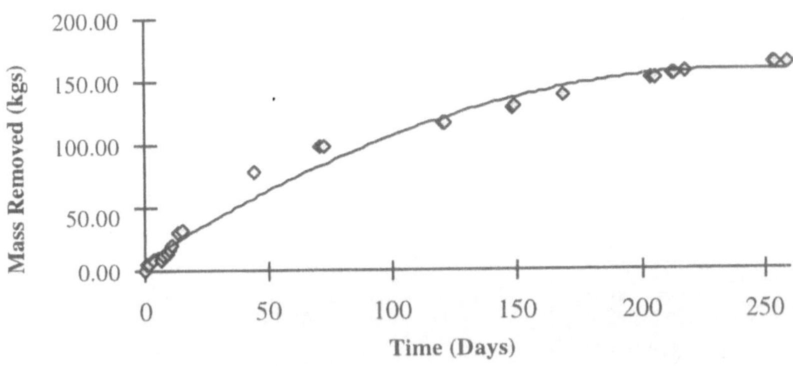

Figure 2. Total VOC mass removed vs. time

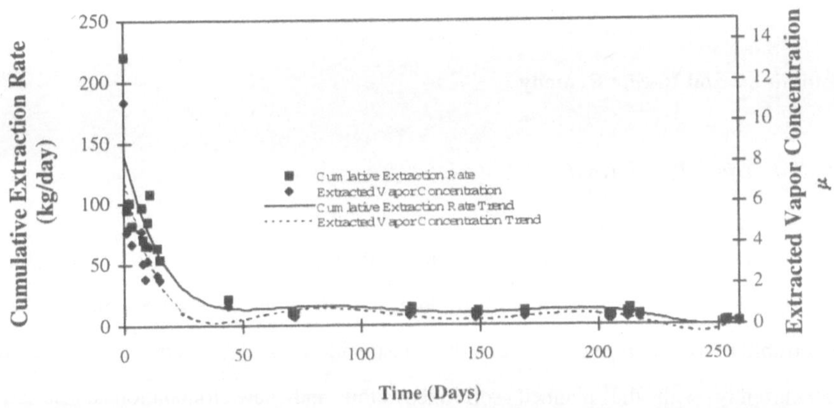

Figure 3. Cumulative VOC extraction rate and extraction vapor concentration vs.time

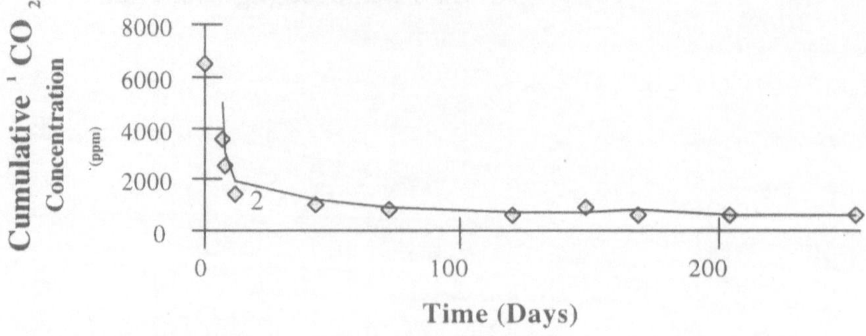

Note 1: Includes Cumulative CO_2 Concentration For Ex-Situ SVE Treatment Units No. 1,2,3,4
Note 2: This point is an averager of five measurements recorded over six days

Figure 4. Cumulative CO_2 concentration vs. time

3. Confirmation of Clean-Up

Monitoring of the system will continue on a monthly basis until VOC concentrations in the extracted vapor indicate the soil has been treated to the desired level. At that time, soil sampling will be performed to confirm soil clean-up levels have been reached.

Building 2 Soil In-Site Remedy

1. Subsurface Pipe Layout

Soil excavation activities associated with the Building 1 modernization were planned throughout the western portion of Building 1. In order to ensure minimal future interruption to the production area, the installation of SVE components occurred concurrently with the planned soil excavation and new foundation construction activities. The subsurface extraction system consisted of perforated horizontal gas extraction pipes installed in five gravel-backfilled trenches. Each of the five trenches is approximately 45 m to 60 m in length. Separation between individual trenches is approximately 15 to 20 m. The perforated piping in the trenches was connected to separate header pipes which extended to the surface. Figure 5 illustrates the general subsurface extraction pipe layout.

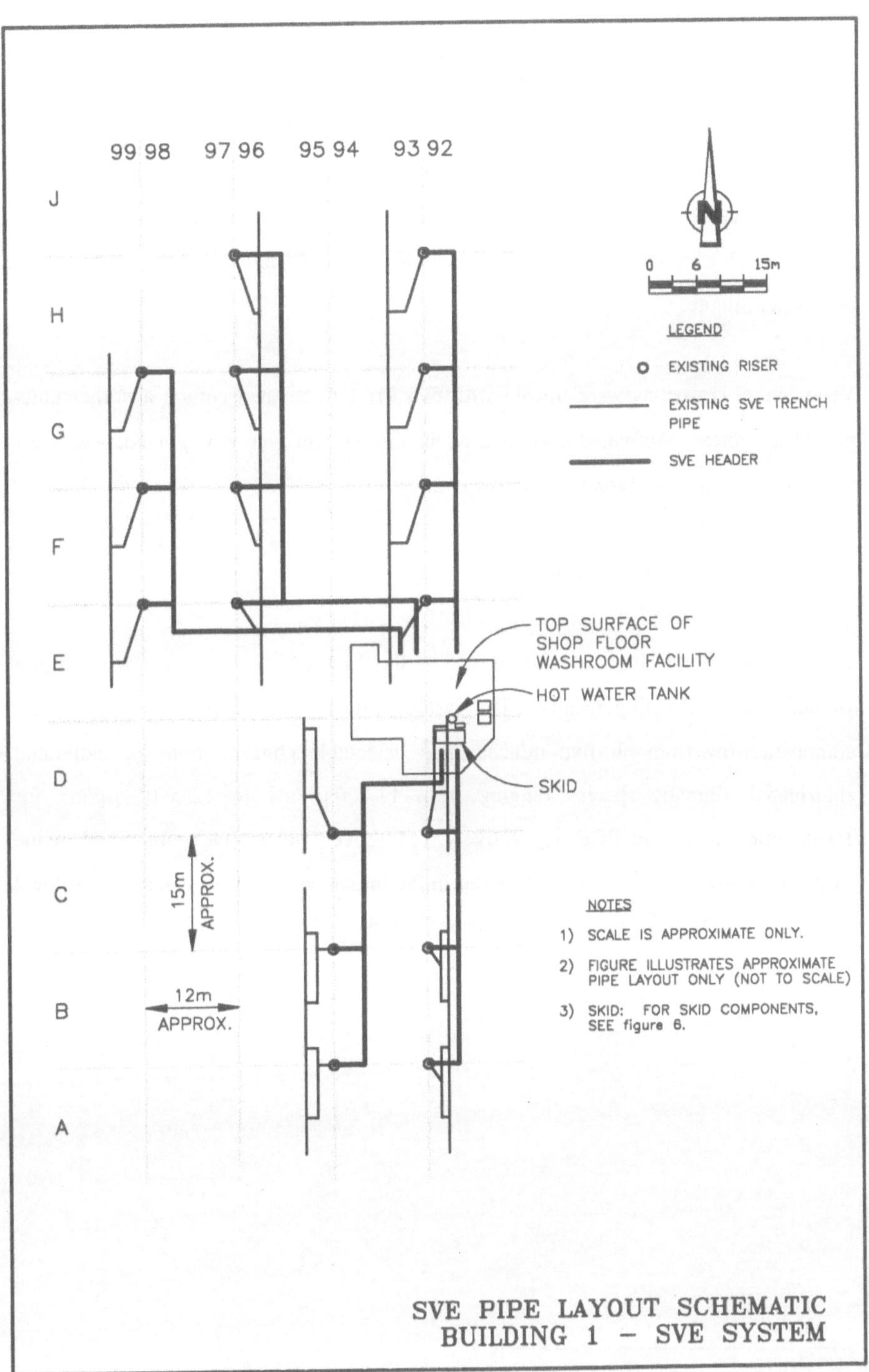
99 98
97 96
95 94
93 92
J
H
G
F
E
D
C
B
A
0
6
15m
LEGEND
EXISTING RISER
EXISTING SVE TRENCH PIPE
SVE HEADER
TOP SURFACE OF SHOP FLOOR WASHROOM FACILITY
HOT WATER TANK
SKID
15m APPROX.
12m APPROX.
NOTES
1) SCALE IS APPROXIMATE ONLY.
2) FIGURE ILLUSTRATES APPROXIMATE PIPE LAYOUT ONLY (NOT TO SCALE)
3) SKID: FOR SKID COMPONENTS, SEE figure 6.
SVE PIPE LAYOUT SCHEMATIC
BUILDING 1 – SVE SYSTEM

Figure 5

2. *ISVE System Blowers, Controls and Vapor Capture*

The ISVE system installed in Building 1 is shown schematically as Figure 6. A general process flowsheet is presented as Figure 7. Each of the five extraction trenches were piped to a central location and equipped with actuated valves to allow each trench to either be vented to the atmosphere or to direct flow to the SVE vacuum blowers. Two parallel ten horsepower blowers were used and were capable of drawing 13 cubic meters per minute.

Vapor-phase emissions were treated with two 816 kg activated carbon adsorber units placed in series. Activated carbon used in this system was a virgin coconut shell product supplied by Calgon Carbon Corporation.

3. *System Performance to Date*

Contaminant monitoring[1] conducted within the first 24 hours of system operation yielded some interesting results. The system influent sample, representative of the combined flow from all five trenches had detectable concentrations of individual chlorinated aliphatic species ranging from 11,000 $\mu g/m^3$ to 1.2×10^6 $\mu g/m^3$ for 1,1-dichloroethane and PCE, respectively. Moreover, other VOCs, unrelated to the suspected source material, were observed in the initial system influent sample. Table 2 summarizes VOC influent concentrations at startup.

---

[1]Samples collected in Summa canisters were analyzed by USEPA Method TO-14.

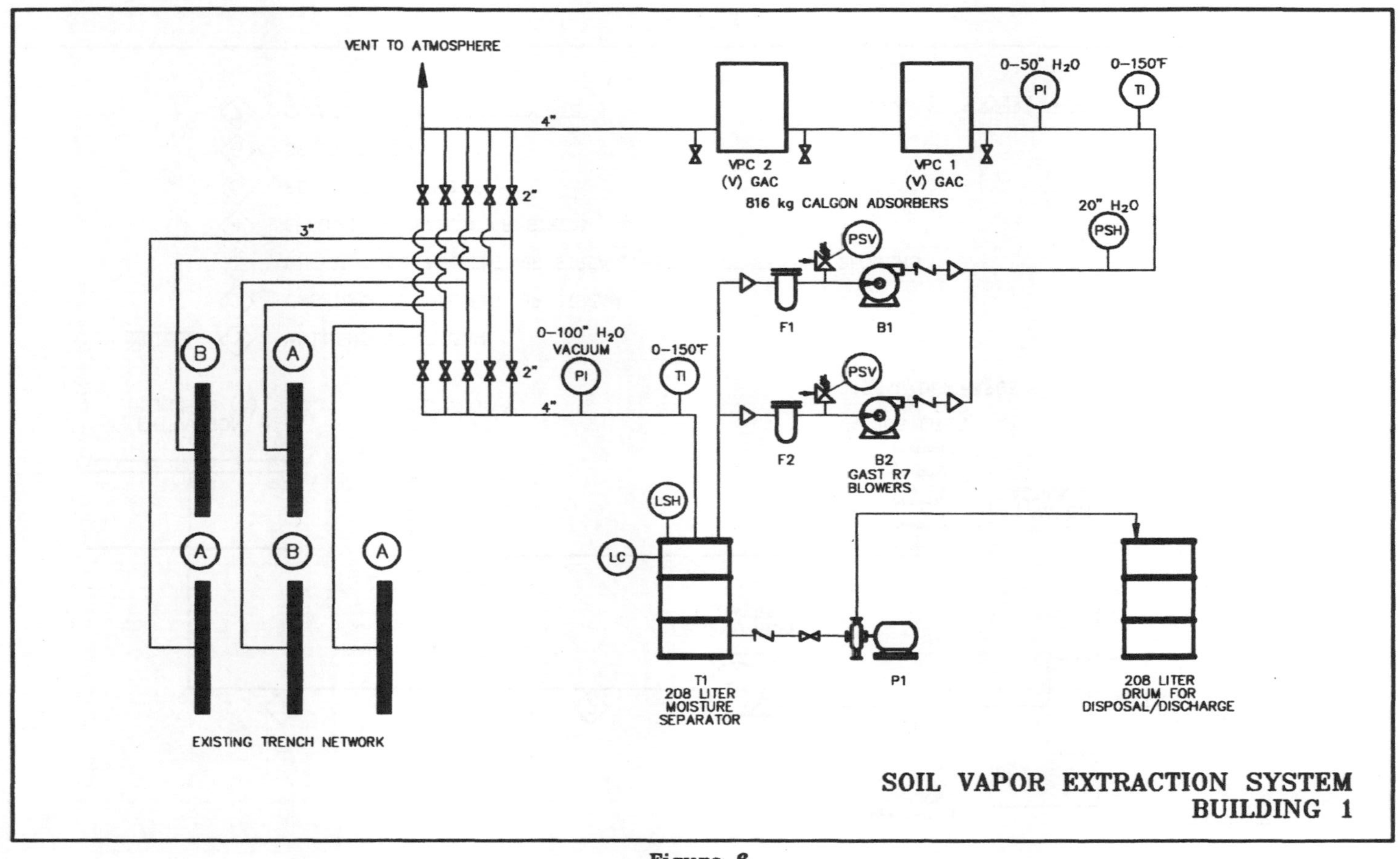
VENT TO ATMOSPHERE
4"
2"
3"
VPC 2
(V) GAC
VPC 1
(V) GAC
816 kg CALGON ADSORBERS
0–50" H2O
PI
0–150°F
TI
20" H2O
PSH
PSV
F1
B1
F2
B2
GAST R7
BLOWERS
0–100" H2O
VACUUM
LSH
LC
T1
208 LITER
MOISTURE
SEPARATOR
P1
208 LITER
DRUM FOR
DISPOSAL/DISCHARGE
A
B
EXISTING TRENCH NETWORK
SOIL VAPOR EXTRACTION SYSTEM
BUILDING 1

Figure 6

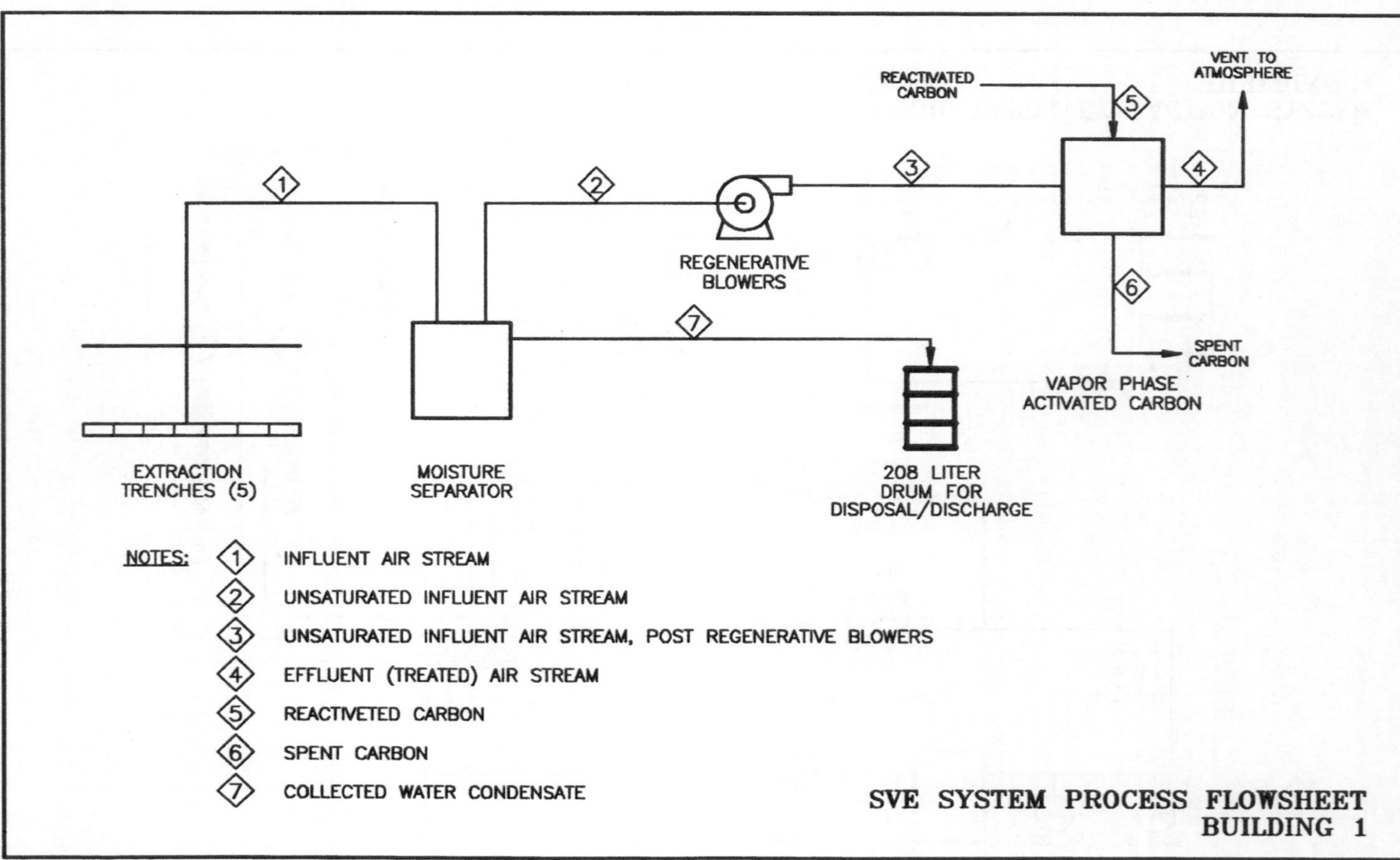
REACTIVATED CARBON
VENT TO ATMOSPHERE
5
1
2
3
4
REGENERATIVE BLOWERS
6
7
SPENT CARBON
VAPOR PHASE ACTIVATED CARBON
EXTRACTION TRENCHES (5)
MOISTURE SEPARATOR
208 LITER DRUM FOR DISPOSAL/DISCHARGE
NOTES:
1 INFLUENT AIR STREAM
2 UNSATURATED INFLUENT AIR STREAM
3 UNSATURATED INFLUENT AIR STREAM, POST REGENERATIVE BLOWERS
4 EFFLUENT (TREATED) AIR STREAM
5 REACTIVETED CARBON
6 SPENT CARBON
7 COLLECTED WATER CONDENSATE
SVE SYSTEM PROCESS FLOWSHEET
BUILDING 1

Figure 7

TABLE 2. VOC ISVE System Influent Concentrations at Startup

Compound	Observed Influent Concentration ($\mu g/m^3$) x 10^3
Vinyl Chloride	70
1,1,2-Trichlorotrifluoroethane	16
trans-1,2-Dichloroethene	15
1,1-Dichloroethane	11
cis-1,2-Dichloroethene	960
Trichloroethene	220
Tetrachloroethene	1,200

Summary and Preliminary Conclusions

On the basis of the experience at this facility, it is clear that SVE is a good candidate technology for situations in which the remedy implementation and ongoing operation can not unduly hinder facility modifications or production. In addition, in cases where a large bulk mass removal of volatile organics is necessary at relatively low cost, SVE is an excellent remedial alternative.

IV.4

DESIGN AND IMPLEMENTATION OF REMEDIAL WORKS AT THE FORMER SOVIET NAVAL BASE AT SWINOUJSCIE, POLAND

RYSZARD KOSLACZ

Zespol Badan i Ochrony Srodowiska EKOKONREM, Sp. z o.o.

A Heidemij Group

Wroclaw, Poland

Introduction

Ekokonrem applied the SOS Selective Skimmer System to start an emergency removal of free floating petroleum products from groundwater at former Soviet naval base at Swinoujscie. Skimming of 80 000 L of free product during the first two years of the operation for 1/4 of the conventional recovery price proved that the recommended recovery method was very efficient and cost effective. When the product thickness reduced to a sheen, a pilot test of bioventing method started in September 1995.

Background

A site investigation at the former Soviet military base in Swinoujscie was performed by State Inspection of Environmental Protection immediately after the Soviet troops' withdrawal and lead to detection of huge contamination of ground and groundwater with various petroleum products. The significant seepage of oil from groundwater to the nearby marine channel and the proximity of municipal groundwater intake located 1.5 km from the site alarmed the City Government of Swinoujscie. An emergency tender

E.A. McBean et al. (eds.) Remediation of Soil and Groundwater, 395–397.

was announced to submit proposals for remedial action. The Ekokonrem proposal of using SOS Selective Skimmer System for removal of thick layer of petroleum product floating on the groundwater table was finally selected by authorities.

Ekokonrem Approach

The main objectives of the contract were to quickly delineate the contamination zone, design and install hydrocarbon recovery system to contain the plume and eventually remove free floating petroleum products from the ground.

Soil gas investigation method was used to determine extent of the plume. Stitz TM sonde and Drager TM field indicators were applied to properly assess maximum concentration areas. Oil seepage points were identified and floating, pneumatic barriers were installed on the channel to stop migration of the oil plume on the surface water.

Based on soil gas investigation results and exploratory drillings a network of recovery wells was drilled. Using pneumatically driven SOS Selective Skimming System recovery of free floating product started within the weeks of the contract award.

After 15 months of operation the free product thickness was reduced to few centimeters and seepage of oil to the nearby water channel stopped. Using SpillCAD^{R} software, a design of final recovery and remediation system was developed. A concept of using horizontal screens and injection wells combined with further skimming resulted in further reduction of contamination levels. A field respiration test conducted in the beginning of 1995 proved that bioventing may be the most effective remedial method.

Performance Highlights

- use of soil gas investigation to delineate extent of contamination significantly shortened the required time schedule

- use of SOS Selective Skimming System for product recovery instead of pumping water and product and subsequent separation on the surface reduced the cost of total operation by 400 %

- use of SpillCAD^R software for design of final remedial system reduced the risk of "overdesigning"

- presentation and educational training of province and local regulators of selective skimming system application as a very cost-effective method

- bioventing pilot test is being recently launched as a most promising remedial technology

Project Features

Project value:	-$280 000
Regulatory Agency:	- Chief Inspectorate of Environmental Protection, Warsaw, - Environmental Department of Szczecin Province Government, - Environmental Section of Swinoujscie City Government
Physical features:	- Surficial sandy aquifer at Baltic Sea coastline - Depth to water 3 m or less - Saturated thickness 15 m
Contaminants:	- Various petroleum products
Techniques/Strategies:	- Soil gas investigation - Selective Skimming - SpillCAD^R modeling - Horizontal screens - Respiration test - Bioventing

IV.5

THE ELIMINATION OF HYDROCARBON AND METAL CONTAMINATION AND ITS RECENT HISTORY IN HUNGARY

ZSOLT FEKETE

Terre-vite KFT

3300 Eger

Maklari u.7. Hungary

On the 30th of May 1995, the Hungarian Parliament passed Act LIII on the general rules of environmental protection. According to this document, a separate act is in force for the protection of agricultural land. The general view points of soil protection, the limit values and the definitions of the concept explained by the law, were also published.

For experts involved in observing, discovering and assessing ground protection, and for those planning, permitting and carrying put damage elimination, this provision of the law promises a lot for the future. However, at present, it cannot be used in daily practice.

Soil contamination as a problem of the environmental protection authorities and engineering has not been among the tasks of daily practice for the recent 10-15 years, apart from some more serious cases of contamination causing nationwide scandals.

E.A. McBean et al. (eds.) Remediation of Soil and Groundwater, 399–403.

Hydrocarbon Contamination

The first serious cases of ground contamination, which occurred regularly and needed engineering damage-eliminating measures by the authorities, were brought about by breakdowns in the pipeline "Friendship" about 7 years ago. The contamination was soon noticed by the populations observation and the pressure-decrease in the pipeline. Being a problem of modern nature, it delayed damage-elimination work to a significant extent, which multiplied the costs of the activity itself. In this period there were no companies having ground-cleaning experience in Hungary.

Four or five companies volunteered to solve the problem; among others, the author's company, Terre-vita KFT, which has the technology (their own development) to eliminate waste containing oil, the University of Agriculture (Godollo) who also bred and applied oil-decomposing cultures of micro-organisms in their technology.

The program to clean barracks after the withdrawal of the Russian troops can be called a milestone in the field of applying ground cleaning technologies. This extensive nationwide program, which produced 230 applications, brought to the surface everything, out of which the two most important factors are as follows:

i) it was difficult for Hungarian environment protection to define the measuring figures of hydrocarbon contamination and the extent of the needed remediation, as well as the laboratory methods needed for definition;and

ii) in the field of environmental protection, a greater number of entrepreneurs appeared, out of which, several American Hungarian, German, Hungarian, Austrian-Hungarian joint ventures, using western technologies, a created a change in standards and brought about a competitive atmosphere.

In the Russian military bases left behind, typically in the ground and the subsoil water, a significant quantity of oil (used for heating or as fuel), kerosene, petrol contamination was discovered and eliminated. At the same time the latter activity has not been carried out at a number of remaining sites .

After the Russian barracks program, as part of the changing of the system, there appeared the selling and liquidating of industrial factories, the assessment of environmental damage during the debtor-consolidation and there was a sharp increase in the quantity of damage-elimination work.

The frequency of road accidents related to the activity of carrying fuel and damage to tankers suffering crashes were unusual in comparison with the figures in the past; the demand for eliminating damage to the environmenthad to be faced.

The examination of the industrial workshops from the point-of-view of environment protection made the scale of the nature of contamination wider. Although it happened rarely, the problem of metal -contaminated ground did appear, mostly in connection with galvan-related sites, and not long ago, lead-contamination, arising from the illegal activity of taking apart batteries killed people.

The increase in the number of private ventures, besides its positive effects, also activate some Mafia-elements, which ended up in unattended contaminating materials, which were supposed to have been eliminated by entrepreneurs in the waste-handling sphere. Eliminating damage done this way is either unsolved or needs control by the state and is often premeditated in the guise of "big business".

There are typical ground cleaning tasks in rebuilding filling stations. The Hungarian Railways also started considerable modernization work, which requires the elimination of soil and subsoil water contamination accumulated during the years.

The Definition of the Extent of Hydrocarbon Contamination and the Efficiency of Damage Elimination Tasks.

The definition of the extent of hydrocarbon contamination is a difficult task: there is no accepted common method that would really show the state of ground contamination, although the laboratories are already aware of the international rules in question. As a result of this, the definition of the efficiency of damage-elimination is vague.

Compared with Dutch, German and Mexican charts of limit values, we can experience significant differences with respect to some components and the selection of characteristic components, which makes the damage-elimination system neither under- or over-planned.

Damage Elimination Methods Applied in Hungary

Compared with hydrocarbon, the measurement of metal contamination is not a problem, but in the respect of limit values to realize these, there are different figures.

Applied Ground Cleaning Technologies

Hydrocarbon

In the beginning almost everywhere in Hungary only-on-site technologies were applied, as well as aerobic (microbiological) methods. As the tasks increased in number, in situ ground-wash methods using enzymes, and/or airing with bacteria became common.

Out of the methods used in the world and mentioned above, only some became implanted, as the standard of costs limit the use of others.

A Hungarian ground-cleaning on-Site method is the Terre-vita procedure, which is the patent of Terra-vita KFT. With organic agricultural waste (straw, sawdust, stable-manure) the Terra-vita activator including soil-bacteria, fungi and radiating fungi is fed into the removed contaminated soil. Airing and homogenization are carried out with the help of machines. Using this method, the total hydrocarbon contents can be decreased to the extent of 100-1000 mg/kg in three to six months. This method can also decrease PAH contents in the soil.

The efficiency of ground-cleaning can be controlled with the frequency of turn-overs. Increasing the frequency of turnover can decrease the period of cleaning, while the costs of engineering will rise.

The enzyme-mix procedure is an in situ one, which carries the oil-decomposing enzymes to the contaminated levels of soil in wet agent. Recycling the water washes the soil again. Contaminated water will be placed in a separate system for cleaning.

Depending on the circumstances, the two methods can be used separately or combined. Soil, with great contamination near the surface, which is easy to remove, is remediated with Terra-vita technology; for eliminating hydrocarbon contamination in deeper layers the enzyme mix method is employed.

Eliminating Metal Contamination

In most cases the elimination of metal contamination is solved by removal and placing the contaminated soil in permanent waste containers.

Suggestions and Conclusions

Preparing and introducing an internationally accepted system of limit value measurement is an urgent tasks for Hungary.

The implementation of the treating technology of some types of metal contamination and starting ventures are the tasks of the near future.

IV.6

APPLICATION OF FLOTATION AND/OR BIOSORPTION FOR THE REMOVAL OF TOXIC METALS FROM DILUTE AQUEOUS SOLUTIONS - GROUNDWATERS

A.I. ZOUBOULIS & A. MATIS
Chemical Technology Division,
Chemistry Dept., Aristotle Univ.,
GR-540 06 Thessaloniki

Introduction

It is recognized that a principal source of contamination of ground water is industrial wastewater impoundments which contain a large variety of wastes, some of them with toxic metals. Acid mine drainage is another source of contamination. This water often contains high concentrations of iron and sulfur compounds and is highly acidic. Contamination of ground water also can occur from the leaching of oxidized minerals and radioactive materials found in tailings piles or slurry ponds and lagoons.

The behavior of heavy metals in a natural aqueous system has attracted researchers because of environmental issues. It is known that hydrous metal oxides, such as those of iron and manganese, are important scavengers of many toxic metals in stream sediment, soil and in the ocean. Examples were given [1] of this phenomenon, which indicate that adsorption, flocculation and precipitation of heavy metals on various substrates seem to be the important processes. This is also the fundamental idea of the so-called adsorbing

E.A. McBean et al. (eds.) Remediation of Soil and Groundwater, 405–423.

colloid flotation. Although "pump and treat" methods suffer from several disadvantages due to a series of subsurface processes such as adsorption/desorption and hydrolysis, they constitute a common way for remediation of groundwater contaminated with toxic metals. It has been shown that remediation depends directly on the physico-chemical properties of the contaminants and the hydrogeology of the site.

The contamination of subsurface soils with non-aqueous phase organic liquids, like petroleum hydrocarbons, is another environmental problem, which is often solved by the use of surface active agents for residual levels of light non-aqueous phase liquid (such as automatic transmission fluid); in order to increase water solubility of these liquids and decrease the interfacial tension. The use of microbubble suspensions of colloidal gas aphrons (i.e. a flotation technique) for in-situ soil flushing was recently reported [2].

Flotation has been extensively reviewed [3]. According to the technique used for the generation of gas (usually air) bubbles which are the transport medium of this gravity separation process, flotation is generally distinguished [4],[5] in two broad categories, while the various less common techniques (like electrolytic flotation, colloidal gas aphrons and so on) may be also classified in are:

i) dispersed-air, and
ii) dissolved-air flotation.

Adsorbing Collid Flotation

1. The Method

This separation method was developed mainly by Wilson et al. [6]. An interesting application is to arsenic removal, for instance from geothermal fluids [7]. Arsenic species were concentrated by adsorbing colloid flotation onto, or precipitate flotation with, colloidal ferric hydroxide, depending on the pH. Sodium lauryl (dodecyl) sulphate was the anionic surfactant used below the isoelectric point of the particles. Similar results are presented in Figure 1, following dispersed-air flotation by sodium oleate. The ferric hydroxo-complexes, because of hydrolysis, were produced in-situ from ferric sulphate by pH adjustment [8],[5]. Aging of the sorbent solution appeared to reduce its activity. Also, aluminum sulphate used instead showed lower arsenic removals. An actual mine wastewater, containing arsenic, chromium and lead, was tested by the same method [9]. The interaction mechanisms between $Fe(OH)^3$-bearing arsenic floc and sodium dodecyl sulphate (the collector) were analyzed and interpreted by means of z-potential, infrared spectra analysis and molecular orbital theory.

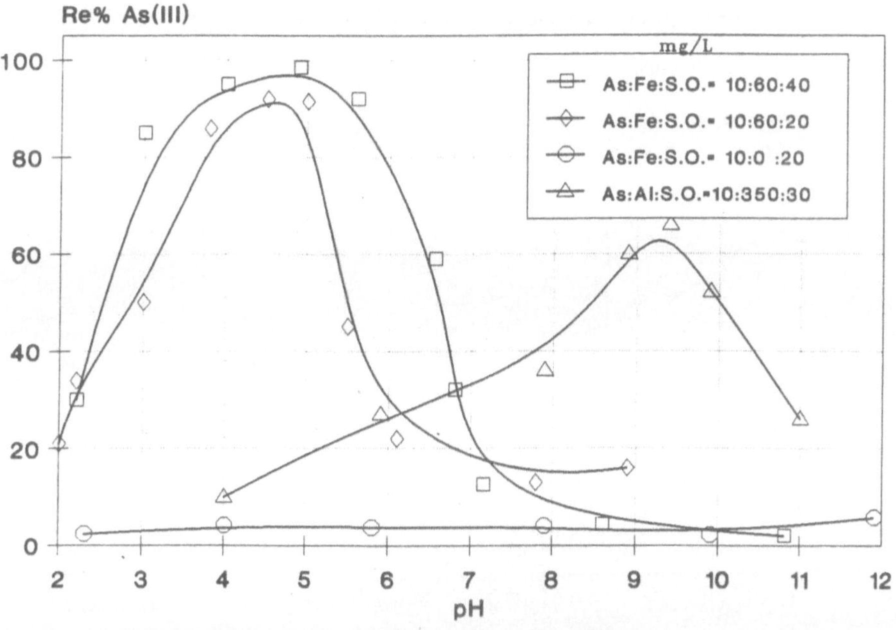

Figure 1. Effect of solution pH in adsorbing colloid flotation of arsenite anions; ferric .or alum adsorbates were used and sodium oleate as collector. Reprinted from ref. [8].

The adsorption of cations, for example from seawater, onto amorphous FeOOH was elsewhere published [10]. The adsorption behavior of Mn, Co, Ni, and Zn from different electrolyte solutions was studied. The properties of hydrous oxides of iron and their applications to both cations and anions removal have been reviewed [11].

Arsenic is also present in molybdenum ores, so the wastewater generated during hydrometallurgical treatment of molybdenite, for example, contains arsenic and molybdenum species. Adsorbing colloid flotation (also ion flotation) was examined [12] for effective metal recovery from dilute aqueous solutions (see Figure 2). The effect of sulphate in high ionic strength was quite detrimental to removals. But, in actual cases, the respective wastewaters contain usually chloride ion and little or no sulphates. For this reason, ferric chloride was applied for the production of the coprecipitant flocs. A 5 min contact time (mixing by magnetic stirring) was adequate in bench scale experiments.

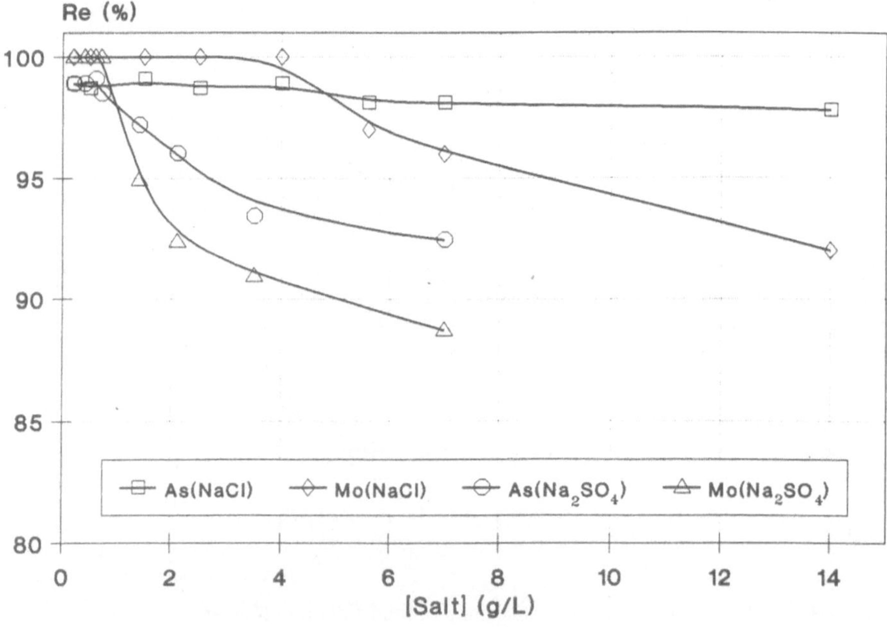

Figure 2. Removal of Mo(VI) and As(V) ions by adsorbing colloid flotation: influence of foreign ions presence. Reprinted with permission from ref. [12]..

Various salts (such as Na_2SO_4, Na_2CO_3 or NaOH) have also been used to treat the collapsed foamate and displace the surfactant in order to recover the latter. Certainly, surface active materials tend to concentrate at the gas-liquid interface. All the surfactants recovered were found [13] to be reusable for foam flotation with qualitatively the same efficiency as that of the freshly-prepared surfactant. Recycling of surfactant, the most expensive chemical used in the process on a cost-per-weight basis, would greatly reduce the chemical cost of foam separation treatment. Additionally, it would reduce also the likelihood of contamination of surface and ground water with surfactant, if the sludge is deposited in a landfill.

Surfactant recovery would markedly improve the economics of the process. Estimated costs (capital costs plus chemical costs) for lead removal from dilute wastewater by adsorbing colloid flotation was reported [13] to be \$ 0.75/1000 gal, while the corresponding costs by lime precipitation was \$ 1.47/1000 gal.

An analysis of the complexed metal treatment by conventional hydroxide precipitation was made [14]; this was aimed to evaluate the mechanism of the ligand-sharing effect of metals which are added to wastewater to ensure effective removal of complexed heavy metals. EDTA, NTA and succinic acid were selected as organic complex formers of different strengths. Theoretical diagrams may be used to predict the favorable conditions.

2. Extensions of the Method

The commonly used inorganic flocculants are by far iron and aluminum compounds; it is known [15] that the hydrolysis products are responsible for any destabilizing effects on particles. An optical monitoring technique was used [16] to investigate the dynamics of coagulation of kaolin suspensions with ferric chloride. In fact, all the development in the application of dissolved-air flotation (following coagulation), in the water and effluent treatment areas, was based on the aforementioned - see for instance, among others, [17],[18].

Another outcome of adsorbing colloid flotation may be the use, in the first stage of the process, as a conventional sorbent (i.e. fine particle), like powdered activated carbon in suspension for heavy metals removal; followed by flotation downstream, to separate the metal-laden sorbent. In the dispersed phase, the larger available surface area of the sorbent and higher kinetics are usually encountered, hence smaller equipment as required. An example for gold removal, this time a valuable product, is presented in Figure 3, where non-cyanide technology was examined leaching gold with thiourea in acidic medium [19].

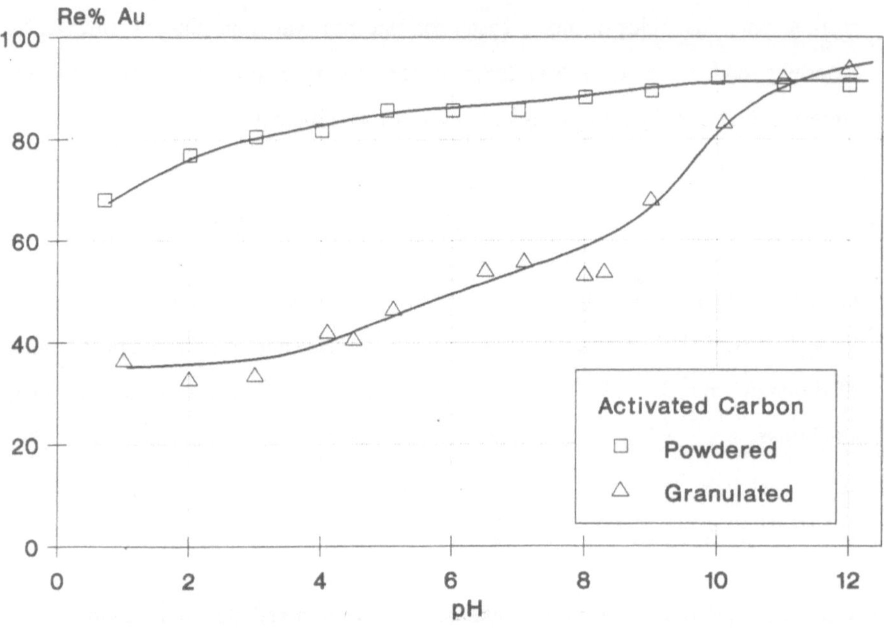

Figure 3. Comparison of powdered with granular activated carbon (both at 0.25 g/L): effect of pH during gold ion removal in thiourea solutions; 20 mg/L Au, 400 mg/L TU, contact time 600 s. Reprinted with permission from ref. [19].

3. Metal Biosorption

Metals can be dissolved from ores during bacterial leaching, by the oxidation of iron and sulfur by bacteria like Thiobacillus ferrooxidans. However, waste microbial biomass produced during many industrial fermentation processes (like antibiotics production), but also in biological wastewater treatment, may well present a suitable metal-sorbing material. Dead and live biomasses have equal biosorptive capacities [20]. Actinomycetes form a large group of gram-positive bacteria, naturally occurring in soil; anionic polymers, including teichoic acids, on their cell wall play an important role and give them favorable properties for metals removal.

When the initial cadmium concentration was varied, metal removal was effectively improved (see Figure 4), reaching almost 100% by increasing the addition of S. griseus biomass. The latter was a streptomyces laboratory grown, initially isolated from contaminated soil. This strain was kindly supplied by I.C. Hancock (Microbiology Department, University of Newcastle upon Tyne, UK). The kinetics of cadmium hydroxide precipitation was studied elsewhere [21].

It was argued [22] that the use of non-immobilized biomass would necessitate an elaborate scheme of well mixed reactors of CSTR type, followed by separators, which is costly and less efficient than other schemes, such as packed or fluidized beds. Acidic mine waters with toxic metals such as copper, cadmium, zinc, lead and mercury, were treated by porous polymeric beads containing biological materials [23].

Flotation, dissolved-air and dispersed-air, has been successfully tested by the authors as the solid/liquid separation method downstream for both actinomycetes and fungi, loaded with metals. The concept of a novel process termed biosorptive flotation was introduced. Figures 5 and 6 present some preliminary results obtained with a fungi industrial sample. A collectorless run is shown for Rhizopus arrhizus (at concentration 0.67 g/L), a filamentous fungi being a fermentation waste [24]. A small amount (0.25% v/v) of ethanol, as a frother agent, had beneficial results. The pH of solution (initially

1x10-4 M Cd) was observed to be critical parameter for biosorption (depending on metals specification) but also flotation. The optimum pH was not necessarily the same in the two cases. A really fast process was realized.

Extracellular polymers of bacterial origin also were reported to be plausible carriers for metals in soil or aquifer systems [25]. Bacterial extracellular polymers occur naturally in groundwaters and some have well established metal (Pb and Cd) binding properties; thirteen bacterial strains were studied.

A strain of a species of Citrobacter was found to accumulate cadmium extensively [26]. The accumulated metal precipitate was identified as cell-bound cadmium phosphate. Alcaligenes eutrophus, a heavy metal resistant strain, was proposed for metal carbonate and hydroxide precipitation [27]. A tubular membrane reactor was further developed in order to keep the cells metabolically active.

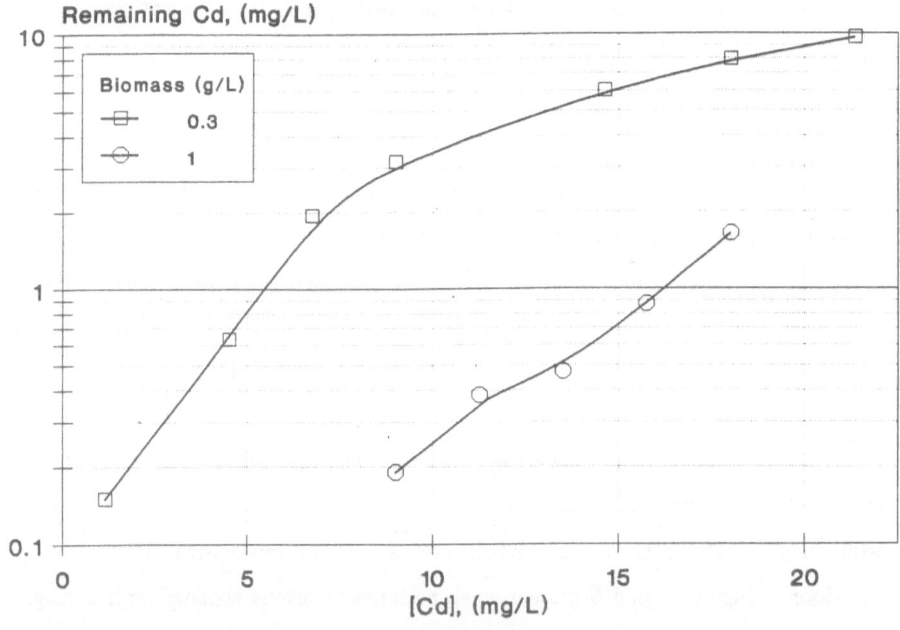

Figure 4. Influence of initial metal concentration in biosorption on Streptomyces griseus; results of remaining cadmium following also dissolved-air flotation (at pH 6.2), at two different biomass concentrations. Reprinted with permission from ref. [20].

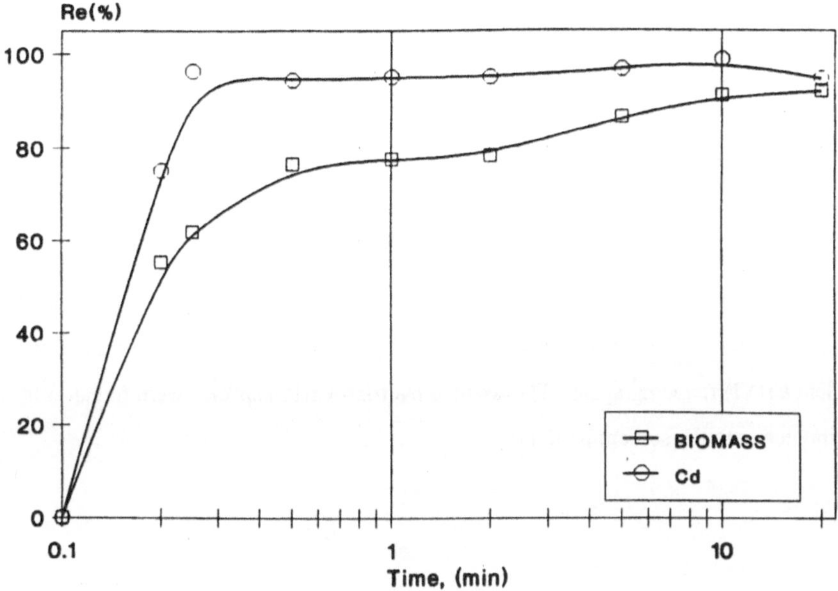

Figure 5. Removal of metal cation by biosorption on Rhizopus arrhizus: effect of flotation time on cadmium removal and biomass recovery; for the latter, dispersed-air flotation was applied as the S/L separation process. Reprinted with permission from ref. [24].

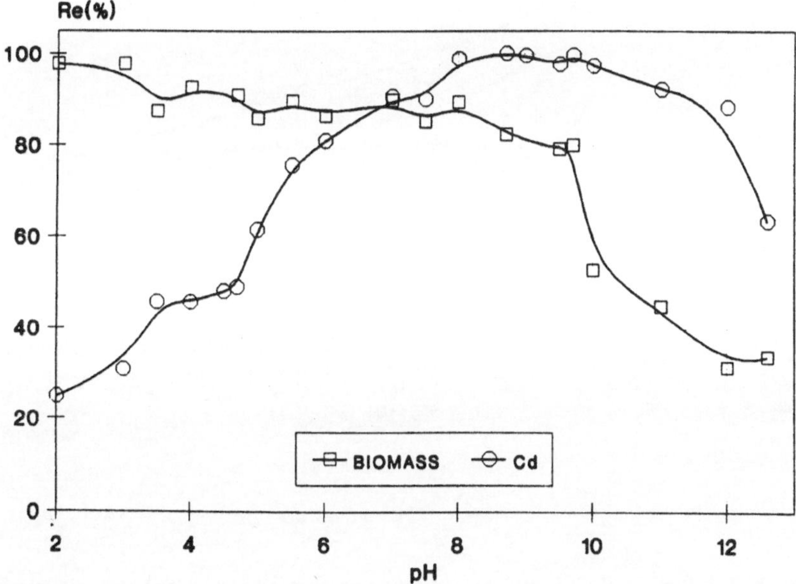

Figure 6. As in Fig. 5: effect of solution pH.

4. *Mineral Fine Particles*

Fine particles, by-products from mineral processing (or even industrial solid wastes), like pyrite (FeS2) may be also applied as alternative sorbents. These materials, compared with conventional sorbents (activated carbon, ion exchangers) present specific advantages such as lower price and higher availability. Figure 7 gives an example; iron sulfide mineral fines (-45 Ãm) proved to separate remarkably hexavalent chromium anions at highly acidic pH values [28]. Pyrite was observed to act as an efficient Cr(VI) reducing agent. The resulted hydroxo-Cr(III) species were found to be precipitated onto the sorbent particles.

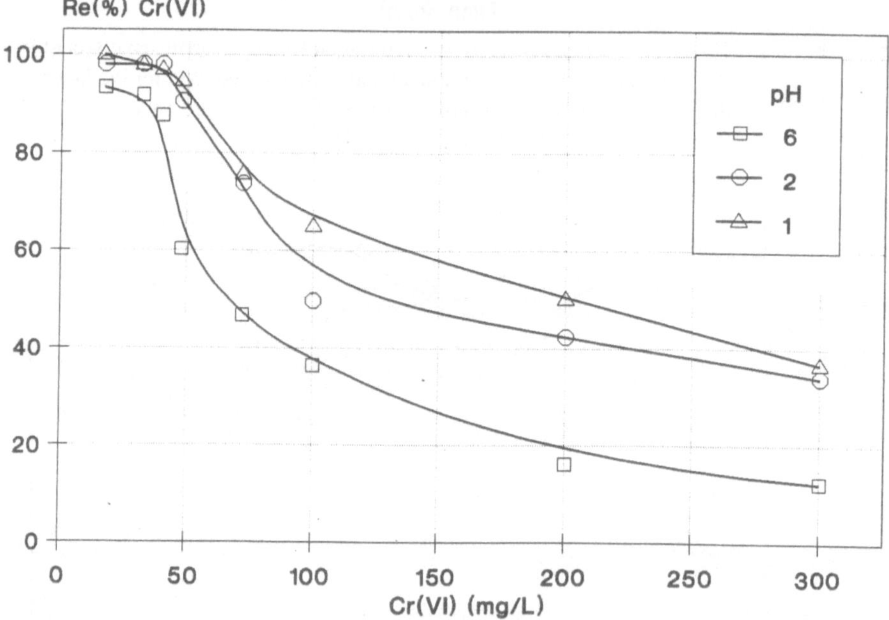

Figure 7. Influence of Cr initial concentration on chromates removal by pyrite applied as sorbent (5 g/L concentration) at different pH mixing conditions, as shown. Reprinted with permission from ref, [28].

Therefore, an interesting aspect of mineral fines utilization was presented in the removal of a priority pollutant. Earlier, chromate adsorption on amorphous iron oxyhydroxide was examined [29], in the presence of common cations and anions present in groundwater (K^+, Mg^{2+}, Ca^{2+}, SO_4^{2-}, CO_2(aq), H_4SiO_4).

The removal of copper ions from dilute aqueous solutions by the addition of pyrite fines was studied following an adsorbing (scavenging) flotation mechanism [30]. The latter was examined by electrokinetic measurements, as Figure 8 shows; a reversal z-potential is apparent. A surface-induced hydrolysis reaction model (i.e. interfacial precipitation) was suggested as a realistic explanation.

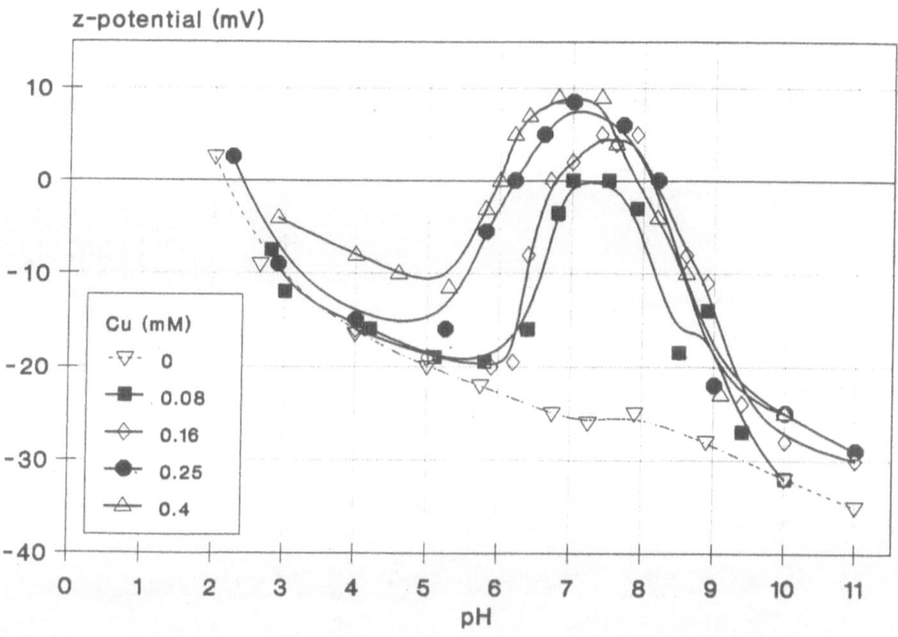

Figure 8. Zeta-potential measurements of pyrite particles in the presence of copper ions at various concentrations. Reprinted with permission from ref. [30].

It is known that a common characteristic generally of sulfide minerals is their instability under atmospheric conditions as they tend to be transformed into more stable oxidized forms; the oxidation of particle surfaces being also a problem in minerals dressing. This surface oxidation of sulfides, which results in a simultaneous formation of a sulfur-rich inner surface and a hydrophilic coating of metal hydroxide, has consequences to the environment and the pollution of groundwaters.

5. *Flotation and Separation Process*

In fact, flotation originated as a unit operation process in mineral processing. There it has been applied to beneficiate and enrich minerals and industrial ores, usually aiming to selectivity [3]. Figure 9 gives an idea from the separation of pyrite/arsenopyrite by cetyl trimetyl-ammonium bromide, from an auriferous bulk concentrate. In this case, unmodified flotation was tested and no selectivity was observed as both minerals were floating.

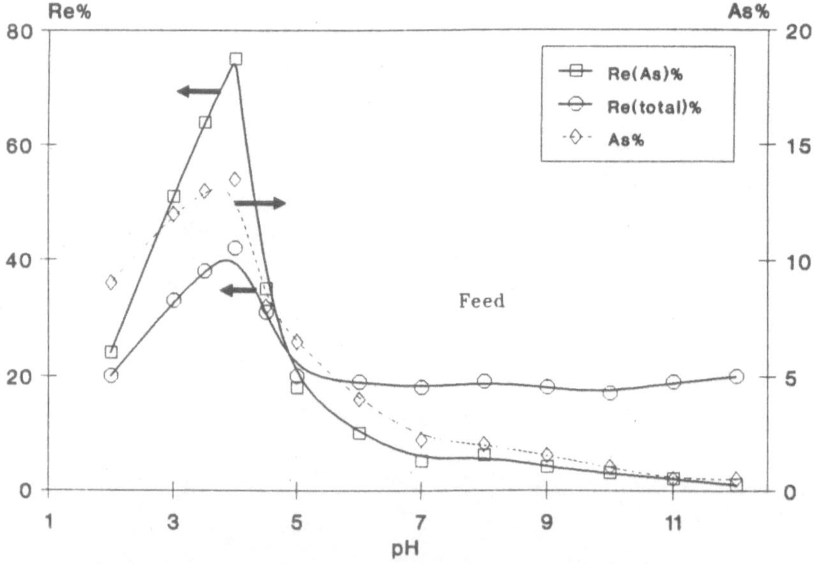

Figure 9. Cationic flotation of industrial pyrite concentrate, for arsenic separ ation, in the presence of ferric ions (5 mg/L): effect of pH (CTMA-Br 0.5 g/L); the dotted curve refer to the right-hand axis. Reprinted with permission from ref. [31].

Pyrite, obviously, is not especially significant from an economic point-of-view, unless it is associated with gold or uranium. It was once exploited for sulfuric acid production, but now it is often stockpiled, usually in the mine area, in tons of waste every year. Nevertheless, pyrite flotation is of particular interest both in sulfide ores and in coal processing. Arsenopyrite constitutes an unwanted admixture of sulfide concentrates (i.e. flotation products), due to its arsenic content, and penalties must be paid in the subsequent pyrometallurgical processing of the concentrates.

A fundamental investigation of cationic flotation was undertaken [31]; the latter being rather unconventional (in the place of xanthate flotation). Pyrite and arsenopyrite due to their surface similarities present almost the same response to sulphydryl collectors, like the xanthates are. It was found that moderate atmospheric oxidation or the presence of ferric ions in solution led to a surface alteration of the two minerals.
Xanthate reagents have also been used for toxic metal removal from wastewaters. The xanthate group is established for its high reactivity towards heavy metals. A comparative study between insoluble and soluble starch xanthate was presented [32]; the soluble xanthate process was found to be relatively less expensive.

In mineral processing, various modifying agents are generally used, such as potassium permanganate and copper sulphate, in order to obtain a differentiation in floatation (see Figure 10). This application is often based on the metal influence on the mineral particle surface; a varying conditioning time sometimes helps. In Figure 10, O-ethyl-dithiocarbonate was again the collector during dissolved-air flotation [33].

Many times the separation can be affected just by choosing another flotation technique instead of the conventional one, for instance electrolytic flotation (see Figure 11); pyrite recoveries were doubled in a broad pH range (4 to 8 approximately) [34]. The improved flotation may be due to many operational reasons and among them, a smaller bubble size with more favorable hydrodynamic conditions. Flocculation of particulate matter prior to flotation separation is another reason. Electrolytic flotation has also been applied in effluent treatment [3].

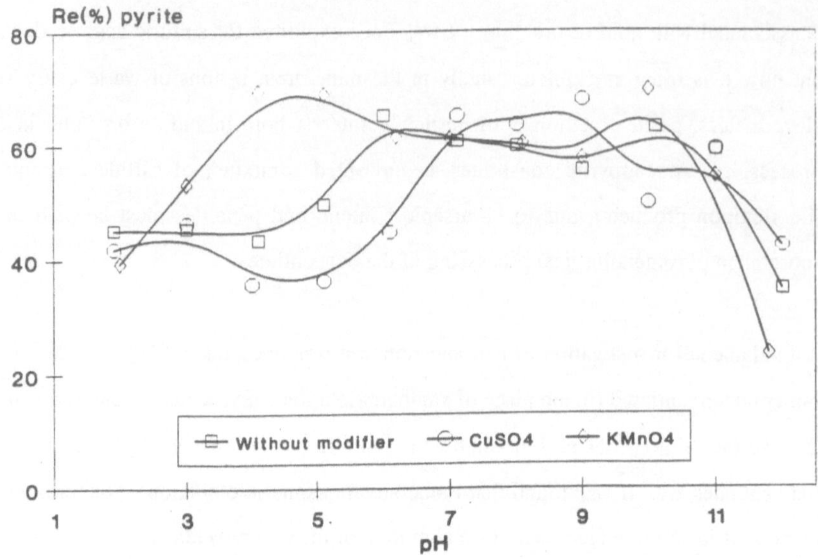

Figure 10. Presence of copper or manganese ions, used as modifiers, during dissolved-air flotation of pyrite by potassium ethylxanthate (15 mg/L): influence of pH; [CuSO4]= 10 mg/L, conditioning time 10 min; [KMnO4]= 100 mg/L, conditioning time 30 min.Reprinted with permission from ref. [33].

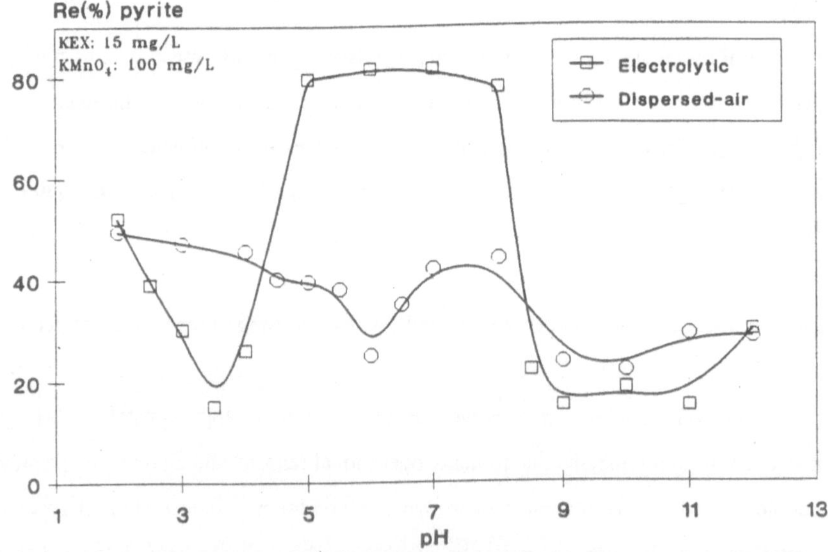

Figure 11. Flotation of pyrite fines by xanthate in presence of permanganate and using lime for pH modification: comparison of flotation techniques. Reprinted with permission from ref. [34].

In the removal of metal ions from dilute aqueous solutions, a selective separation among the metals may lead to subsequent recovery and hence, metal recycling. Flotation can be applied for this reason, as Figure 12 shows. It was found that Ni^{2+} floats only in the pH region where its hydroxide forms, so it can be separated from ferric ions. When xanthate, however, was introduced as collector in the Ni-Fe system, similar behavior of the ionic species was noticed, with total removal of both for pH values over 7, i.e. no selectivity [35]. It was demonstrated that dissolved-air flotation may achieve two goals: the concentration and separation of metal ions.

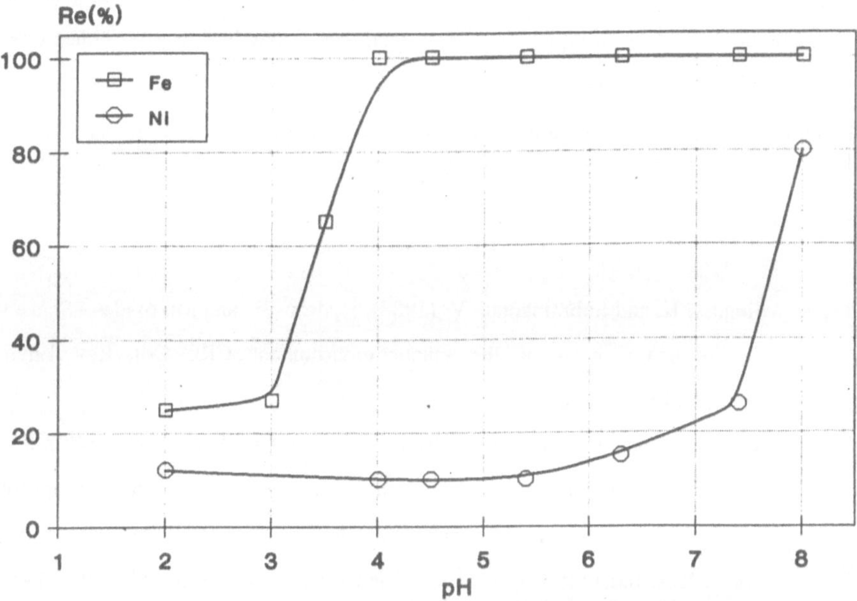

Figure 12. Dissolved-air flotation of nickel-ferric solution (each 50 mg/L), in absence of xanthate. Reprinted with permission from ref. [35].

An efficient liquid-solid separation was suggested to be the key to the successful pretreatment of industrial wastewaters [36]. Objectives of the former is to eliminate or reduce harmful matter in water. A correlation was presented between sedimentation and flotation under the same physical and chemical boundary conditions.

Concluding, modern environmental control is initially trying to avoid the formation of pollution, possibly by substituting hazardous chemicals or raw materials by safer matter and changing the process to minimize emissions (gaseous, liquid or solid). A reliable and efficient process solution is generally researched and a good separation system is always required. Flotation, in many cases, can be the answer to the problem.

Flotation presents some advantages over other, more conventional processes (such as sedimentation), because of its better effluent quality and mainly, lower residence time, which implies a lowering of plant costs, including the need for less floor space. Flotation science and engineering seems also to have a definite role in bioseparation.

References

1. Singh, S.K. and Subramanian, V. (1988), Hydrous Fe and Mn oxides - Scavengers of heavy metals in the aquatic environment, CRC Crit. Rev. Envir. Control 14 (1), 33-90.
2. Roy, D., Kommalapati, R.R., Valsaraj K.T., and Constant, W.D. (1995), Soil flushing of residual transmission fluid: application of colloidal gas aphron suspensions and conventional surfactant solutions, Wat. Res. 29 (2), 589-595.
3. Matis, K.A. (ed.) (1994), Flotation Science and Engineering, Marcel Dekker, New York.
4. Zouboulis, A.I., Matis, K.A., and Stalidis, G.A. (1992), Flotation in wastewater treatment, in P. Mavros and K.A. Matis (eds.), Innovations in Flotation Technology, Kluwer Academic Publishers, Dordrecht, pp. 475-498.

5. Zouboulis, A.I., Matis, K.A., and Stalidis, G.A. (1990), Parameters influencing flotation in the removal of metal ions, Int. J. Envir. Studies (Section B) 35, 183-196.6. Clarke, A.N. and Wilson D.J. (1983), Foam Flotation - Theory and Applications, Marcel Dekker, New York.

7. De Carlo, E.H. and Thomas, D.M. (1985), Removal of arsenic from geothermal fluids by adsorptive bubble flotation with colloidal ferric hydroxide, Environ. Sci. Technol. 19, 538-544.

8. Stalidis, G.A., Matis, K.A., and Zouboulis, A.I. (1986), Flotation techniques for the separation of trace pollutants, Chim. Chron. (New Ser.) 15 (3), 133-146.

9. Peng, F.F. and Di, P. (1994), Removal of arsenic from aqueous solutions by adsorbing colloid flotation, Ind. Eng. Chem. Res. 33, 922-928.

10. Kanungo S.K. (1994), Adsorption of cations on hydrous oxides of iron, J. Coll. Interface Sci. 162, 86-92.

11. Dzombak, D.A. and Morel, F.M.M. (1990), Surface Complexation Modelling: Hydrous ferric oxides, J. Wiley & Sons, New York.

12. Zhao Youcai, Zouboulis, A.I., and Matis (1995), K.A., Removal of molybdate and arsenate from aqueous solutions by flotation, Sep. Sci, Tech. to be published.

13. Huang, S.-D. (1983), Recovery of dodecylbenzene sulfonate from adsorbing colloid flotation foamates, Sep. Sci. Tech. 18 (11), 1017-1022.

14. Tunai, O. and Kabdasli, N.I. (1994), Hydroxide precipitation of complexed metals, Wat. Res. 28 (10), 2117-2124.

15. Gregory, J. (1992), Flocculation of fine particles, in P. Mavros and K.A. Matis (eds.), Innovations in Flotation Technology, Kluwer Academic Publishers, Dordrecht, pp. 101-124.

16. Ching, H.-W., Tanaka, T.S., and Elimalech, M. (1994), Dynamics of coagulation of kaolin particles with ferric chloride, Wat. Res. 28, 559-569.

17. Bunker, D.Q., Edzwald, J.K., Dahlquist, J., and Gillberg, L. (1994), Pretreatment considerations for dissolved-air flotation: water type, coagulants

and flocculation, in Flotation Processes in Water and Sludge Treatment, Conference preprints, IAWQ, Orlando, pp.35-44.

18. Schneider, I.A.H., Manera Neto, V., Soares, A., Rech, R.L., and Rubio, J. (1995), Primary treatment of a soybean protein bearing effluent by dissolved-air flotation and by sedimentation, Wat. Res. 29 (1), 69-75.

19. Zouboulis, A.I., Kydros, K.A., and Matis, K.A. (1994), Flotation of powdered activated carbon with adsorbed gold(I)-thiourea complex, Hydrometallurgy 36, 39-51.

20. Matis, K.A., Zouboulis, A.I., and Hancock, I.C. (1994), Biosorptive flotation in metal ions recovery, Sep. Sci. Tech. 29 (8), 1055-1071.

21. Luo, B., Patterson, J.W., and Anderson, P.R. (1992), Kinetics of cadmium hydroxide precipitation, Wat. Res. 26 (6), 745-751.

22. Tsezos, M. (1990), Engineering aspects of metal binding by biomass, in H.L. Ehrlich and C.L. Brierley (eds.), Microbial Mineral Recovery, McGraw-Hill, New York, pp. 325-339.

23. Jeffers, T.H., Ferguson, C.R., and Bennett, P.G. (1991), Biosorption of metal contaminants from acidic mine waters, in R.W. Smith and M. Misra (eds.), Mineral Bioprocessing, TMS, Wrrendale.

24. Zouboulis, A.I. and Matis, K.A. (1993), Flotation as a bioseparation process for fungi removal, Biotech. Techniques 7, 867-872.

25. Chen, J.-H., Lion, L.W., Ghiorse, W.C., and Shuler, M.L. (1995), Mobilization of adsorbed cadmium and lead in aquifer material by bacterial extracellular polymers, Wat. Res. 29 (2), 421-430.

26. Macaskie, L.E., Dean, A.C., Cheetham, A.K., Jakeman, R.J.B., and Skarnulis, A.J. (1987), J. Gen. Microbiol. 133, 539-544.

27. Diels, L., Van Roy, S., Taghavi, S., Doyen, W., Leysen, R., and Mergeay, M. (1993), The use of Alcaliggenes eutrophus immobilized in a tubular membrane reactor for heavy metal recuperation, in A.E. Torma, M.L. Apel and C.L. Brierley (eds.), Biohydrometallurgical Technologies, TMS, Warrendale.

28. Zouboulis, A.I., Kydros, K.A., and Matis, K.A. (1995), Removal of hexavalent chromium anions from solutions by pyrite fines, Wat. Res. 29, 1755-1760.

29. Zachara, J.M., Glrvin, D.C., Schmidt, R.L. and Resch, C.T. (1987), Chromate adsorption on amorphous iron oxyhydroxide in the presence of major groundwater ions, Environ. Sci. Technol. 21, 589-594.

30. Zouboulis, A.I., Kydros, K.A., and Matis, K.A. (1992), Adsorbing flotation of copper hydroxo-precipitates by pyrite fines, Sep. Sci. Tech. 27 (15), 2143-2155.

31. Kydros, K.A., Matis, K.A., and Stalidis, G.A. (1993), Cationic flotation of pyrites, J. Coll. Interface Sci. 155, 409-414.

32. Tare, V., Chaudhari, S., and Jawed, M. (1992), Comparative evaluation of soluble and insoluble xanthate process for heavy metal removal from wastewaters, Wat. Sci. Tech. 26 (1-2), 237-246.

33. Matis, K.A., Mavros, P., and Kydros, K.A. (1991), A dissolved-air flotation micro-cell for floatability tests with particulate systems, Separ. Technol. 1, 255-258.

34. Kydros, K.A., Gallios, G.P., and Matis, K.A. (1994), Electrolytic flotation of pyrite, J. Chem. Techn. Biotech. 59, 223-232.

35. Lazaridis, N.K., Matis, K.A., Stalidis, G.A., and Mavros, P. (1992), Dissolved-air flotation of metal ions, Sep. Sci. Tech. 27 (13), 1743-1758.

36. Mihopoulos, J. and Hahn, H.H. (1994), Concepts for efficient liquid-solid separation - The key to successful pretreatment of industrial wastewaters, Wat. Sci. Tech. 29 (9), 347-350.

IV.7

THE CENTRE FOR INTERNATIONAL PROJECTS (CIP): OPPORTUNITIES IN RUSSIA

SERGEI TIKHONOV
PO Box 46, CIP
Moscow, Russia 11792

The Center for International Projects (CIP) was established in 1980 to facilitate UNEP projects on the territories of the former USSR. At present, the CIP is responsible for the implementation of ecological projects in Russia. In 1993 at the third session of the Intergovernmental Ecological Council (IEC), the CIP was appointed to serve as the focus for information and coordination activities for ecological body assessments of CIS countries with UNEP.

Being a state organization, CIP operates under the auspices of the Russian Ministry of the Environment Protection and Natural Resources (MEDNR) and at the same time it is independent financially, having its own accounts for hard currency and rubles. The CIP has certain tax and custom privileges in Russia.

During the period 1980-1995 CIP managed the following main projects: Problems of Aral Sea; decertification and soil degradation; International Program if Chemical Safety (IPCS) in cooperation with WHO; Hygienic aspects of the environment; publication activities in the field of environment and dissemination of environmental information.

E.A. McBean et al. (eds.) Remediation of Soil and Groundwater, 425–427.

The CIP mow maintains three main directions of CIP, namely:

i) Activities according to plans of international and other organizations such as: UNEP (Oceans and Seas Program: Black Sea, Caspian Sea, Baltic Sea, North-West of Pacific Ocean); WHO (Global Environmental, Monitoring System (GEMS)/WATER Program; ILO (Major Hazard Control); Protection of Arctic Environment (Arctic Data-Base from the Russian sources. Data base for Protection of Arctic Marine Environment); Translation into Russian and Publication of the Materials and Documents for the following Conventions: Climate change. Decertification Control, Convention for the International Trade of Endangered Species; and Publication in Russian and dissemination among CIS countries "Environment News Digest" based on the information from UNEP Headquarters.

ii) Participation in the Programs of the Russian Ministry of the Environment, namely: Ecological Safety of Russia; preparation, edition and publication "State Report of the Environmental in Russian Federation" in 1992, 1993 and 1994; and reviews of the "State of the Environment in different Regions of Russia".

iii) Coordination activities in the Intergovernmental Ecological Council (IEC) according to the Agreement between IEC and UNEP.

This involved: biodiversity conservation (forest resources, water-swamp territories, natural reserves in different parts of the country), the conducting of conferences, seminars, workshops and training courses. During the last 15 years of CIP activities we conducted four International Congresses and more than 400 conferences, seminar, workshops, etc., in which approximately 15,000 participants took part; transfer and disseminate information; and consulting services to the Ecological Bodies and Institutions in CIS countries.

The staff of CIP includes highly qualified Russian scientists, experts and interpreters acting as temporary advisers to International Organizations (UNEP, WHO, UNIDO, VRO and some international institutions (AMAP, HELCOM, etc.)

CIP delivers assistance in solving scientific and technical tasks in Russia and CIS countries for International Organizations (mentioned above), programs, commissions, institutions and implementation of their Projects.

Part V:

SUMMARY ASPECTS

V.I

WHERE DO WE GO FROM HERE? AN OUTLINE FOR A PROGRAM OF ACTION

J. BALEK
ENEX Tabor
Czech Republic

Introduction

Specific tasks were given to the Workshop. We were to discuss the soil and groundwater remediation in polluted areas of central and eastern Europe and other states once under communist rule. In addition to the technical problems, we were also to discuss institutional and socio-economic issues.

The quality of both soil and groundwater needs to be judged as a complex system involving quality and quantity regimes.

Humans are an important part of any ecosystem; humans are responsible for damage to the ecosystem and must be responsible for the restoration of former natural conditions. A complex chain of interactions also exists among human beings. Therefore, when assisting in the restoration of natural ecosystems, we must be concerned with social and economic problems. In the course of our discussions, we discovered how very specific issues of groundwater flux and soil physics are linked with political, economic, social and health aspects because the deterioration of the soil and groundwater is not a

E.A. McBean et al. (eds.) Remediation of Soil and Groundwater, 431–444.

scientific problem per se but first if all it has to be considered as a health hazard with serious social and economic consequences.

Political and Socio-Economic Situation

Before summarizing our conclusions we should realize how immense are the spatial dimensions of the area of concern. The region of concern stretches from the borders of Germany to the borders of China and includes more that 20 states. More than 200 million people live in the region. The political climate and socio-economic conditions vary greatly from state to state and within each state itself, and so does the state of environment. As in many countries of the world, the environmental conditions are generally better and more sustainable in the states and regions which have been less advanced industrially and agriculturally and where soil and water remain more or less undisturbed. However, a good environment itself does not guarantee economic stability. Whether a sustainable environmental accompanied by the economic retardation can be considered as a good or bad omen is not a simple question. At least, we can conclude that a different remedial program will be necessary for a small country such as, for example, rural Albania with three million inhabitants, than for vast industrial regions found in the Russian Federation with a population of nearly 150 million.

Whether we, as technically-oriented professionals, like it or not, at some stage we have to be concerned with the political situation in the region under consideration. At least we have to be aware that central and eastern Europe countries formerly under the communist system are passing through a period of profound change. Rather sudden, rapid and uncontrolled political developments have been followed by striking changes in the economic structure and this process is far from completion. As a result, in many places the situation remains rather chaotic. Economic problems accumulated from the past have been solved by trial and error as opposed to utilizing a systematic approach. During the communist period, political and economic injustices to large percentages of the population remained hidden and authorities pretended they did not exist. Examples can be given of the problems of national minorities, nationalistic tensions, slow or even stagnant economic development, low standards and poor health conditions of the population. Under the new situation, and a more democratic atmosphere these once-accumulated problems can easily explode in the form of the clashes, civil strife and even wars. Unfortunately, in some countries such events have already been experienced.

A decline of moderate social benefits available in the past, to levels experienced in the 19th century has occurred for unprepared populations in the early capitalist system . Structural unemployment has emerged and living standards have decreased. Income differentials have widened and many people on low incomes are being exposed to poverty and reduced social security. The gross national product has dropped substantially making it increasingly difficult to maintain reasonable social and environmental standards. Budgetary constraints have starved the health and education services of adequate funding. Sudden and uncontrolled political freedom and democracy has contributed to the unethical competition of the private sector and industrial conflicts.

Over-developed administrations and the expansion of new bureaucracies in the central and eastern countries represents another major obstacle to the development process. The number of employees in the governmental administration have increased markedly. Important posts at all levels of the political and economic sphere are still controlled by former bureaucrats once actively supporting and promoting the communist system. On the contrary, in the governmental and parastatal sectors, important for further development and the application of new technologies, innovation, remedial and health programs, the number of employees and allocated budgets have been rapidly decreasing. Health, education, science, technology and worker's protection are the areas most affected. As well, a strong decline of cultural background has been observed elsewhere.

Environmental Issues

Environmental issues can be added to the list of neglected spheres. Before the political changes occurred most of the environmental issues in many states of central and eastern Europe had belonged to the unmentionable topics, many facts and actual data relevant to the damaged environment were considered as classified. Therefore, these days an expert tracing the history of regional soil and groundwater pollution can face serious difficulties due to the lack of solid data and relevant information.

After the introduction of the democratization process, other extremes could be observed. First, the environmental issues have become not only a popular and fashionable matter of concern of the politicians but also a dangerous toy in their hands. This is because most of the new politicians understand the environmental problems only in general terms, but use them as a political tool. Second, both international and local environmentally-oriented NGO's have spread throughout the region. Many of them choose easy, visible environmental targets as significant issues in their propagation campaigns. Unfortunately, soil and groundwater quality do not belong to such topics because they are much less visible than the smoke from chimneys, the cooling towers of nuclear power plants, and so on. Thus, respective ministries of environment have to work under the frequent and often unjustified pressure of the Greens.

For the above reasons a solid engineering approach is not very favored these days and the environment has ceased to be a matter of reliable concern supported by scientific evidence, experiment and observation.

As well, political, social, economic and legislative trends have not been very favorable for the general working conditions of the ministries of environment and their executive agencies. The utmost difficulties can be observed, first, in the preparation of general concept and strategy and their implementation and second, in the formulation, administration and enforcement of supporting legal measures. Within governmental structures, the ministries of environment are of secondary importance. Actual decision-makers on important environmental issues are ministries concerned with the industrial development, planning, finance and defense. As a result some of the newly-established enterprises, which in the future may significantly contribute toward the deterioration of soil, groundwater and complex ecosystems are built (sometimes with the participation of western companies) without consent and compensation of the communities likely to be influenced.

Such decision are made even if environmental impact assessment studies clearly signal the possibility of serious damage to the ecosystem. One of the reasons is that some of the responsible officers with decision-making authority are exposed to all kinds of political power-plays and corruption. Toward such chaotic situations, a lack of an effective legal system contributes which demonstrates not only the inability but also the unwillingness of responsible legislators to introduce and to enforce reasonable legislative measures. A pressure for the adoption of the legislation, comparable with the legislation in western countries, may be found sooner or later successfully in the states which intend to join the European Union; however, the enforcement of such legislation will be a much more difficult task because of the lack of effective control.

Groundwater and Soil Remediation Activities

Among many factors, our concern about remediation alternatives focused in the soil and water quality cannot be seen as isolated events. First of all, in an attempt to improve the quality of soil and water we have to focus our attention on one of the most important environmental issues which is human health.

The health problems and conditions vary throughout the region. For instance, according to the statistics of the Chartered Institute of Environmental Health in the Netherlands, the health problems in the Baltic States are related to poor technology and the management of the industry, while central and eastern European regions suffer from diseases developed by high levels of lead in air, cobalt in soil, high levels of nitrates and hydrocarbons in water and a food chain contaminated with heavy metals. The Central Asian Republics experience deficiencies in adequate nutrition, safe water supply, sanitation and basic medical care. Groundwater and soil become saline due to non-effective irrigation.

To identify priorities in the remedial programs implemented at the regional scale one will need the results of solid environmental monitoring. Present environmental networks are uniformly spread throughout the region. For some states it is difficult to purchase costly monitoring equipment. Thus, it can be difficult to obtain data in terms of groundwater and soil quality. To develop effective monitoring systems, capable of identifying the most significant sources of soil and groundwater pollution and relevant consequences they may impose on the human health, is an important initial step in remediation programs.

A serious technical problem of interest is how to monitor the status of health and water/soil pollution in the vast areas of the eastern sphere, and how to identify the most endangered zones to establish the priorities for remediation. For instance, in Kazakhstan, a significant part of the population in nomadic and the assessment of the relationship between the soil and water pollution on one hand, and human health on the other is difficult. This can also be a serious constraint when implementing remediation programs in places like the immense Aral Sea basin, whether the groundwater is saline from irrigation, and chemically and bacteriologically-polluted by fertilizers, pesticides, and defoliants and where assistance should be implemented as an international program.

In the Ukraine, a solution of different groundwater-related problems can be found as deserving priority. Some 80% of the Ukranian residents draw water from shallow wells contaminated with heavy metals and pesticides. In the Ukraine and in the Russian Federation serious pollution is found around the military establishments. Certain parts of the country have experienced pollution from nuclear tests and accidents.

Even if some of the formerly polluted sites along the western borders have been partly cleaned, some western organizations in cooperation with obscure local firms try to use them as dumping sites of their own hazardous and toxic wastes.

It can be said that different problems will need different approaches for national and regional needs, depending on the type of environmental problems. Economic development in respective countries, their capability to mobilize local manpower and willingness to cooperate are all potentially important. As many environmental features remain unknown, more attention has to be paid to the environmental monitoring, including acquisition, processing and analysis, and the distribution and publication of the results. Significant achievements have been made under special programs such as Tacis or Phare, but may gaps still exists.

A common feature for large parts of the region under consideration is the lack of well organized environmental management. Neither the environmental impact assessment practices nor the auditing are applied regularly. Also, general environment awareness remains very low.

A support for the mobilization of human resources is also important . In some countries there are professional with solid theoretical backgrounds. Difficulties can be expected when the theory is to be applied to field applications. For example, shortly after the revolution numerous offers for technical assistance were given by some western institutions to the former Czechoslovakia. Unfortunately, most of them were given by consulting companies capable of producing one or another form of general report but not the know-how and practices of technology. Some of them recruited local experts to accomplish the work for a fragment of the fee they had obtained from various international support programs. Very soon the Czechoslovak private sector, which had been slowly but steadily developing, recognized a big gap in the relatively new but attractive environmental market. This year, no less than 418 private firms in both the Czech and Slovak Republics offer services for the liquidation of hazardous and toxic waste, remedial works, and reconstruction of the waste deposit sites. (Anon. 1994). Some of the applied technologies, such as the washing the polluted aquifer, stripping, development of the monitoring facilities, etc., have been developed by the local manufacturers without any external assistance. Perhaps the main constraint experienced by those companies is heavy taxation imposed by the government, regardless of the significance of this type of business for public health and environment.

When discussing technology transfer, we should also examine whether the understanding of soil and/or water contamination is identical in both political spheres. As generally understood and postulated by the NATO Committee on the Challenges of Modern Society (Denner 1991) "...the contaminated land is a land which contains substances, when present in sufficient quantities or concentrations, likely to cause harm, directly or indirectly, to man, to the environment or on other occasions to other targets".

Of course such definition is generally valid, but when using it we have to check whether the terms "sufficient" and "harm to the environment" have the same meaning in different parts of the world. In other words, we have to find out what kind of risk in relation to the environment the countries at different stage of development are willing to accept. When performing such an exercise, we immediately tackle a wide spectrum of relevant problems, such as the water quality standards, health policy and environmental awareness at high administrative levels. At this stage we should be aware that in many parts of the eastern world, it is not always the polluter who pays, but often the public pays the toll.

Any transfer of technology and "know how" from west to east is usually based on the presumption that remedial problems are much more extensive in eastern than in the western countries. As stated by Wood (1994), this is not always true, since every major town and city in the UK possesses large tracts of polluted wasteland abandoned by industry. We can expect that with few exceptions the situation is similar in other western countries (Balek 1992). This is because, regardless of the present environmental awareness, mines, brickworks, tanneries, railway sidings, chemical factories and other workplaces have been built many years ago and later on were replaced by urban development or new industries. Limited remedial activities were applied years ago. Only recently the potential hazards were recognized and these days a reuse of such land without careful consideration of the possible consequences both on and off the site have become unthinkable.

Two remedial approaches can be identified in the countries where environmental awareness is more advanced. Under a rigid approach some environmentalists support the original Dutch philosophy of multifunctionality which requires any site, however badly contaminated, to be made fit for any purpose. As an example, consider the massive spending on clean-up operations in the United States where a levy was imposed on problematic industries in 1976 (Griffiths, 1992). This however, led to large proportions of the financial resources earmarked for remediation being diverted to the costs of establishing responsibility by litigation.

Under more liberal approaches, some western experts feel that there is considerable overreaction to the risk presented by contamination in general and that large sums of money are being spent unnecessarily on remediation practices and are being unreasonably reclaimed.

Even in the western world it is a rather difficult to determine the correct boundaries between economic viability on one hand, and proper standards of environmental safety on the other. Such a type of the decision-making is even more difficult in most of the eastern European countries. From a practical and economic point-of-view perhaps a reasonable approach can be followed with the aim of achieving a fitness of the site for a specified purpose. This, of course, requires a sensible commercial judgment not only of the polluted site and its future utilization but also possible impacts to neighboring properties and hydrologic and hydrogeologic catchments. In other words, some acceptable standards need to be introduced for the time being.

Such an approach is unthinkable without intensive monitoring of the polluted sites and areas likely to be influenced in the future. In addition, there is a need to introduce solid theoretical methods for the determination of the flux of pollution and the identification of the boundary conditions. While some eastern companies generally agree to establish basic monitoring networks and to finance the remedial works in situ, they are much less in favor of supporting any wider theoretical approach such as, for instance, the mathematical modeling of the complex processes involved and the monitoring of pollution beyond the factory boundaries in time and space. In other words, often a very pragmatic approach based on trial and error is practiced.

A reluctance to accept the results of mathematical modeling studies stem from the fact that the new generation of environmentalists in the administrative sector tend to be biological as opposed to engineering backgrounds, and as such are not inclined toward advanced mathematical studies. "...I have received some mathematical description supporting the proposed alternative but I do not understand it..." This is a quotation from a conclusion of one biologically-oriented decision-maker who, without nay further discussion, rejected the alternative proposed by an engineer. Prior to monitoring, a preliminary inventory of the site polluted above some pre-defined standards should be emphasized, based on all sources of information including verbal reports, remote sensing and mapping and analysis of the history of industrial development. One of the primary aims of such an exercise is to determine the priorities for the regional remedial program. Again, a question can be given whether the same standards and priorities as considered in the western world are valid for the eastern countries. Perhaps the highest priority should be given to the protection of potable water supplied, because they are essential for the support of human health. However, we should bear in mind that local economists and decision-makers may have a different opinion about the priorities. This occurs because, traditionally, human health has been of secondary importance to the former political regimes and in some states the situation may not have changed to a large extent.

When discussing the priorities in relation to human health, another problem can be identified. Many accumulated impacts resulting from continuous agricultural and industrial activities can be detected and proven only after a prolonged time. Unaware of the future hazard and risk, political decision-makers may consider the invisible soil pollution and relevant remedial programs to be of secondary importance. Therefore, it is absolutely necessary to emphasize as any part of the know-how and technology transfer in remedial programs, an assessment of hazard and determination of pathways through which the hazard can reach the population as a target. The term "risk assessment" is generally known in most eastern countries, but much less is known about the difference between hazard and risk as a degree of damage that can be accepted. Therefore, the relation between the level of hazard, sensitivity of the target, risk as a probability of acceptable damage, and appropriate and economically feasible level of protection has to be emphasized, and the concept of risk estimation based on the probability analysis introduced whenever possible.

Considering limited budgets and minimal data availability for the remediation of contaminated sites, the above approaches can facilitate an introduction of simple, robust and small scale remediation systems which can provide an effective barrier between contamination on the one hand and the user of that site, be the construction worker or the public.

Another factor to be considered is that the remedial works must remain effective for a considerable length of time. Therefore, to impose effective measures an assessment should be made on how, and when both soil and/or water can be considered as "clean" enough. This is because the cleaner the contaminated area, the more expensive is further clean-up and the remedial activities should be stopped once they have reached certain standards specified in advance.

Any institution responsible for the remedial program should be aware that even rather simple remedial methods which have been considered as locally sufficient for the time being, may have to be upgraded to comply with rather rigid EC legislation. To satisfy both requirements can be a difficult task for some countries. Therefore, differentiated standards should be introduced in the legislative system, these being based on realistic values that can be met by responsible institutions at present and in the future. Otherwise, by introducing unreasonable standards at the initial stage we are at risk that they may not be respected at all.

Priorities in the Remediation Program

Based on the discussions held at this Workshop, the following priorities of the remedial program can be specified:

- identification of the sites subject to remediation activities should be prioritized in relation to human health risks. This importance exists because many socio-economic and environmental problems are solved in relation to the improvement of general health conditions;

- improvements in systematic monitoring system of water and soil contamination are needed. Data processing, analysis and retrieval have to become an integrated part of the monitoring program. The monitoring of human health and living conditions should also be introduced;

- support the provision and transfer of long and short term knowledge necessary for an improvement of the legal and institutional structure and capacity building. This is particularly important when introducing remediation activities in an environmentally-sound and sustainable manner;

- any financial support for a specific remediation program should have a follow-up component to ensure that the budget has been efficiently utilized for the planned purpose;

- support should be made to ensure the participation of local and external NGO's which have shown an interest to assist directly in the above activities (instead of those preferring to increase their own publicity through environmental issues);

Some ambitious politicians in the eastern world would try to accelerate the moment when their respective countries will join the NATO alliance. These politicians already see themselves on numerous committees as equal partners to their western colleagues. Often the problems of the compatibility of weapons, airplanes, tanks and military structures are raised, discussed and seen as basic conditions for any type of such participation. However, from the scientific point-of-view the compatibility of weapons can be considered of secondary importance. Much more important are problems of the compatibility of health and environmental structures, comparable background supporting further scientific development in the above fields and, last but not least, decent living conditions of the population comparable with the living conditions as existing in the western countries. In other words, eastern states entering the western structures should reach beforehand a certain stage of development for which comparable environmental, health and socio-economic standards are set up in advance as prerequisite conditions for any kind of the partnership.

Conclusion

We as the earth cleaners have a difficult job. We are invited to clean and restore the areas once polluted by the industry, agriculture or military. After making the area acceptable for living, we have to leave for another filthy place. This does not mean that our work is less important than that of the economists, politicians or army officers. Actually our job can be considered as very satisfactory and some of the advanced remediation programs as a real challenge.

Even if it has not been officially stated, this Workshop, as a meeting place of soil cleaners from all parts of the world, is another type of event accomplished in the spirit of the NATO West-East program known as the "Partnership for Peace". Actually, scientists involved in one or another field of the water management and development had practiced such partnerships long before the cold war was at an end. Numerous meetings, symposia and seminars organized under the umbrella of UNESCO and other international organizations were organized with the aim to facilitate contracts and the exchange of information between eastern and western scientists. And all of us, regardless whether coming from eastern or western spheres, have always been aware that there is no place for peace without political and economic stability. Sustainable development, a sound environment, healthy populations and reasonable socioeconomic conditions for living are important factors which contribute to stability and peace. Both soil and water are important parts of the environment and health conditions. Our attempts to keep this world clean and inhabitable can be considered perhaps as a small but highly relevant and important contribution toward the global prosperity.

Literature

1. Anon., 1994 SOS Waste. Catalogue of the organizations concerned with then waste management (In Czech), ENZO Prague, 1994, 354 p.
2. Anon., 1994 Two workshops held in Siberia. Newsletter, NATO Science & Society, No. 42. p. 2.
3. Anon., 1994 Building the foundations for a sustainable future. Info Phare No. 7, p. 7-8.
4. Anon., 1994 Environmental progress in Eastern Europe. European Environmental Extra, No. 3, p. 2-5.
5. Balek, J., 1992 The environment for sale. Carlton Press, New York, 268 p.
6. Denner, J., 1991 Contaminated land: policy development in the UK. Proc. conf. on contaminated land, IWM, Birmingham, 1991.
7. Griffiths C.M., Approaches to the assessment and remediation of polluted land in Board N.P., 1992 Europe and American. J.Inst. Wat. and Env. Mgmt, 1992 6, No. 6, Dec., p. 720-725.
8. MacArthur I., 1995. Environmental health service in Europe. Europ. Bull on Environ . and Health, Vol. 3, No. 3, p. 5-6.
9. Wood A. A., Griffiths C. M., 1994. Contaminated sites engineered. Proc.Instn. Civ. Engrs, Civ. Engng, 1994, 102, Aug., p. 97-105.

V.2

A CONSCENSUS OF THE CURRENT SITUATION AND THE NEEDS AND OPPORTUNITIES FOR SOIL AND GROUNDWATER REMEDIATION IN CENTRAL AND EASTERN EUROPE

EDWARD A. MCBEAN

Conestoga-Rovers & Associates Limited

651 Colby Drive

Waterloo, Ontario, Canada N2V 1C2

The need for soil and groundwater remediation in central and eastern Europe is immediate. The legacy of soil and groundwater pollution inherited by the new governments in these countries not only creates direct health and ecological costs but is also likely to affect future industrial development and investment. However, in comparison with many western economies, the economies in eastern Europe are weaker, environmental problems much greater, and regulatory institutions are less developed. The result is that the soil and groundwater contamination problems are only going to be resolved at great economic and political cost and with considerable difficulty. With these considerations in mind, the conscensus of the current situation and the needs and opportunities apparent from discussions at the Advanced Research Workshop include the following:

E.A. McBean et al. (eds.) Remediation of Soil and Groundwater, 445–450.

1) Remediation Alternatives Currently Being Utilized are Primarily Bio-Remediation , Pump and Treat, and Excavation/Landfarming

For many sites in central and eastern Europe, there are no ongoing remediation efforts due to cost considerations. However, for those sites that are being remediated, the preponderance of remediation alternatives currently being utilized are the following: bioremediation, pump and treat, and excavation/landfarming. Bioremediation is the most frequently adopted because of its low cost. Pump and treatment systems are being utilized to a lesser extent but still relatively often and excavation/landforming is being utilized for soil remediation where volatiles are the major problem. Low labor rates make these options attractive. The more energy-intensive technologies (e.g. soil washing and excavation/incineration) are not in widespread use due to their expense.

2) Increased Efforts Are Needed in the Development of Modeling and Design Criteria for the Examination of Remediation Alternatives fot Site-Specific Applications

Improvements are needed in the mathematical modeling and design criteria employed in the implementation of remediation alternatives. The effectiveness of some remediation procedures is, as yet, unproven. Given the short time during which the various remediation procedures have gone from development to implementation, only a limited experience base exists. Many of the technologies that have been classified as standard (as opposed to emerging) remediation technologies are just now showing performance problems.

3) Haste May Make Waste Indicating the Need for Careful Ordering of the Sequence of Activities at a Site

In the desire to quickly remediate soil and groundwater problems, there is a concern that activities aren't always being initiated in the most effective order. For example, procedures to prevent the source from releasing future contaminants should be put in place before the soil and groundwater is remediated. However, this is not always occurring. Instead, the emphasis in some situations is being placed on the remediation of the groundwater which may become contaminated again due to the absence of source controls.

4) Uniformity of Legislation Between Countries Is Highly Desirable but Currently Does Not Exist

From one country to another, there are substantial variations in environmental legislation. Individual countries have signed agreements with other countries with the intent of making the legal atmosphere similar, but such endeavors are not uniform (e.g. the Latvian Ministry of the Environment has signed an agreement with the Polish Ministry of the Environment).

5) Environmental Issues Must be a Central Part of the Decision-Making at Governmental Levels and Not Subservient to Other Ministries

Some politicians understand environmental issues only marginally and are attempting to utilize these issues to accelerate the moment when their respective countries will join the European Union. These politicians already see themselves on numerous committees as equal partners to their western colleagues. Often the problems of compatibility of weapons, airplanes, tanks and military structures are raised, discussed and seen as basic conditions for any type of participation. However, from a scientific point-of-view, the compatibility of weapons can be considered of secondary importance. Much more important are problems of the compatibility of health and environmental structures. For example, in Poland, the Ministry of Environmental Protection dictates the law and this law cannot be superseded by other ministries. However, in numerous other countries, the environmental laws are old, very general, and departments are utilizing loopholes in the legislation to circumvent important environmental considerations.

Any new laws should be implemented reflecting European Union directives and contain a sufficient basis and power to require strict environmental levels. In addition, the institutional enforcement powers must be present to ensure that environmental legislation is implemented. The desire for countries to join the European Union appears to be a very effective mechanism to ensure that any legislation is adequate, and implemented.

7) The Assessment of Environmental Risk is an Effective Mechanism to Set the Priorities for Soil and Groundwater Remediation

There are insufficient funds available from existing and foreseeable institutional systems to clean up to European Union standards. As an alternative, risk assessment procedures represent an apparent mechanism for setting the priorities to select the oreder in which sites should be remediated (as opposed to establishing a uniform cleanup standard). As a result, since the current priority system for determining how available funds should be distributed is not clear, environmental risk assessment represents an effective mechanism. The highest priority should be given to the protection of potable water supplies. However, an unresolved question appears to be what kind of risk in relation to the environment that countries at different stages of remediation are willing to accept.

8) Environmental Audits are the Most Effective Mechanism to Separate Old Versus New Pollution and Much Greater Utilization of Such Audits Should be Undertaken. Environmental Monitoring Programs Must Be Greatly Increased

Environmental audits are an effective mechanism to allow separation of old versus new pollution, and such audits are strongly encouraged. Specific and extensive environmental monitoring and data management systems should be emplaced. Neither the environmental impact assessment practices nor the auditing approaches are applied regularly or consistently. Data processing, analysis and retrieval have to become an integral part of the monitoring programs. The monitoring of human health and living conditions must be more widely utilized.

9) Funding Availability is a Major Difficulty Associated with Environmental Remediation, Privatization Provides an Important Opportunity to Collect Some of the Needed Revenue

In many parts of central and eastern Europe, it is not always the polluter who pays, but often the public where this payment may be in the form of general revenue being expended but, more often than not, in terms of decreased environmental quality and the health implications this entails. In some countries there is almost no experience with remediation and the rate of privatization (which provides an opportunity to raise revenue) is very slow. Water utilities in some situations have paid for some remediation to imminent threats to drinking water sources but this is site-specific and occurs only in immediately threatening situations.

The degree of privatization in a country is directly related to the availability of money for remediation. An effective model for the funding of remediation is some type of National Fund for Environmental Protection such as is functioning in Poland and the Czech Republic in which user fees and fines are utilized as revenue generators. In the Czech Republic for cases of privatization (takeover), the past damages (liabilities) are covered (financed) by the National Property Fund and a new owner applies to the fund for such revenue. Remediation for State-owned companies (water utilities, etc.) is financed from the National Environmental Protection Fund.

Some interesting alternatives to the above exist but although desirable, will be difficult to envision as being widely available. For example, regional support from Germany is available to Western Bohemia to big business (polluters) on the River Elbe with a view toward increasing the quality in the rivers subsequently flowing into Germany. As well, in Germany the costs are paid by the government in that new investors in privatization are not considered as responsible for prior contamination (to prevent deterrence of investment). However, it is the strength of the economy of Germany that is the reason this approach is functioning but is not universally available.

Alternatively, Russia is establishing an Eco-Fund which will collect fines and penalties. This Eco-Fund will transfer money directly to various sectors. The World Bank is going to provide a loan for support of this system.

Privatization and the private sector is highly relevant in relation to environmental quality since this approach forces a better characterization of existing problems (for what may the company be ultimately responsible?), forcing the directors to think in ecological terms.

10) Any Financial Support to a Country or an Industry Should Be Considered as Temporary and Marginal

Any financial support for a specific remedial program should have a follow-up component to ensure that the allocated budget has been efficiently utilized for the planned purpose. Very specific controls on any expenditures should be mandated. As well, any financial assistance to a country should be considered as temporary and marginal. Sooner or later each country should become independent and self-supporting in the remedial programs that match generally respected standards.

11) Need for Separate Environmental Ministries

An important need exists to develop environmental ministries as separate entities within governments. When the ministry does not exist as a separate entity, there is a tendency toward subservience to other ministries such as those of economic development and defence.

Acknowledgments: This paper represents an attempt by the author to assemble the most important points of discussion at the Workshop. Consequently, the content of this paper relies heavily upon the inputs from the delegates.

List of Participants

Jirí Balej
Czech Inspection of the Enviornment
Vystupní 1644
CZ-400 07 Ustí nad Labem
Czech Republic

Jaroslav Balek
ENEX
P.O. Box 8
CZ-391 56 Tábor 4
Czech Republic

Walter Buydens
ERM
Huybrechtsstraat 3
B-2060 Antwerp
Belgium

Joseph Capka
HEIDEMIJ
Utrechtseweg 68
6800 AG Arahem
The Netherlands

Martin Carville
Aspinwall & Company Ltd.
Walford Manor
Baschurch
Shrewsbury SY3 9EG
U.K.

Milena Cislerová
Faculty of Civil Engineering
Czech Technical University
Thákurova 7
CZ-166 29 Prague 6
Czech Republic

Bruce Clegg
Conestoga-Rovers & Associates
8615 West Bryn Mawr Avenue
Chicago, Illinois
60631, USA

E.A. McBean et al. (eds.) Remediation of Soil and Groundwater, 451–456.

Zsolt Fekete
TERRA VITA KFT
Maklári u. 7
H-3300 Eger
Hungary

M. Talha Gönüllü
Dept. of Environmental Engineering
Yildiz Technical University
80750 Yildiz
Turkey

Dzidra Hadonina
Ministry of the Environmental Protection
and Regional Development
Peldu Iela 25
LV-1494 Riga
Latvia

Don Haycock
Conestoga-Rovers and Associates Limited
651 Colby Drive
Waterloo, Ontario
N2V 1C2
Canada

Zuzana Horická
Dept. of Hydrobiology
Charles University
Vinicná 7
CZ-128 44 Prague 2
Czech Republic

Krasimir Ivanov
Technical University
Suedinenie Str. 4
BG-1111 Sofia
Bulgaria

Janusz Jasinski
Mining and Shelter Company "Boleslav"
Kolejowa 37
32-332 Bukowno
Poland

Ryszard Koslacz
A Heidemij Group
EKOKONREM Ltd.
Ujejskiego 4/3
51-141 Wroclaw
Poland

Josef Krecek
Institute of Applied Ecology
Czech Agricultural University
Jungmannova 11
CZ-110 00 Prague 1
Czech Republic

Andrzej Krzyzowski
Law Office
Mazowiecka 25/501-503
30-019 Krakow
Poland

Arthur Kurzydlo
Conestoga-Rovers & Associates
8615 W. Bryn Mawr Avenue
Chicago, Illinois
60631, USA

Ryszard Maraszek
Government of Legnica Province
Pl. Skowiánski 1
59-220 Lagnica
Poland

Edward A. McBean
Conestoga-Rovers & Associates Limited
651 Colby Drive
Waterloo, Ontario
N2V 1C2

Viktors Melbardis
Baltec Assoc. Inc.
Kr. Garona Iela 88/9
LV-1001 Riga
Latvia

Marek Nawalany
Warsaw University of Technology
Nowawiejska 20
00-653 Warsaw
Poland

Sirekan Oganian
Government of Armenia
Urban Planning Dept.
House of Government 1
Republic Square
375010 Yerevan
Armenia

Zhora Oganian
Ministry of Economy
Dept. of Natural Resources and Ecology
House of Government
Republic Square
375010 Yerevan
Armenia

Pavol Pospisil
EKOKONZULT
Racianska 95/28
83 102 Bratislova
Slovakia

Silvia Reppe
Federal Environmental Agency
Bismarckplatz 7
Berlin, Germany
74793

Vialit Rezepov
Centre for International Projects
P.O. Box 46
117292 Moscow
Russia

Magda Sass
State of Illinois Hungary Office
East-West Business Center
Rákoczi út 1-3
H-1088 Budapest
Hungary

Youri Sedloukho
Polotsk State University
Blokhin Str. 29
BY-211440 Novopolotsk
Belarus

Julius Sivickis
Ministry of Environmental Protection
A. Juozapaviciaus 9
LT-2600 Vilnius
Lithuania

Krysztof Strynkovski
Government of Legnica Province
Pl. Skowia´nski 1
59-220 Legnica
Poland

Jan Svoma
AQUATEST
Geologická 4
CZ-150 00 Prague 5
Czech Republic

Neil R. Thomson
Department of Civil Engineering
University of Waterloo
Waterloo, Ontario
N2L 3G1
Canada

Sergei E. Tikhonov
Centre for International Projects
P.O. Box 165
117292 Moscow
Russia

Gabriel Constantin Tomescu
National Institute of Meteorology and Hydrology
Sos. Bacuresti-Ploiesti 97 sect.1
Bucharest
Romania

Miroslav Tyl
Ministry of the Environment
Vrsovická 65
CZ-100 10 Prague 10
Czech Republic

György Várallyay
Research Institute for Soil Science and Agricultural Chemistry
of the Hungarian Academy of Sciences
Herman Otto15
H-1022 Budapest
Hungary

Igor S. Zektser
Institute of Water Problems
Novo-Basmannaya 10
107078 Moscow
Russia

Anastasios I. Zouboulis
Department of Chemistry
Aristotle University
GR-540 06 Thessaloniki
Greece

INDEX

E.A. McBean et al. (eds.) Remediation of Soil and Groundwater, 457–462.